RECOMBINANT TECHNOLOGY IN HEMOSTASIS AND THROMBOSIS

RECOMBINANT TECHNOLOGY IN HEMOSTASIS AND THROMBOSIS

Edited by

Leon W. Hoyer
and

William N. Drohan

Jerome H. Holland Laboratory
American Red Cross Blood Services
Rockville, Maryland

PLENUM PRESS • NEW YORK AND LONDON

Library of Congress Cataloging-in-Publication Data

American Red Cross Scientific Symposium (21st : 1990 : Washington,
    D.C.)
     Recombinant technology in hemostasis and thrombosis / edited by
   Leon W. Hoyer and William N. Drohan.
         p.   cm.
     "Proceedings of the American Red Cross Twenty-first Annual
   Scientific Symposium: impact of recombinant technology in hemostasis
   and thrombosis, held May 15-16, 1990, in Washington, DC"--T.p.
   verso.
     Includes bibliographical references and index.
     ISBN 0-306-43893-3
     1. Blood--Coagulation, Disorders of--Congresses. 2. Recombinant
   proteins--Congresses. 3. Blood--Coagulation--Congresses. 4. Blood
   coagulation factors--Biotechnology--Congresses.   I. Hoyer, Leon W.
   II. Drohan, William. III. Title.
     [DNLM: 1. Blood Coagulation Disorders--diagnosis--congresses.
   2. Blood Coagulation Disorders--drug therapy--congresses. 3. Blood
   Coagulation Factors--congresses. 4. Recombinant Proteins-
   -congresses. 5. Thrombosis--drug therapy--congresses.   WH 322
   A215r 1990]
   RC647.C55A53   1990
   616.1'57--dc20
   DNLM/DLC
   for Library of Congress   91-3009
                                                                 CIP

Proceedings of the American Red Cross Twenty-first Annual Scientific
Symposium: Impact of Recombinant Technology in Hemostasis and
Thrombosis, held May 15–16, 1990, in Washington, D.C.

ISBN 0-306-43893-3

PREFACE

Recent progress in molecular biology has led to a rapid expansion of our understanding of the proteins that are essential for hemostasis and thrombosis. The goal of the XXI Annual Scientific Symposium of the American Red Cross was to provide a forum to explore and document the impact of recombinant DNA technology in this field. The speakers described the essential features of the genes responsible for key plasma proteins important in hemostasis, including procoagulant Factors VIII and IX and anticoagulant proteins, Antithrombin III and Protein C. They emphasized the advances in recombinant DNA technology that have led to the cloning of these genes. Careful examination of the gene sequence has then provided a clearer understanding of the structure of the encoded proteins, and has given additional insight into their functional domains and their interactions in hemostasis.

At the same time, these advances have made it possible to better characterize hemostatic disorders. A large number of published studies have shown that the mutations affecting biological activity are clustered in areas that define functional domains. They have led to fundamental advances in our understanding of specific diseases and they have made it possible to develop more accurate and sensitive diagnostic tests for the detection of the disease states.

Recombinant DNA technology is also beginning to be used to produce coagulation proteins for clinical use. The first example of this is the outstanding technical accomplishment of producing human coagulation Factor VIII in mammalian tissue culture cells, and the subsequent evaluation of the recombinant-produced protein in clinical trials. During this symposium, the advantages and limitations of protein production by recombinant technology were carefully examined by the participants. The rapid progress during the past decade provides strong support for the view that even more remarkable advances will soon be available to improve the treatment of diseases affecting hemostasis and thrombosis.

The success of the symposium was due to the efforts of the Program Committee that planned and chaired the scientific sessions: Morris A. Blajchman, M.D., William N. Drohan, Ph.D., Leon W. Hoyer, M.D., Kenneth G. Mann, Ph.D., and Frederick J. Walker, Ph.D. The rapid publication of these proceedings has been facilitated by the excellent editorial assistance of Debbie Wilder.

Leon W. Hoyer, M.D.

William N. Drohan, Ph.D.

CONTENTS

CHARACTERIZATION OF GENE AND PROTEIN STRUCTURE

MOLECULAR DEFECTS AFFECTING HEMOSTASIS

CHARACTERIZATION OF GENE AND PROTEIN STRUCTURE

BIOSYNTHESIS AND ASSEMBLY OF THE FACTOR VIII-VON WILLEBRAND FACTOR
COMPLEX

Jan A. van Mourik, Anja Leyte, Harm B. van Schijndel,
Martin Ph. Verbeet, Jan Voorberg, Ruud D. Fonteijn,
Hans Pannekoek, Koen Mertens

Central Laboratory of the Netherlands Red Cross
Blood Transfusion Service
P.O. Box 9190, 1006 AD Amsterdam
The Netherlands

INTRODUCTION

Factor VIII and the von Willebrand factor (vWF) are plasma
proteins that serve an essential role in the hemostatic proces;
factor VIII functions as a cofactor in the intrinsic coagulation
pathway (1,2) whereas vWF is hemostatically important in the mediation
of platelet-vessel wall interactions at sites of vascular injury
(3,4). In blood, factor VIII and vWF are not present as distinct proteins
but rather circulate as a linked complex. Several lines of evidence
indicate that this phenomenon is of physiological significance. For
instance, it now seems clear that vWF functions as a carrier protein
and as such has a stabilizing effect on factor VIII. This view stems
from the observation that reduced or absent synthesis of vWF (as
seen in von Willebrand's disease) is associated with markedly reduced
concentrations, or absence, of plasma factor VIII (5). Similarly, a rise
in vWF concentration, as observed in disorders associated with acute-phase
reactions, is accompanied with concomitant rises in plasma factor VIII
concentrations (6). Taken into account that the half life of factor VIII
infused in animals is determined by the presence of endogenous vWF (7),
and vWF also stabilizes factor VIII in vitro (8), these observations
clearly illustrate that vWF not only binds to factor VIII but also
confers stability to factor VIII. As vWF protects factor VIII from proteolytic attack by proteases including thrombin and activated factor X (9,10),
it seems likely that limited proteolysis is a factor that determines
the stability of factor VIII in vitro and in vivo. The importance of the
apparent stabilizing effect of vWF on factor VIII is underscored by recent
observations which show that an aberrant interaction between factor VIII
and vWF predisposes to a bleeding diathesis (11,12) Taken together,
these observations clearly document the physiological importance of
the factor VIII-vWF complex formation.

The cloning of the factor VIII- and vWF gene and the expression of
recombinant proteins and mutants thereof, together with immunochemical

studies, have provided the basis for significant advancements in the understanding of the structure-function relationship of factor VIII and vWF and of the nature of the interaction between these molecules.

BIOSYNTHESIS AND ASSEMBLY OF VON WILLEBRAND FACTOR

In the early 1970's, when antibodies to the factor VIII-vWF complex became available, immunofluorescence studies revealed that vascular endothelial cells of a variety of human tissues contain immuno-reactive material (13,14). Soon it became clear that most normal endothelial cells synthesize and secrete vWF, including endothelial cells isolated from large and smaller veins, capillaries, aorta, and arteries (15,16). Besides megakaryocytes (17,18), the endothelial cell is the only cell type that synthesizes vWF.

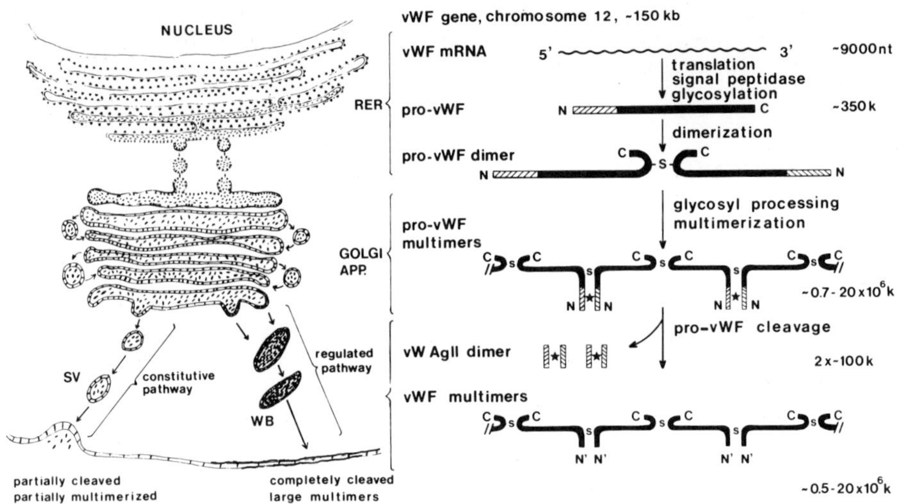

Fig. 1. Schematic representation of the processing steps involved in the biosynthesis of vWF. The bars on the right represent the vWF protein precursor and the products derived from it. The hatched area represents the propeptide (vWAg II) of vWF and the dark area mature vWF. WB = Weibel-Palade body; SV = secretory vesicles; * = noncovalent interaction; S = disulfide bonds.

Endothelial cells or megakaryocytes do not synthesize factor VIII (see below). VWF distinguishes itself from many other endothelial proteins in that it can be secreted by the cell by more than one pathway. VWF is either released directly by the constitutive pathway or is released into the medium upon treatment with endothelial cell agonists that trigger release of previously synthesized vWF from storage vesicles (regulated pathway). The constitutive secretory nature of the endothelium is reflected by the observation that soon after synthesis vWF, and other proteins such as fibronectin or thrombospondin, accumulate extracellularly in the absence of a stimulus (15,19). No external stimulus or trigger is required for this type of secretion. On the other hand, if endothelial cells are exposed to stimuli such as thrombin, the Ca-ionophore A23187 or phorbol-esters, vWF rapidly (within minutes) accumulates outside the cell (19,

20-22). These observations indicate that vWF, released when the endothelial cell receives an appropriate signal, originates from a storage pool. Several cell biological- and immuno-histochemical studies indicated that the endothelial cell-specific Weibel-Palade bodies serve as a storage vesicle for vWF (23-25).

VWF externalized by the constitutive route is structurally and, most likely also functionally different from vWF secreted by the regulated pathway. These differences can most readily be explained if one considers our present knowledge of the synthesis, processing and assembly of vWF (Fig. 1).

Molecular cloning of the full-length vWF cDNA has revelead that the vWF mRNA is translated as a pre-pro-polypeptide, composed of a signal peptide (22 amino acids), a pro-polypeptide (741 amino acid residues, also

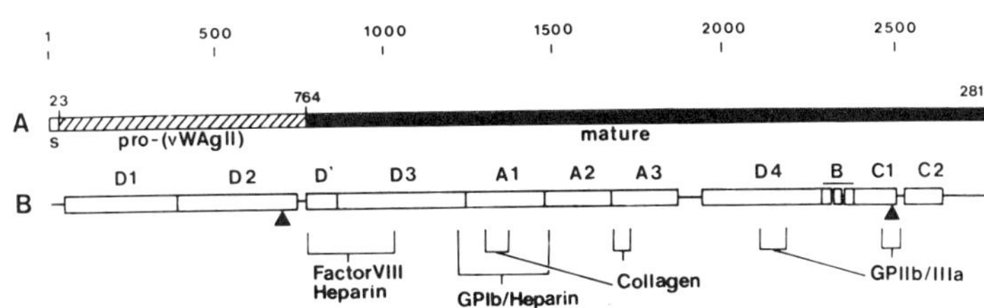

Fig. 2. Structure of vWf precursor. A, Precursor of vWF includes a signal peptide (S, amino acids 1-22), a pro-polypeptide (pro-vWF Ag II, 23-763), and a mature subunit (mature, 764-2813). B, Internal homologous domains and the structure-function relation-ship are depicted as are the positions of RGD peptides (triangles) known to be involved in cell attachment. GP = glycoprotein.

known as vW Antigen II) and mature vWF (2050 amino acids) (26,27). First, pro-vWF undergoes N-linked glycosylation and dimerization in the rough endoplasmic reticulum. Dimerization of pro-vWF monomers occurs by inter-molecular disulfide bridge formation, mediated by sulfhydryl groups located in the carboxy-terminal part of pro-vWF (28,29). These dimers serve as promotors for multimerization, a process that occurs during the travel through the Golgi compartments (30). During transfer through the Golgi complex several other post-translational processes occur, including sulfation processing of the mannose-type carbohydrate to the complex type and proteolytic cleavage of the pro-polypeptide. The latter process results in the formation of mature, multimeric and biologically active vWF. Also multimer assembly involves intermolecular disulfide bonding. The cysteine residues involved are located in the amino-terminal part of mature vWF (29). Only the largest and most biologically active vWF multimers are stored in the Weibel-Palade bodies and are released after stimulation. In contrast, all multimeric species are secreted in a constitutive way (22,24). More than 90% of the pro-vWF molecule consists of four types of repeated domains, denoted A, B, C and D, respectively (Fig. 2, refs. 26,31-33).

Studies on the fate of pro-vWF and mutants that lack one or more of
these domains, transiently expressed in Cos-1 cells, provided insights
into the individual role of these domains in the assembly of vWF multimers.
First the multimerization of vWF is directed by the vWF pro-polypeptide
(domains D1 and D2) (34,35). For instance, upon deletion of pro-vWF
only dimers are formed. Similarly, purified vWF dimers which lack the
pro-polypeptide are not able to multimerize in vitro, whereas pro-vWF does
form multimers under these conditions (36). Also premature cleavage leads
to aberrant multimerization (37). As multimerization of vWF protomers
involves intermolecular disulfide linkages in the region spanning the
amino acid residues 283 to 695 of the amino-terminal portion of the
mature vWF subunit (29), it seems possible that the pro-peptide could
serve to recognize and align the amino-terminal segments of vWF protomers
to permit the formation of appropriate disulfide links (35). Second,
proteolytic processing of vWF to generate the pro-polypeptide is not
required for multimer formation (37). The introduction of a point
mutation at the cleavage site at position 763 (at gly substitution),
preventing proteolytic processing, does not interfere with
multimerization of pro-vWF. In addition to the pro-polypeptide, both the
D' and D3 domain are required for multimer assembly (Fig. 3, ref. 38). As
intermolecular disulfide bridges are located in the 3D domain in fully
assembled multimers (29), the role of this domain in multimerization is
not unexpected. In contrast no intermolecular disfulfide bridges were
found in domain D' (29). It seems likely, therefore, that D'
provides for non-covalent interactions between pro-vWF dimers.
Presumably, such interactions are a continuation of the non-covalent
association initiated by the pro-polypeptide (39). Both the amino-
terminal and carboxy-terminal region of mature vWF serve an autonomous
role in the assembly of vWF multimers. A mutant lacking the A, B, C,
and D4 domains through the carboxy-terminus (Fig. 2) is able to form
dimers (Fig. 3, ref. 38). Similarly, a carboxy-terminal 151 aminoacid
residue may undergo dimerization (Voorberg, submitted for publication).
Taken together, these data suggest that dimerization and multimerization are processes that may occur independently. Recent studies
suggest that the pro-peptide also directs the folding and disulfide
formation within the amino-terminus of mature vWF that is required
for factor VIII-binding (Leyte, submitted for publcation).

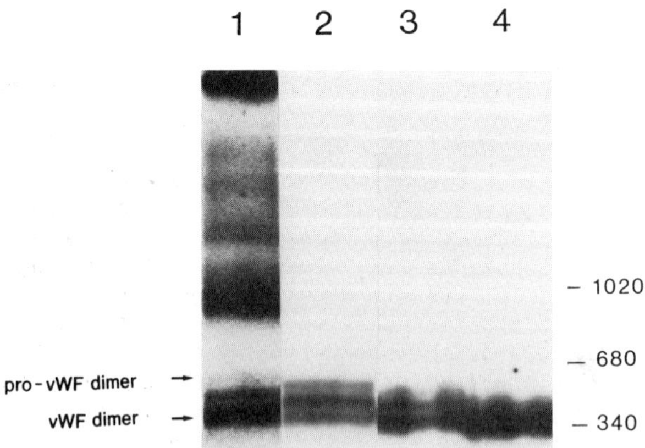

Fig. 3. Multimeric analysis of wild-type vWF and mutant proteins
expressed in Cos-1 cells. Medium was analyzed with respect to
multimeric composition under non-reducing conditions on 2% SDS-agarose
gels. Lane 1, wild-type vWF; lane 2, vWF lacking D'; lane 3, vWF
lacking D3; and lane 4, vWF lacking both D' and D3. (From ref. 38.)

BIOSYNTHESIS OF FACTOR VIII

Unlike the site of synthesis of vWF, the major production site of factor VIII is less clear and insight into processes that regulate its processing and release form the tissue is limited. Using monoclonal antibodies to factor VIII, immuno-histochemical studies revealed that this protein is located within the sinusoidal endothelium of the liver (40), supporting the view that the liver is an important site for production of factor VIII (41-43). Whether sinusoidal endothelial cells produce factor VIII is still a matter of debate, however. Using the same monoclonal antibodies employed by Stel and coworkers (40), Zelechowska and colleagues showed that these antibodies recognized factor VIII determinants, not only in the sinusoidal endothelial cells but also in the parenchymal cells (44). It is possible that differences in the fixation of the liver tissue examined may explain the apparent discrepant findings.

Similarly, factor VIII mRNA has been demonstrated in parenchymal cells but not in sinusoidal cells (45). In addition, the latter study demonstrated that a variety of other tissue, including that of the kidney, spleen and lymphnodes, contain factor VIII mRNA and together with the liver should be considered as potential sites of factor VIII synthesis. Although purified sinusoidal endothelial cells do not contain detectable amounts of factor VIII mRNA, production of factor VIII by these cells is still not ruled out. Recently it has been shown that apparently pure rat liver endothelial cells maintained in culture and incubated with 35S-methionine, produced radiolabeled factor VIII (46). In situ mRNA hybridisation studies could possibly answer the question whether factor VIII is produced by these cells in its natural environment.

ASSEMBLY OF THE FACTOR VIII-VON WILLEBRAND FACTOR COMPLEX

One of the most intringuing aspects of the biology of factor VIII and vWF is the observation that these proteins circulate in the blood as stable non-covalently linked complexes. At high

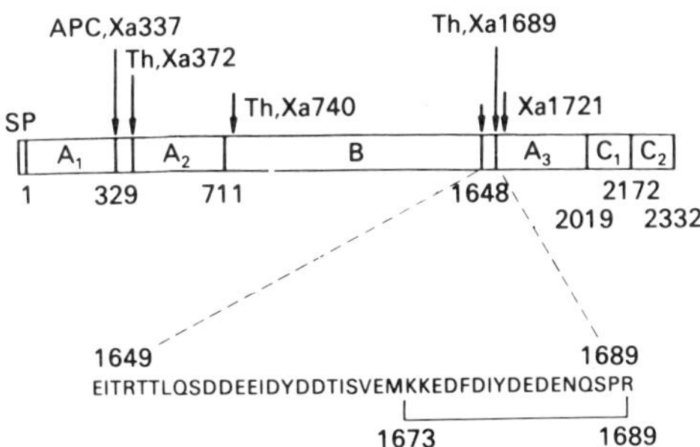

Fig. 4. Schematic representation of the factor VIII protein and the amino acid sequence of the N-terminal acidic region of the light chain. At the top the factor VIII protein is represented by the open bar; the signal peptide (SP) and the various domains (A1, A2, B, A3, C1, C2) are indicated. The arrows mark the known cleavage sites for thrombin (Th), factor Xa (Xa) and activated protein C (APC). The bottom shows the acidic region of the light chain; the bar indicates a synthetic peptide containing the epitope of a monoclonal antibody (CLB-CAg 69) that inhibits vWF binding (from ref. 48).

ionic strength the factor VIII-vWF complex dissociates, suggesting
that hydrophobic intramolecular interactions contribute to the stability
of the factor VIII-vWF complex, although electrostatic interactions may
also play a role in maintaining the stability of the complex (47).
It is still unclear how factor VIII and vWF interact at the molecular
level. The binding sites on factor VIII and vWF that are involved
in the mutual interaction have now been identified (48-51) and it is
expected that detailed information on the mode of interaction level will
soon become available.

Precise epitope mapping of monoclonal antibodies which inhibit the
factor VIII-vWF interaction have provided insight into the indentification
of regions of both factor VIII and vWF which are required for complex
formation. Using this approach it has been shown (48,50) that the
sequence Lys1673-Arg1689 of the acid region of the factor VIII light
chain is involved in the high-affinity interaction with vWF (fig. 4).
Similarly, mutant factor VIII that lacks this sequence is not able to
bind to vWF (Leyte, submitted for publication) Interestingly,
synthetic peptides presenting this sequence are recognized by the inhibiting
antibodies, but they do not compete with the factor VIII-vWF interaction
(48,50). This observation suggests that this sequence does not contain
the complete vWF binding site. The inability of the peptide Lys1673-Arg1689
to compete with the factor VIII-vWF complex formation may relate to
the absence of other sequences located close to the peptide in the intact
molecule. Alternatively, it is possible that post-translocational modifications are essential for proper interaction. In this respect, it is noteworthy that the residue Tyr1680 is flanked by acidic residues at position -2,
-4 and -5, and, therefore, meets the structural requirements to serve as a
substrate for tyrosylprotein sulfotransferase, an enzyme that catalyzes the
sulfation of certain Tyr residues (52). Indeed, Cos-1 cells transfected with
factor VIII cDNA produced factor VIII that was less effective in vWF binding
when the transfected cells were incubated with chlorate, an inhibitor of
protein sulfation.

Substitution of Tyr1680 for Phe also abolished vWF binding. Although
it remains to be proven that Tyr1680 of recombinant factor VIII is indeed
sulfated, it seems likely that sulfation is a post-translational event that
is essential for complex-formation. Tyrosine sulfation of proteins is a
rather common process in eukaryotes (52). However, insights into the biological significance of protein sulfation is limited. It has been reported that
inhibition of tyrosine sulfation affects the transport of proteins to the
cell surface to some extent (53), or decreases the activity of the fourth
component of complement (54). The finding that sulfation of factor VIII
seems essential for the factor VIII-vWF interaction is another example
in which protein sulfation influences the functional property of a protein.
The physiological significance of Tyr1680 as a residue involved in vWF
binding is underscored by a recent observation showing that the factor VIII
deficiency of a patient with hemophilia A is associated with a Tyr1680-Phe
substitution (55).

On vWF, a major factor VIII binding site has been assigned to an amino-
terminal fragment of mature vWF spanning the amino acid residues
1-272 (49). This fragment competes with factor VIII-vWF complex formation.
Also, a monoclonal antibody to this fragment inhibits factor VIII-vWF
interaction. Similarly, antibodies directed to more distinct stretches
on this peptide, including residues 1-106 (ref. 56) and 78-96 (ref. 57)
appear to inhibit complex formation. Previous studies have shown that
the factor VIII-vWF complex not only dissociates at high ionic strength
but also under reducing conditions (58, 59).

     Similarly, reduced and alkylated vWF does not bind factor VIII (49)
and reduction of the vWF peptide 1-272 abolishes its factor VIII binding
properties. The sequence 1-272 contains 24 cysteine residues (26, 31),
all of which are involved in intrachain disulfide bonds (29). It is likely,
therefore, that intrachain disfulfide bond formation is essential
for proper factor VIII binding. This view is supported by recent findings
showing that the pro-peptide of vWF not only directs the folding and
inter-chain disulfide bond formation required for multimerization
(vide infra), but also seems to direct the folding of the amino terminal
portion of vWF to allow factor VIII binding. The latter hypothesis is
based on the observation that a vWF mutant that lacks the pro-peptide
not only fails to polymerize but is also defective in factor VIII binding
(60)

CONCLUDING REMARKS

     It is well established that vWF serves an important role in mediating
the adhesion of platelets at sites of vascular injury. More recently
it has become clear that vWF has a stabilizing effect on factor VIII,
another important physiological function. The latter is most clearly
demonstrated by observations showing that defective factor VIII-vWF
interaction, either caused by genetically determined defects at the
factor VIII level or at the vWF level, may predispose to a bleeding
diathesis. The availability of sophisticated biochemical tools,
including monoclonal antibodies and recDNA techniques, have provided
detailed insights into the factor VIII-vWF interaction at the molecular
level. Although factor VIII and vWF may readily form complexes in vitro,
it is not clear how and where these proteins interact under physiological
conditions. Do these proteins form complexes intracellularly (e.g. the
sinusoidal endothelial cell in the liver) or in the blood stream? Which
structural features on either factor VIII or vWF determines the
stoichiometry of the factor VIII-vWF complex (1 mole factor VIII binds
approximately 100 moles of vWF) and the complex mode of interaction?
These, and several other questions raised by the biochemical studies
discussed above remain to be answered and will certainly be the focus
of future studies.

REFERENCES

1. Van Dieijen G, Tans G, Rosing J, Hemker H C: The role of
    phospholipid and factor VIIIa in the activation of bovine factor X.
    J Biol Chem 256:3433, 1981.
2. Mertens K, van Wijngaarden A, Bertina R M: The role of
    factor VIII in the activation of human blood coagulation factor X
    by activated factor IX. Thromb Haemostas 54:654, 1985.
3. Tschopp T B, Weiss H J, Baumgartner H J: Decreased adhesion of
    platelets to subendothelium in von Willebrand's disease. J Lab
    Clin Med 83:206, 1974.
4. Sakariassen K S, Bolhuis P A, Sixma J J: Human platelet
    adherence to artery subendothelium is mediated by factor VIII-
    von Willebrand factor bound to the subendothelium. Nature 279:
    636, 1979.
5. Ruggeri Z M, Zimmerman T S : Von Willebrand factor and von
    Willebrand disease. Blood 70:895, 1987.
6. Bloom A L: The biosynthesis of factor VIII. Clin Haemat 8:53, 1979.
7. Brinkhous K M, Sanberg H, Garvis J B, Mattsson C, Palm M, Griggs T,
    Read M S: Purified human factor VIII procoagulant protein:
    comparative hemostatic response after infusion into hemophilic
    and von Willebrand disease dogs. Proc Natl Acad Sci USA 82:8752, 1985.

8. Weiss H J, Sussman I I, Hoyer L W: Stabilization of factor VIII in plasma by the von Willebrand factor. Studies on post-transfusion and dissociated factor VIII and in patients with von Willebrand's disease. J Clin Invest 60:390, 1977.

9. Koedam J A, Meijers J C M, Sixma J J, Bouma B N: Inactivation of human factor VIII by activated protein C. Cofactor activity of protein S and protective effect of von Willebrand factor. J Clin Invest 82: 1236, 1988.

10. Hamer R J: in: FVIII: isolation, characterization and interaction with von Willebrand factor. Doctoral Thesis, University of Utrecht, ICG Printing BV, Dordrecht 1986.

11. Nishino M, Girma J P, Rothschild C, Fressinaud E, Meyer D: New variant of von Willebrand disease with defective binding to factor VIII. Blood 74:1591, 1989.

12. Mazurier C, Dieval J, Jorieux S, Delobel J, Goudemand M: A new von Willebrand factor (vWF) defect in a patient with factor VIII (FVIII) deficiency but with normal levels and multimeric patterns of both plasma and platelet vWF. Characterization of abnormal vWF/FVIII interaction. Blood 75:20, 1990.

13. Bloom A L, Gidding J C, Wilks C J: Factor VIII on the vascular intima: possible importance in haemostasis and thrombosis. Nature New Biol 241:217, 1973.

14. Hoyer L W, de los Santos R P, Hoyer J R: Antihemophilic factor antigen. Localization in endothelial cells by immuno-fluorescence microscopy. J Clin Invest 52:2737, 1973.

15. Jaffe E A, Hoyer, L W, Nachman R L: Synthesis of von Willebrand factor by cultured human endothelial cells. Proc Natl Acad Sci USA 71:1906, 1974.

16. Folkman J, Haudenschild C C, Zetter B R: Long-term culture of cappillary endothelial cells. Proc Natl Acad Sci USA 76:5217, 1979.

17. Nachman R L, Levine R, Jaffe E A: Synthesis of factor VIII antigen by cultured guinea pig megakaryocytes. J Clin Invest 60: 914, 1977.

18. Sporn L A, Chavin S I, Marder V J: Biosynthesis of von Willebrand protein by human megakaryocytes. J Clin Invest 76: 1102, 1985.

19. Reinders J H, de Groot Ph G, Dawes J, Hunter N R, van Heugten H A A, Zandbergen J, Gonsalves M D, van Mourik J A: Comparison of secretion and subcellular localization of von Willebrand protein with that of thrombospondin and fibronectin in cultured human vascular endothelial cells. Biochim Biophys Acta 844:306, 1985.

20. Loesberg C, Gonsalves M D, Zandbergen J, Willems Ch, van Aken W G, Stel H V, van Mourik J A, de Groot Ph G: The effect of calcium on the secretion of factor VIII-related antigen by cultured human endothelial cells. Biochim Biophys Acta 763:160, 1983.

21. Levine J D, Harlan J M, Harker L A: Thrombin-mediated release of factor VIII antigen from umbilical vein endothelial cells in culture. Blood 60:531, 1982.

22. Sporn L A, Marder V J, Wagner D D: Inducible secretion of large biologically potent von Willebrand factor multimers. Cell 46:185, 1986.

23. Reinders J H, de Groot Ph G, Sixma J J, van Mourik J A: Storage and secretion of von Willebrand factor by endothelial cells Haemostasis 18:246, 1988.

24. Ewenstein N M, Warhol M J, Handin R I, Pober J S: Composition of the von Willebrand factor storage organelle (Weibel-Palade body) isolated from cultured human umbilical vein endothelial cells. J Cell Biol 104:1423, 1987.

25. Wagner D D, Oluisted J B, Marder V J: Immunolocalization of von Willebrand factor protein in Weibel Palade bodies of human endo-thelial cells. J Cell Biol 95:355, 1982.

26. Verwey C L, Diergaarde P, Hart M, Pannekoek H: Full-length
    von Willebrand factor (vWF) cDNA encodes a highly repetitive
    protein, considerably larger than the mature vWF subunit. EMBO
    J 5:1839, 1986.
27. Fay P F, Kawai Y, Wagner D D, Ginsburg D, Bonthron D, Ohlsson-
    Wilhelm B M, Chavin S I, Abraham G N, Handin R I, Orkin S H,
    Montgomery R R, Marder V: Pro-polypeptide of von Willebrand
    antigen II. Science 232:995, 1986.
28. Fretto L J, Fowler W E, McCaslin D R, Erickson H P, McKee P A:
    Substructure of human von Willebrand factor. J Biol Chem 261:15679,
    1986.
29. Marti T, Rösselet S J, Titani K, Walsch K A: Identification of
    disulfide-bridged substructures within human von Willebrand factor.
    Biochemistry 26:8099, 1987.
30. Wagner D D, Marder V J: Biosynthesis of von Willebrand protein
    by human endothelial cells: processing steps and their intracellular
    localization. J Cell Biol 99:2123, 1984.
31. Titani K, Kumar S, Takio K, Erisson L H, Wade R D, Ashida K,
    Walsh K A, Chopek M W, Sadler E, Fujikawa K: Amino acid
    sequence of human von Willebrand factor. Biochemistry 25:3171, 1986.
32. Bonthron D T, Handin R I, Kaufman R J, Wasley L C, Orr E C, Mitsock L M,
    Ewenstein B, Loscalzo J, Ginsburg D, Orkin S H: Structure of pre-
    pro-von Willebrand factor and its expression in heterologous cells,
    Nature 324:270, 1986.
33. Shelton-Inloes B B, Titani K, Sadler J E: cDNA sequences for human
    von Willebrand factor reveal five types of repeated domains and
    five possible protein sequence polymorphisms. Biochemistry 25:
    3164, 1986.
34. Verwey C L, Hart M, Pannekoek H: Expression of variant
    von Willebrand factor (vWF) cDNA in heterologous cells: requirement
    of the propolypeptide in vWF multimer assembly. EMBO J 6:2885, 1987.
35. Wise R J, Pittman D D, Haudin R I, Kaufman R J, Orkin S H: The
    propolypeptide of von Willebrand factor independently mediates the
    assembly of von Willebrand multimers. Cell 52:229, 1988.
36. Mayadas T, Wagner D D: In vitro multimerization of von Willebrand
    factor is triggered by low pH. J Biol Chem 264:13497, 1989.
37. Verwey C L, Hart M, Pannekoek H: Proteolytic cleavage of the
    precursor of von Willebrand factor (pro-vWF) is not essential for
    multimer formation. J Biol Chem 263:7921, 1988.
38. Voorberg J, Fontijn R, van Mourik J A, Pannekoek, H: Domains
    involved in multimer assembly of von Willebrand factor (vWF):
    multimerization is independent of dimerization. EMBO J 9:797,
    1990.
39. Wagner D D, Fay P J, Sporn L A, Sinha S, Lawrence S O, Marder V J:
    Divergent fates of von Willebrand factor and its propolypeptide
    (von Willebrand antigen II) after secretion from endothelial
    cells. Proc Natl Acad Sci USA 84: 1955, 1987.
40. Stel H V, van der Kwast Th H, Veerman E C I: Detection of
    factor VIII/coagulant antigen in human liver tissue. Nature 303:
    530, 1983.
41. Owen Ch A, Bowie E J W, Fass D N: Generation of factor VIII
    coagulant activity by isolated, perfused neonatal pig
    livers and adult rat livers. Brit J Haematol 43:307, 1979.
42. Shaw E, Giddings J C, Peake I R, Bloom A L: Synthesis of
    procoagulant factor VIII, factor VIII releated antigen and other
    coagulation factors by the isolated perfused rat liver. Brit J
    Haematol 41:585, 1979.
43. Lewis J H, Bontempo F A, Spero J A, Ragni M V, Starzl T E:
    Liver transplantation in a hemophiliac. N Engl J Med 312:1189, 1985.

44. Zelechowska M G, van Mourik J A, Brodniewics-Proba T: Ultra-
    structural localization of factor VIII procoagulant antigen in
    human liver hepatocytes. Nature 317:726, 1985.
45. Wion K L, Kelly D A, Summerfield J A, Tuddenham E G D, Lawn R M:
    Distribution of factor VIII mRNA and antigen in human liver
    and other tissues, Nature 317:726, 43, 1985.
46. Hellman L, Smedröd B, Sandberg H, Petterson U: Secretion
    of coagulant factor VIII activity and antigen in vitro
    cultivated rat liver sinusoidal endothelial cells. Brit J Haematol
    73:348, 1989.
47. Owen W G, Wagner R H: Antihemophilic factor: Separation of
    an active fragment following dissociation by salts or detergents.
    Thromb Diath Haemorrh 27:502, 1972.
48. Leyte A, Verbeet M Ph, Brodniewicz-Proba T, van Mourik J A, Mertens, K:
    The interaction between human blood-coagulation factor VIII
    and von Willebrand factor. Biochem J 257:697, 1989.
49. Foster P A, Fulcher C A, Marti T, Titani K, Zimmerman T S:
    A major factor VIII binding domain resides within the amino-
    terminal 272 amino acid residues of von Willebrand factor. J Biol
    Chem 262:8443, 1987.
50. Foster P A, Fulcher C A, Houghton R.A., Zimmerman T S: An
    immunogenic region within amino acid residues Val1670-Glu1684
    of the factor VIII light chain induces antibodies which inhibit
    binding of factor VIII to von Willebrand factor. J Biol Chem 263:
    5230, 1988.
51. Takahashi Y., Kalafatis M, Girma J-P, Sewerin K, Andersson L-O,
    Meyer D: Localization of a factor VIII binding domain on a
    34 kilodalton fragment of the N-terminal portion of von Willebrand
    factor. Blood 70: 1679, 1987.
52. Huttner W B, Baeuerle P A. Protein sulfation on tyrosine.
    Modern cell Biology 6:97, 1988.
53. Friederich E, Fritz H-J, Huttner W B. Inhibition of Tyrosine
    sulfation in the trans-Golgi retards the transport of a constitutively
    secreted protein to the cell surface. J Cell Biol 107:1655, 1988.
54. Hortin G L, Farrier T C, Graham J P, Atkinson J P: Sulfation
    of tyrosine residues increases activity of the fourth component
    of complement. Proc Natl Acad Sci USA 86:1338, 1989.
55. Higuchi M, Traystman M, Wong C, Olek K, Kazazian H H, Antonarakis S E:
    Detection of point mutations in hemophilia A using PCR
    amplification of selected regions of the factor VIII gene. Thromb
    Haemostas 62:201, 1989 (Abstract).
56. Pietu G, Ribba A S, Meulie P, Meyer D: Localization within
    the 106 N-terminal amino acids of von Willerband factor (vWF) of
    the epitope corresponding to a monoclonal antibody which inhibits
    vWF binding to factor VIII. Biochem Biophys Res Comm 613:618, 1989.
57. Bahou W F, Ginsburg D, Sikkink R, Litwiller R, Fass D N:
    A monoclonal antibody to von Willebrand factor (vWF) inhibits
    factor VIII binding. J Clin Invest 84:56, 1987.
58. Counts R B, Paskell S L, Elgee S K: Disulfide bonds and
    the quarternary structure of factor VIII/von Willebrand factor.
    J Clin Invest 62:792, 1978.
59. Vehar G A, Davie E W (1980) Preparation and properties of bovine
    factor VIII (antihemophilic factor). Biochemistry 19:401, 1980.
60. Leyte A, Voorberg J, van Schijndel H B, Duim B, Pannekoek H,
    van Mourik J A: The pro-polypeptide of von Willebrand factor is
    required for the formation of a functional factor VIII binding
    site on mature von Willebrand factor: Biochem J, in press.

FACTOR IX: GENE STRUCTURE AND PROTEIN SYNTHESIS

D.B.C. Ritchie, D.L. Robertson and R.T.A. MacGillivray

Department of Biochemistry
University of British Columbia
Vancouver, B.C. V6T 1W5

INTRODUCTION

The development of recombinant DNA techniques during the past ten years has led to an explosion of the field of molecular genetics. Using these techniques, DNA fragments can be cloned and characterized at the molecular level. This in turn led to the discovery of intervening sequences in some eukaryotic genes, the identification of promoter elements, and to the production of recombinant proteins. As with many other fields, the field of thrombosis and hemostasis has also been changed by the application of recombinant DNA techniques. In this manuscript, we will review the structure and expression of the human factor IX gene, and discuss the various approaches to producing recombinant factor IX as a pharmaceutical. The molecular genetics of factor IX deficiency (hemophilia B) are discussed elsewhere in this book. The molecular biology and molecular genetics of blood coagulation have been reviewed recently[1,2].

HUMAN FACTOR IX cDNA STRUCTURE

Cloned factor IX cDNAs have been isolated by several groups. Choo et al.[3] synthesized two sets of degenerate oligonucleotides; these mixtures coded for part of the amino acid sequence of bovine factor IX[4]. Because of the low abundance of factor IX mRNA in the liver, one of the oligonucleotide mixtures was used to prime liver cDNA synthesis using reverse transcriptase. The resulting cDNA (enriched for factor IX cDNA sequences) was cloned into pBR322, and used to transform *Escherichia coli*. The enriched cDNA library was screened for factor IX cDNA sequences by using the second oligonucleotide mixture as a hybridization probe. DNA sequence analysis of the single positive colony showed that a partial bovine factor IX cDNA had been isolated.

A fragment of the bovine factor IX cDNA was then used as a hybridization probe to screen a human genomic phage library. Because of nucleotide differences between the two species, the hybridization was performed under conditions of reduced stringency. Again, DNA sequence analysis revealed that a recombinant phage had been isolated that contained part of the human factor IX gene. Brownlee's group extended these studies to include the isolation of a full-length cDNA clone[5], and the isolation of recombinant phage containing the complete human factor IX gene[5].

A similar approach was used by Jaye et al.[6] Based on the codon usage in the bovine prothrombin and fibrinogen cDNAs, these investigators designed a unique oligonucleotide (52 nucleotides in length) that coded for part of bovine factor IX as predicted by the amino acid sequence. This oligonucleotide was then used as a hybridization probe to screen a human liver cDNA library. A full-length cDNA was obtained. Jagadeeswaran et al.[7] also screened a human liver cDNA library but with a mixture of synthetic oligonucleotides (17-mers) coding for the bovine factor IX sequence. A partial cDNA was isolated that was 917bp in length.

A novel approach was used by Kurachi and Davie[8] to isolate a human factor IX cDNA. These investigators tried two different approaches to enrich for liver factor IX mRNA sequences. Initially, a baboon was treated with affinity-purified goat anti-human factor IX antibodies such that the plasma factor IX levels decreased to less than 1% of the normal level. Subsequent studies showed that this depletion treatment resulted in liver factor IX mRNA levels being increased by five fold. The factor IX mRNA levels were further increased by specific immunoprecipitation of polysomes and chromatography with oligo(dT) cellulose. In this way, an RNA preparation was obtained of which factor IX mRNA constituted about 2%. This enriched mRNA preparation was then used as a hybridization probe; in addition, Kurachi and Davie used a second hybridization probe that was comprised of a mixture of synthetic oligonucleotides that were 14 nucleotides in length. By using these two probes, Kurachi and Davie were able to isolate a full-length factor IX cDNA.

McGraw et al.[9] also screened a human liver cDNA library but used a unique oligonucleotide (an 18-mer) as a hybridization probe; this 18-mer was directed against the highly conserved region surrounding the active site serine residue. The resulting partial clone was then used to rescreen a liver cDNA library, and a full-length factor IX cDNA clone was obtained.

DNA sequence analysis of these clones allowed the nucleotide sequence of factor IX mRNA to be determined. In turn, this allowed the amino acid sequence of the primary translation product of the mRNA to be predicted. Excluding the poly (A) tail, human factor IX mRNA is 2802 nucleotides in length[5], and is comprised of 29 nucleotides of 5' untranslated sequence, 1,383 nucleotides of coding sequence, a UAA stop codon, and 1,390 nucleotides of 3' untranslated sequence. The location of the 5' end of the mRNA was determined by two different approaches, nuclease S1 protection assay using a cloned genomic DNA fragment and human liver RNA and primer extension analysis using human liver poly(A) RNA[5]. These results suggested that there was a major transcription start site located 29 nucleotides upstream of the initiator AUG codon. In addition, two minor start sites were identified. Recently, Salier et al.[10] have investigated the transcription start site of a chimeric gene consisting of the human factor IX gene promoter linked to the gene for chloramphenicol acetyl transferase. This construct was transfected into HepG2 cells, and poly(A) RNA prepared. The 5' end of the mRNA was determined by using a ribonuclease protection assay and antisense RNA transcribed from a cloned genomic DNA fragment spanning from -416 to the CAT gene. These results indicated that the transcription start site was located much further upstream from the AUG codon, at position -150. In addition, there were other transcription start sites at positions -247, -79 and -62. The reason(s) for the discrepancy in the location of the transcription start site is not known at present, but could reflect differences between human liver and HepG2 cells or differences between the expression of the native factor IX gene and the chimeric gene.

STRUCTURE OF THE FACTOR IX POLYPEPTIDE CHAIN

From the structure of the mRNA, the amino acid sequence of factor IX could be predicted (there were some minor discrepancies between the cDNA sequences determined by the different groups but these have now been shown to be sequencing errors or polymorphisms) - see Fig. 1. The cDNA sequence predicted that factor IX is synthesized as a precursor having an amino-terminal extension. The conversion from the precursor form to the plasma form of factor IX required the cleavage of an Arg-Tyr bond. As this cleavage is not typical of signal peptidase (which has an elastase-like specificity[11]), it was proposed that factor IX is synthesized as a prepro-protein. The cDNA sequences of prothrombin, factor X, factor VII, protein C and protein S showed that these vitamin K-dependent proteins were also synthesized as prepro-peptides (see [1,2] for a review). Interestingly, the signal peptides were not homologous between the various vitamin K-dependent proteins; however, the pro-peptides did share sequence homology suggesting a possible function for this region.

The amino-terminal region of the prepro-peptide constitutes the hydrophobic signal peptide that is involved in transfer of secreted proteins across the rough endoplasmic reticulum membrane[12]. Several recent studies have now shown that the pro-peptide of factor IX constitutes at least in part the recognition sequence for the vitamin K-dependent carboxylase. Initially, two independent cases of hemophilia B (Oxford-3 and Cambridge) were described in which cleavage of the signal peptide had occurred but cleavage of the pro-peptide had not occurred[13,14]. The mutations in these cases involved Arg(-4) to Gln, and Arg(-1) to Ser, respectively. Many other mutations at the -4 position have now been reported[15]. Mutations at these two positions results in the abnormal factor IX circulating as a larger protein, caused by the presence of the extra 18 amino acid residues of the pro-peptide. These abnormal proteins appear to have impaired carboxylation with a corresponding decrease in factor IX activity[14,15].

The propeptide has been mutated *in vitro* by Jorgensen et al.[16] These investigators had previously described an expression system for the production of recombinant human factor IX in chinese hamster ovary cells[17]. The secreted, recombinant factor IX was characterized by various physico-chemical methods (see later section). Jorgensen et al.[16] then introduced mutations into the pro-peptide of factor IX including the deletion of the pro-peptide, and mutation of two highly conserved residues

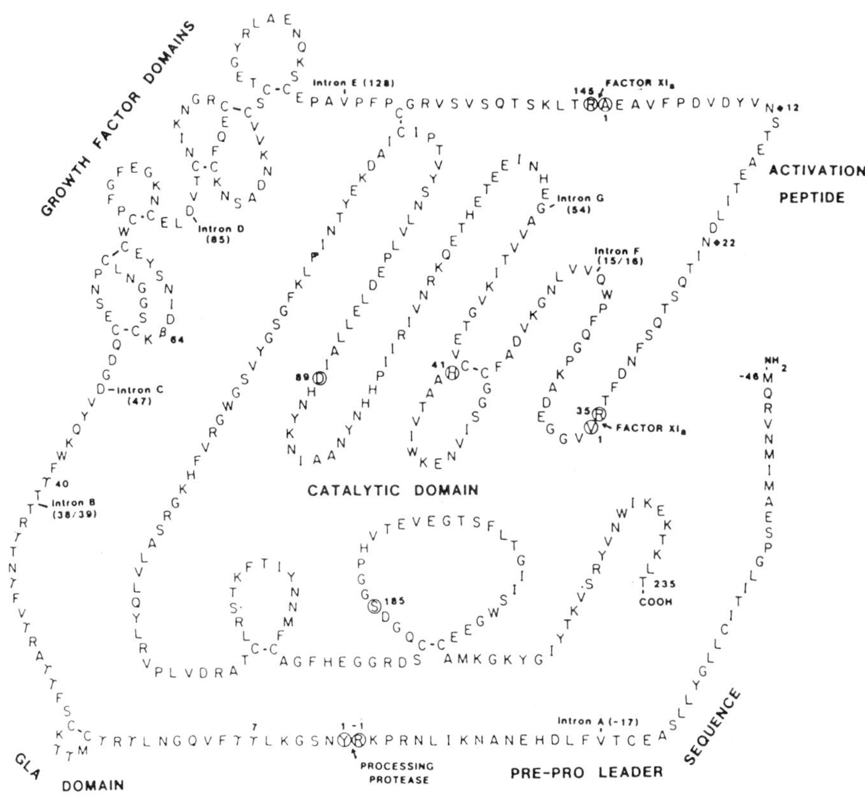

Fig.1. Amino acid sequence of prepro-factor IX showing the location of the seven introns. The amino acid sequence is given in the single letter code. γ stands for γ-carboxyglutamic acid; β stands for β-hydroxyaspartic acid. The positions where introns interrupt the coding region in the factor IX gene are indicated. The three attachment sites for carbohydrate are indicated by the black diamonds. The three amino acids that comprise the active site (H-41, D-89 and S-185) are numbered. Taken from Yoshitake et al.[26].

in the pro-peptide (Phe-16 to Ala and Ala-10 to Glu). The mutant factor IX cDNAs were expressed in CHO cells and analyzed with two antibodies; one recognized the calcium-dependent epitope found in native factor IX while the other antibody measured total factor IX regardless of carboxylation status. While each of the mutations resulted in the secretion of factor IX antigen, the amounts of native (calcium-dependent) factor IX antigen varied from undetectable (the deletion) to 6.1% and 2.4% of wild-type for the mutations at -16 and -10, respectively. Similar studies[18,19] have demonstrated the importance of the pro-peptide in the carboxylation of recombinant protein C.

By comparison with other proteins, the plasma factor IX polypeptide chain can be divided into several regions or domains (Fig. 1). The amino-terminal 40 residues contain the 12 glutamic acid residues that are converted to γ-carboxyglutamic acid (Gla) by the vitamin K-dependent carboxylase[20,21]. This region binds calcium ions and induces a conformational change that allows factor IX and factor IXa to function efficiently in the blood coagulation cascades[22]. This region is followed by a short aromatic amino acid-rich region of unknown function and two epidermal growth factor-like regions. Brownlee's laboratory has recently shown that the first epidermal growth region contains a high affinity calcium-binding site[23,24]. Initially, Rees et al.[23] developed a system for the expression of recombinant human factor IX in dog kidney cells in tissue culture. Expression of the factor IX cDNA was driven by the SV40 early promoter. Several point mutations were then introduced into the first EGF domain including the mutation of Asp-64 to Lys, Val and Gly; in plasma derived factor IX, approximately 30% of Asp-64 is found as β-hydroxy aspartic acid[25]. These mutations resulted in low factor IX coagulant activity suggesting that the EGF domain is essential for factor IX function. Rees et al. suggested that calcium ions bind to carboxylate ions in the EGF domain and stabilize a conformation necessary for the correct interaction of factor IXa with factor VIIIa, factor X and phospholipid[23]. In further experiments, Handford et al.[24] expressed the first EGF domain in yeast and demonstrated a direct binding of calcium ions to the isolated domain. A high affinity binding site for calcium was detected having a Kd of 200-300μM. Because yeast did not hydroxylate Asp-64, Handford et al. concluded that β-hydroxyaspartic acid is not required for calcium-binding. However, Asp-64 must be essential in some other role as mutation of this residue can cause hemophilia B[15].

The two EGF domains are followed by the activation peptide region and the protease region. Extensive homologies with other serine proteases at the activation site and in the protease region have suggested that factor IX is part of a large gene family of proteases that have serine at their active site[1].

HUMAN FACTOR IX GENE STRUCTURE

The complete factor IX gene has been isolated by two different groups. Anson et al.[5] isolated four recombinant phage that together spanned approximately 40kbp. The factor IX gene mapped to a 34kbp region within the cloned sequences. Anson et al. further characterized the transcription start and stop sites together with the intron-exon junctions. Yoshitake et al.[26] extended these studies by determining the complete nucleotide sequence of the factor IX gene. These authors screened phage libraries constructed from human fetal liver DNA and from a human fibroblast cell line containing five copies of the X chromosome. Five phage were isolated that contained the complete factor IX gene. Taken together, these two studies showed that the factor IX gene was about 33,500bp in length and consisted of eight exons interrupted by seven introns (Fig. 2).

As has been found with many other genes, the introns do not occur randomly within the factor IX gene. Instead, introns tend to separate DNA coding for discreet regions or domains within the factor IX molecule[5,26] (see Fig. 1). Thus, the first exon encodes the signal- or pre-peptide, the second exon encodes the pro-peptide and the majority of the Gla-region, exon 3 encodes the short aromatic-rich region, exons 4 and 5 each encode an EGF domain, exon 6 encodes the activation peptide while exons 7 and 8 encode the protease domain. This organization is very similar to the organization of the genes for factor X[27], factor VII[28], and protein C[29,30] (Fig. 2) suggesting that these genes have evolved by means of a relatively recent gene duplication event. Subsequent divergence has given rise to the genes found today; this divergence is reflected by the limited substrate specificity of the respective proteases. The first three exons of the factor IX gene are very similar to the first three exons of the prothrombin gene[31,32]. However, the organization of the remainder of the genes is different including the protease regions (Fig. 2), even though sequences of the protease regions of factor IXa and thrombin are homologous. It is likely that several ancient gene duplications gave rise to several serine protease genes. After divergence, these genes then gave rise to further gene families by recent gene duplications[32].

CHROMOSOMAL LOCATION OF THE HUMAN FACTOR IX GENE

Because of the X-linked nature of hemophilia B, it was long known that the factor IX gene resided on the human X chromosome, where it was linked to color blindness[33]. With the cloning of the factor IX gene, chromosomal localization by molecular hybridization techniques became possible. These studies showed that the gene was on the long arm of the X chromosome at Xq26-ter[34-37]. This is different from the location of the genes for factor X (chromosome 13[38-41]), factor VII (chromosome 13[41]), and protein C (chromosome 2[38,42,43]). These studies suggest that although the gene duplications giving rise to the factor IX-like gene family occurred relatively recently, there has still been enough time for the genes to become dispersed throughout the genome.

FACTOR IX PROMOTER

As discussed earlier, the factor IX gene was characterized by Anson et al.[5] in 1984 and Yoshitake et al.[26] in 1985 (the numbering system of Yoshitake is quoted here as it is more complete). The regulation of gene expression commonly occurs at the level of transcription. In eukaryotes, unlike bacteria, a number of DNA binding proteins known as transcription factors first bind to the DNA template near the site where RNA synthesis begins (the transcription start site). RNA polymerase then binds to the complex. DNA consensus sequences in the 5' end of the gene represent sites of DNA-protein interaction. These consensus sequences include the transcription start site, a TATA box at -26 to -34 where transcription factors localize RNA polymerase to the DNA template, and an upstream regulatory sequence such as a CCAAT box at -60 to -119 bases[44-46]. The latter sequence is highly polymorphic, variable in position, and of

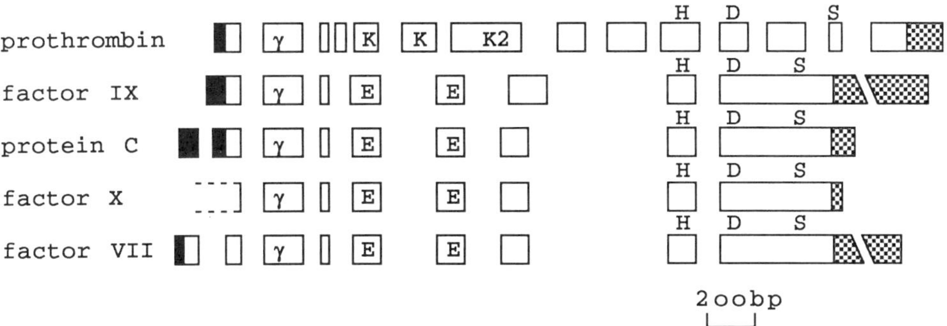

Fig.2. Comparison of the exon organizations of the human genes for prothrombin, factor IX, factor X, factor VII and protein C. The 5' untranslated regions are denoted by solid bars, the protein-coding regions by the open bars and the 3' untranslated regions by the slashed bars. The γ-carboxyglutamic acid region is denoted by γ, the kringles by K, the EGF-like regions by E, and the codons for the active site residues by H,D and S. The bar at the bottom represents 200bp. References are given in the text.

unclear importance to transcription[46]. Other, poorly characterized sequences are important to tissue specific expression of genes[47].

As discussed earlier, the factor IX transcription start site was first mapped by Anson et al. using nuclease S1 protection and primer extension experiments[5]. They found three start sites using both techniques (see Fig. 3, positions a,b, and c), and concluded that the most 5' site (+1) represents the start site. This is consistent with a weak consensus sequence[44]. These primer extension data and more recent consensus sequence comparisons[46], however, suggest that the second site at +4 (b below) is more likely. Anson et al. found a plausible TATA box[44,46] (TGTA) at -27 bases from the putative start site, although a better match was found outside the usual 26-34 bases from the start site at -41 (GTAAATA). The transcriptional start site is 29 bases from the probable translational start site which begins with a typical signal sequence[11] as described earlier. A CCAAT box[44] was not identified by these authors[5,26]. More recently, Crossley et al.[48] identified a CCAAT/enhancer binding protein binding site at +13 in the sequence CACAAT.

Yoshitake et al.[26] proposed two other possible TATA boxes at positions -256 (GATGAA) and -411 (TATATAA). Both of these are well outside the 26-34 bases from the putative start site, but the first is intriguing in that it is part of a sequence that shares strong homology (12 out of 13 base pairs) with the proposed TATA sequence for factor VIII[50]. Reitsma et al.[49] noted an inverted CCAAT box at -96. More recently, Salier et al.[10] mapped the transcriptional start site for a factor IX-chloramphenicol acetyltransferase chimeric gene to -150 by ribonuclease protection and primer extension assays. This is close to the transcription start site mapped for canine factor IX (-170bp from the transcription start site)[51]. Using a chloramphenicol acetyl transferase assay,

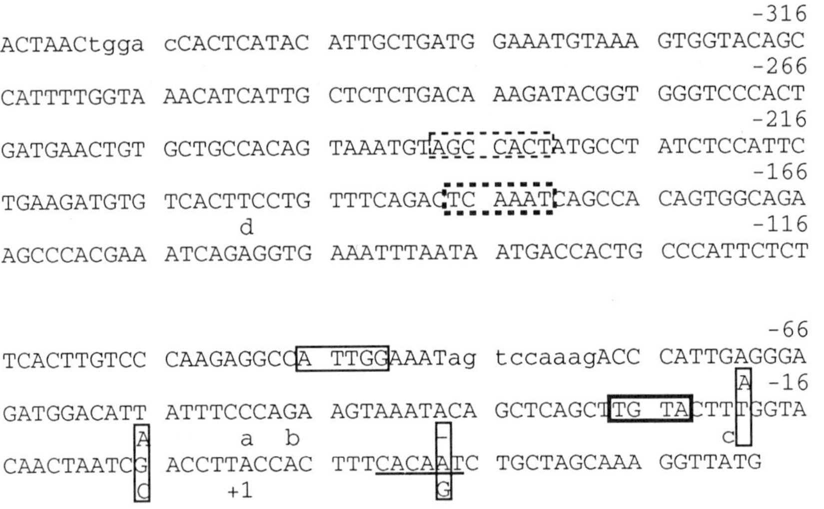

Fig. 3.   The 5' sequence of the factor IX gene. The proposed transcriptional start sites are labeled in small letters above the sequence. a, b, and c, represent the start sites proposed by Anson et al[1]. b represents the most likely major start site. c represents the probable translational start site. d represents the start site mapped by Salier et al.[7] using a chimeric gene made up of the 5' end of the factor IX gene spliced to the chloramphenicol acetyl transferase coding sequence. The TATA box proposed by Anson et al.[5] is boxed in bold lines and the inverted CCAAT box noted by Reitsma et al.[49] is boxed in thin lines. The CCAAT/enhancer protein binding site is underlined. The TATA and CCAAT sequences proposed by Salier et al.[10] for the chimeric gene are in dashed boxes. The proposed liver specific elements noted by Salier et al.[10] are in small letters. Mutations at -20, -6, and +15 causing the hemophilia B-Leyden phenotype are boxed[49,52,53].

Salier et al.[10] found cis acting elements at -170 to -270 and proposed TCAAAT at -187 as a TATA box (37 bases from the newly proposed transcription start site) and AGCCACT at -238 as a CCAAT box. These investigators also found a strong promoter in the opposite orientation between -700 and -800 bases and a silencer element at -1.6 kb to -1.9 kb. They noted homology to known liver-specific elements (LF-A1 and PR-1) at positions -349 and -79, although they have been unable to demonstrate any liver-specific transcription within 2.2 kb of the putative start site.

There are a number of proposed regulatory elements which are in reverse orientation including the promoter and liver specific element PR1 noted by Salier et al.[10], and the CCAAT box noted by Reitsma et al.[49]. These may represent functional regulatory elements for another as yet unidentified gene that is located 5' to the factor IX gene (Fig. 3).

The hemophilia B-Leyden phenotype has been instructive in sorting out the upstream regulatory elements. In these individuals, severe hemophilia B occurs at birth with undetectable factor IX coagulant and antigen levels. Near the time of puberty, however, the factor IX levels begin to rise and the symptoms begin to diminish and eventually disappear[52] The changes are persistent and progressive, and are typical of changes of secondary sexual development at puberty. This phenotype has now been described and the molecular lesion characterized in a number of unrelated families[15,49,53]. The remarkable finding is that there are at least three different regions where point mutations occur: positions -20, -6, and +13. The T to A change at -20 bases occurred in two Dutch families[53] and a German family[15]. The G to A change at the -6 position occurred in two families and a G to C in a third[15]. Finally, an A to G change occurred at the +13 position in three unrelated families[15,49] and deletion of the same A in a fourth family[49]. The occurrence of similar mutations in unrelated families with the same phenotype and the occurrence of different mutations at the same position suggest that all of these nucleotides are important.

Using HepG2 cells transfected with the factor IX-chloramphenicol acetyltransferase chimeric gene described above, Kurachi et al.[54] were able to show diminished promoter activity caused by the -20 and -6 mutations *in vitro*, and that the activity could be restored by adding androgens to the tissue culture medium. More recently, Crossley et al.[48] have reported that the +13 mutation disrupts the binding site for the CCAAT/enhancer binding protein. The three sites are close enough that they could interact with a single regulatory protein. Alternatively, multiple regulatory proteins might act cooperatively on these sites as seen when steroid receptors interact to activate steroid responsive genes[55,56]. The mechanism by which the factor IX gene is turned on at puberty remains unexplained, but likely holds important clues to developmental gene regulation in general.

EXPRESSION OF RECOMBINANT HUMAN FACTOR IX

Several groups have engineered the factor IX cDNA into expression vectors and produced recombinant factor IX or its fragments in *E. coli*[57], yeast[24] and tissue culture cells[17,58-60].

Lin et al.[57] were interested in expressing regions of factor IX in *E. coli* such that they could be used as antigens for production of anti-factor IX antibodies. Lin et al. placed a human factor IX cDNA and 13 different subfragments under the transcription and translation control signals from bacteriophage T7. The 13 subfragments covered the complete coding sequence of plasma factor IX plus residues -40 to -19 of the prepro-peptide. The resulting expression plasmids were introduced into *E. coli*, where they directed the synthesis of fusion proteins made up of the major capsid protein of T7 followed by the factor IX sequence. All of the fusion proteins were insoluble which facilitated their subsequent purification. The fusion proteins were recognized by polyclonal antiserum that had been raised against plasma factor IX. In addition, the fusion proteins were used to produce polyclonal antisera in rabbits against specific regions of factor IX. As discussed previously[24], Brownlee's group have expressed the first EGF-like domain of factor IX in yeast. They placed DNA coding for residues 46-84 of factor IX into the shuttle vector pMA91 that was able to replicate in both *E. coli* and yeast cells. The EGF domain was fused in frame with the yeast α-factor leader sequence which directed secretion of the EGF domain into the medium. The EGF domain was purified from the culture supernatent by batch adsorption onto C18 beads and final purification was achieved by reverse-phase chromatography. Three protein peaks were identified; as judged by peptide sequencing and NMR analysis, two of the peaks corresponded to incorrectly-folded EGF domains while the third peak corresponded to the correctly folded EGF domain. Handford et al. then investigated the effect of pH and calcium on the conformation of the domain. As judged by ^{1}H-NMR analysis, probably a single calcium-binding site was observed that had a Kd in the 250μM range. In addition, a lower affinity site for either calcium or magnesium was observed.

In 1985, three groups reported the expression of biologically-active human factor IX in tissue culture cells. Such recombinant factor IX could potentially be used in the treatment of individuals with hemophilia B. Anson et al.[58] constructed an expression vector by using a factor IX cDNA plus the 5' untranslated region from a genomic DNA clone. This factor IX DNA was placed under the transcriptional control of the long terminal repeat of the Moloney murine leukemia virus and the SV40 small-t-antigen intron and early polyadenylation signal. The expression vector also contained the *neo* gene as a dominant selectable marker in eukaryotic cells. Initially, the expression vector was introduced into a mouse fibroblast cell line and resulted in the secretion of immunologically-reactive factor IX; however, the factor IX failed to adsorb to barium sulfate suggesting that the fibroblast cell line was deficient in the enzymes for γ-carboxylation. However, when the expression vector was introduced into the rat hepatoma cell line H4-11-E-C3, most of the immunologically-reactive factor IX bound to barium salts suggesting that it was γ-carboxylated. In addition, the recombinant factor IX was active in a one-stage clotting assay. The amount of secreted factor IX was relatively low (about 6ng per 10^7 cells per 24 hours).

De la Salle et al.[59] used a vaccinia expression system to produce biologically-active factor IX in human hepatoma cells. In this case, the human factor IX cDNA was placed under the control of the vaccinia early 7.5K promoter and inserted into the N7 strain of vaccinia virus by recombination *in vivo*. HepG2 cells infected with the recombinant virus secreted fully active human factor IX that appeared to be γ-carboxylated in a vitamin K-dependent manner. Mouse fibroblast cells secreted factor IX antigen but even in the presence of high levels of vitamin K, the protein was less active than that produced by HepG2 cells suggesting that γ-carboxylation was less efficient in the mouse fibroblast cell line. The maximum level of recombinant factor IX expressed in HepG2 cells was about 3.1μg per 2 x 10^7 cells per 24 hours.

Busby et al.[60] placed a factor IX cDNA under the transcriptional control of the adenovirus-2 major late promoter, the tripartite leader sequence, a splice set comprised of the adenovirus-2 third leader 5' splice site and an immunoglobulin 3' splice site, and the early SV40 polyadenylation signal. After transfection into baby hamster kidney cells, factor IX was assayed in the cells and in the tissue culture medium by using a clotting assay based on the correction of factor IX deficient plasma and by an enzyme linked immunosorbent assay. Most of the factor IX (70-80%) was secreted into the medium and about 50% of the secreted factor IX was biologically active. Western blot analysis showed that the factor IX migrated with the same molecular weight as plasma derived factor IX.

As mentioned earlier, Kaufman et al.[17] have also produced recombinant factor IX. The expression vector p91023-IX contained the SV40 origin of replication, the adenovirus major late promoter (including the adenovirus tripartite leader and 5' splice site plus a 3' splice site from the immunoglobulin gene) driving expression of the factor IX cDNA and a dihydrofolate reductase (DHFR) cDNA followed by the SV40 early poly(A) signal, the adenovirus-associated genes, and pBR322 genes for propagation of the plasmid in *E. coli.* The vector p91023-IX together with the selectable DHFR gene on another plasmid (pAdD26SVpA#3) were used to transfect DHFR deficient Chinese hamster ovary cells. Cells were selected for the DHFR phenotype; these cells also produced immunoprecipitable factor IX. The transformants were then pooled and grown in increasing concentrations of methotrexate. This treatment resulted in the selection of cells in which the integrated DHFR gene (and co-transfected factor IX DNA) had been amplified. Cloned cell lines were obtained that secreted over 100µg/mL of recombinant factor IX antigen; however, by using antibodies specific for the calcium-dependent conformation, only 0.2% to 4.4% of the recombinant factor IX was fully carboxylated. The well-carboxylated factor IX was purified by immunoaffinity chromatography using the calcium-dependent antibodies. The recombinant factor IX was eluted from the affinity column with EDTA giving a 6250 fold purification from the tissue culture medium. The purified factor IX migrated on SDS-polyacrylamide gels with the same mobility as plasma-derived factor IX. Amino-terminal sequence analysis showed that the pre- and pro-peptides had been correctly processed. Amino acid analysis of the purified factor IX material showed that it contained 6-7moles of γ-carboxyglutamic acid per mole of factor IX.

Recombinant factor IX has also been produced in transgenic mice[61]. However, the expression in transgenic mice required the construction of an engineered cDNA which included seven base pairs of genomic sequence 5' to the putative transcription start site and 142 bases of genomic sequence past the poly (A) addition site, including the apparent termination site. The addition of these short pieces of genomic sequence were critical to the successful expression of factor IX, since previous experiments with only the full length cDNA were unsuccessful. Using this engineered cDNA fragment downstream of the metallothionein promoter, Choo et al. were able to induce the synthesis of normally γ-carboxylated, functional human factor IX in the plasma of a transgenic mouse in levels equivalent to that seen in man ($7.5\,\mu g/ml$).

To date, attempts at gene replacement therapy have met with limited success. Two groups have produced recombinant factor IX in fibroblasts using a retroviral expression system and have then implanted these cells into experimental animals[62,63]. Fibroblasts were chosen because they are easily obtained by biopsy of the skin, they are easily cultured in tissue culture, they are efficiently infected by retroviruses, they are efficient cells for expression of recombinant proteins and they can be easily grafted and removed. The retroviral vectors contain only the 5' and 3' long terminal repeats (LTR) and none of the other elements needed for retroviral replication. Accordingly, the vectors must be passaged through cells which provide the missing elements necessary for phage production before infection of the target fibroblasts. The cDNA is inserted following either the 5' LTR or another promoter and is expressed in conjunction with the *neo* gene which confers antibiotic resistance for selection purposes. Using similar systems, both groups were able to produce cells which produced appreciable quantities of recombinant factor IX in tissue culture. Upon implantation in either subcutaneous or intraperitoneal sites, active recombinant human factor IX could be demonstrated in the plasma of mice and rats. However, in both cases, the levels produced were low (2-6% of that expected from the tissue culture experiments) and transient (40 days maximum) when using normal diploid cells. Using transformed cells (NIH3T3) implanted in immunocompromised mice, more prolonged expression was obtained but also resulted in the development of tumors. Antibodies to the foreign protein developed in most immunocompetant animals in relation to the level of protein produced, but it appears that the main problem is that the promoters become inactive with time. The reasons for this remain unclear but pose the major hurdle to gene replacement therapy at present.

ACKNOWLEDGEMENTS

Research in the authors' laboratory is supported by the Medical Research Council of Canada and the Alberta Heritage Foundation for Medical Research.

REFERENCES

1. MacGillivray RTA, Cool DE, Fung MR, Guinto ER, Koschinsky ML, Van Oost BA: Structure of the genes encoding proteins involved in blood clotting. In Setlow JK (ed): Genetic Engineering. Principles and Methods, vol 10. New York, NY, Plenum, 1988, p 265.
2. Furie B, Furie BC: The molecular basis of blood coagulation. Cell 53: 505, 1988.
3. Choo KH, Gould KG, Rees DJG, Brownlee GG: Molecular cloning of the gene for human anti-haemophilic factor IX. Nature 299: 178, 1982.
4. Katayama K, Ericsson LH, Enfield DL, Walsh KA, Neurath H, Davie EW, Titani K: Comparison of amino acid sequence of bovine coagulation factor IX (Christmas Factor) with that of other vitamin K-dependent plasma proteins. Proc Natl Acad Sci USA 76: 4990, 1979.
5. Anson DS, Choo KH, Rees DJG, Gianelli F, Gould K, Huddleston JA, Brownlee GG: The gene structure of human anti-haemophilic factor IX. EMBO J 3: 1053, 1984.
6. Jaye M, de la Salle H, Schamber F, Balland A, Kohli V, Findeli A, Tolstoshev P, Lecocq JP: Isolation of a human anti-haemophilic factor IX cDNA using a unique 52-base synthetic oligonucleotide probe deduced from the amino acid sequence of bovine factor IX. Nucleic Acids Res 11: 2325, 1983.
7. Jagadeeswaran P, Lavelle DE, Kaul R, Mohandas T, Warren ST: Isolation and characterization of human factor IX cDNA: identification of TaqI polymorphism and regional assignment. Somatic Cell Mol Genet 10: 465, 1984.
8. Kurachi K, Davie EW: Isolation and characterization of a cDNA coding for human factor IX. Proc Natl Acad Sci USA 79: 6461, 1982.
9. McGraw RA, Davis LM, Noyes CM, Lundblad RL, Roberts HR, Graham JB, Stafford DW: Evidence for a prevalent dimorphism in the activation peptide of human coagulation factor IX. Proc Natl Acad Sci USA 82: 2847, 1985.
10. Salier J-P, Hirosawa S, Kurachi K: Functional characterization of the 5'-regulatory region of human factor IX gene. J Biol Chem 265: 7062, 1990.
11. Watson MEE: Compilation of published signal sequences. Nucleic Acids Res 12: 5145, 1984.
12. Blobel G, Walter P, Chang CN, Goldman BM, Erickson, AH, Lingappa R: Translocation of proteins across membranes: the signal hypothesis and beyond. In Hopkin CR, Duncan CJ (Eds) Secretory Mechanisms, vol 33. London, Cambridge Univ Press, 1979, p 9.
13. Bentley AK, Rees DJG, Rizza C, Brownlee GG: Defective propeptide processing of blood clotting factor IX caused by mutation of arginine to glutamine at position -4. Cell 45: 343, 1986.
14. Diuguid DL, Rabiet M-J, Furie BC, Liebman HA, Furie B: Molecular basis of hemophilia B: a defective enzyme due to an unprocessed propeptide is caused by a point mutation in the factor IX precursor. Proc Natl Acad Sci USA 83: 5803, 1986.
15. Gianelli F, Green PM, High K, Lozier DP, Lillicrap DP, Ludwig M, Olek K, Reitsma PH, Goossens M, Yoshitake A, Sommer S, Brownlee GG: Haemophilia B database of point mutations and short additions and deletions. Nucleic Acids Res *in press*.
16. Jorgensen MJ, Cantor AB, Furie BC, Brown CL, Shoemaker CB, Furie B: Recognition site directing vitamin K-dependent γ-carboxylation resides on the propeptide of factor IX. Cell 48: 185, 1987.
17. Kaufman RJ, Wasley LC, Furie BC, Furie B, Shoemaker CB: Expression, purification, and characterization of recombinant γ-carboxylated factor IX synthesized in chinese hamster ovary cells. J Biol Chem 261: 9622, 1986.
18. Foster DC, Rudinski MS, Shach BG, Berkner KL, Kumar AA, Hagen FS, Sprecher CA, Insley MY, Davie EW. Propeptide of human protein C is necessary for γ-carboxylation. Biochemistry 26: 7003, 1987.
19. Suttie JW, Hoskins JA, Engelke J, Hopfgartner A, Ehrlich H, Bang NU, Belagaje RM, Schoner B, Long GL: Vitamin K-dependent carboxylase: possible role of the substrate 'Propeptide' as an intracellular recognition site. Proc Natl Acad Sci USA 84: 634, 1987.
20. Suttie JW: Vitamin K-dependent carboxylase. Annu Rev Biochem 54: 459, 1985.

21. Furie B, Furie BC: Molecular basis of vitamin K-dependent γ-carboxylation. Blood 75: 1753, 1990.
22. Bajaj SP: Cooperative Ca^{2+} binding to human factor IX. J Biol Chem 257: 4127, 1982.
23. Rees DJG, Jones IM, Handford PA, Walter SJ, Esnouf MP, Smith KJ, Brownlee GG: The role of β-hydroxyaspartate and adjacent carboxylate residues in the first EGF domain of human factor IX. EMBO J 7: 2053, 1988.
24. Handford PA, Baron M, Mayhew M, Willis A, Beesley T, Brownlee GG, Campbell ID: The first EGF-like domain from human factor IX contains a high-affinity calcium binding site. EMBO J 9: 475, 1990.
25. Fernlund P, Stenflo J: β-hydroxyaspartic acid in vitamin K-dependent proteins. J Biol Chem 258:12509, 1983.
26. Yoshitake S, Schach BG, Foster DC, Davie EW, Kurachi K: Nucleotide sequence of the gene for human factor IX (antihemophilic factor B). Biochemistry 24:3736, 1985.
27. Leytus SP, Foster DC, Kurachi K, Davie EW: Gene for factor X: a blood coagulation factor whose gene organization is essentially identical with that of factor IX and protein C. Biochemistry 25: 5098, 1986.
28. O'Hara PJ, Grant FJ, Haldeman BA, Insley MY, Murray MJ: Nucleotide sequence of the gene coding for human factor VII, a vitamin K-dependent protein participating in blood coagulation. Proc Natl Acad Sci USA 84: 5158, 1987.
29. Plutzky J, Hoskins JA, Long GL, Crabtree GR: Evolution and organization of the human protein C gene. Proc Natl Acad Sci USA 83: 546, 1986.
30. Foster DC, Yoshitake S, Davie EW: The nucleotide sequence of the gene for human protein C. Proc Natl Acad Sci USA 82: 4673, 1985.
31. Degen SJF, Davie EW: Nucleotide sequence of the gene for human prothrombin. Biochemistry 26: 6165, 1987.
32. Irwin DM, Robertson KA, MacGillivray RTA: Structure and evolution of the bovine prothrombin gene. J Mol Biol 200: 31, 1988.
33. Whittaker DL, Copeland DL, Graham JB: Linkage of color blindness to hemophilias A and B. Am J Hum Genet 14: 149, 1962.
34. Chance PF, Dyer KA, Kurachi K, Yoshitake S, Ropers H-H, Wieacker P, Gartler SM: Regional localization of the human factor IX gene by molecular hybridization. Hum Genet 65: 207, 1983.
35. Camerino G, Grzeschik KH, Jaye M, de la Salle H, Tolstoshev P, Lecocq JP, Heilig R, Mandel JL: Regional localization on the human X chromosome and polymorphism of the coagulation factor IX gene (hemophilia B locus). Proc Natl Acad Sci USA 81: 498, 1984.
36. Purello M, Alhadeff B, Esposito D, Szabo P, Rocchi M, Truett M, Masiarz F, Siniscalco M: The human genes for hemophilia A and hemophilia B flank the X chromsosome fragile site at Xq27.3. EMBO J 4: 725, 1985.
37. Mattei MG, Baetman MA, Heilig R, Oberle I, Davies K, Mandel JL, Mattei JF: Localization by in situ hybridization of the coagulation factor IX gene and of two polymorphic DNA probes with respect to the fragile X site. Hum Genet 69: 327, 1985.
38. Rocchi M, Roncuzzi L, Santamaria R, Sparra D, Mochi M, Archidacono N, Covone A, Cortese R, Romeo G: Mapping of coagulation factor protein C and factor X on chromosome 2 and 13, respectively. Cytogenet Cell Genet 40: 734, 1985.
39. Scambler P, Williamson R: The structural gene for human coagulation factor X is located on chromosome 13q34. Cytogenet Cell Genet 39: 231, 1985.
40. Royle NJ, Fung MR, MacGillivray RTA, Hamerton JL: The gene for clotting factor 10 is mapped to 13q32-qter. Cytogenet Cell Genet 41: 185, 1986.
41. Gilgenkrantz S, Briquel M-E, Andre E, Alexandre P, Jalbert P, LeMarec B, Pouzol P, Pommereuil M: Structural genes of coagulation factor VII and factor X located on 13q34. Ann Genet 29: 32, 1986.
42. Kato A, Miura O, Sumi Y, Aoki N: Assignment of the human protein C gene (PROC) to chromosome region 2q14-q21 by *in situ* hybridization. Cytogenet Cell Genet 47: 46, 1988.

43.   Long GL, Marshall A, Gardner JC, Naylor SL: Genes for human vitamin K-dependent plasma proteins C and S are located on chromosomes 2 and 3, respectively. Somat Cell Molec Genet 14: 93, 1988.

44.   Breathnach R, Chambon P: Organization and expression of eukaryotic split genes coding for proteins. Annu Rev Biochem 50: 349, 1981.

45.   Workman JL, Roeder RG: Binding of transcription factor TFIID to the major late promoter during *in vitro* nucleosome assembly potentiates subsequent initiation by RNA polymerase II. Cell 51: 613, 1987.

46.   Bucher P, Trifonov EN: Compilation and analysis of eukaryotic POL II promoter sequences. Nucleic Acids Res 14: 10009, 1986.

47.   Maniatis T, Goodbourn S, Fischer JA: Regulation of inducible and tissue-specific gene expression. Science 236: 1237, 1987.

48.   Crossley M, Brownlee GG: Disruption of a C/EBP binding site in the factor IX promoter is associated with hemophilia B. Nature 345: 444, 1990.

49.   Reitsma PH, Mandalaki T, Kasper CK, Bertina RM, and Briet E: Two novel point mutations correlate with an altered developmental expression of blood coagulation factor IX (hemophilia B Leyden phenotype). Blood 73: 743, 1989.

50.   Gitschier J, Wood WI, Goralka TM, Wion KL, Chen EY, Eaton DH, Vehar GA, Capon DJ, Lawn RM: Characterization of the human factor VIII gene. Nature 312: 326, 1984.

51.   Evans JP, Watzke HH, Ware JL, Stafford DW, High KA: Molecular cloning of a cDNA encoding canine factor IX. Blood 74: 207, 1989.

52.   Briet E, Bertina RM, Van Tilburg NH, Veltkamp JJ: A sex-linked hereditary disorder that improves after puberty. N Eng J Med 306: 788, 1982.

53.   Reitsma PH, Bertina RM, Ploos van Amstel JK, Riemens A, Briet E: The putative factor IX gene promoter in hemophilia B Leyden. Blood 72: 1074, 1988.

54.   Kurachi K, Hirosawa S, Fahner JB, Wu CT, Salier JP: Regulation of human factor IX gene. Thromb Haemostas 62: 154, 1989.

55.   Mulvihill ER, Palmiter RD: Relationship of nuclear estrogen receptor levels in induction of ovalbumin and conalbumin mRNA in chick oviduct. J Biol Chem 252: 2060, 1977.

56.   Lin Y-S, Carey M, Ptashne M, Green MR: How different eukaryotic transcriptional activators can cooperate promiscuously. Nature 345: 359, 1990.

57.   Lin S-W, Dunn JJ, Studier FW, Stafford DW: Expression of human factor IX and its subfragments in *Escherichia coli* and generation of antibodies to the subfragments. Biochemistry 26: 5267, 1987.

58.   Anson DS, Austen DEG, Brownlee GG: Expression of active human clotting factor IX from recombinant DNA clones in mammalian cells. Nature 315: 683, 1985.

59.   De la Salle H, Altenberger W, Elkaim R, Dott K, Dieterle A, Drillien R, Cazenave J-P, Tolstoshev P, Lecocq J-P: Active γ-carboxylated factor IX expressed using recombinant DNA techniques. Nature 316: 268, 1985.

60.   Busby S, Kumar A, Joseph M, Halfpap L, Insley M, Berkner K, Kurachi K, Woodbury R: Expression of active human factor IX in transfected cells. Nature 316: 271, 1985.

61.   Choo KH, Raphael K, McAdam W, Peterson MG: Expression of active human blood clotting factor IX in transgenic mice: use of a cDNA with complete mRNA sequence. Nucleic Acids Res 15: 871, 1987.

62.   St.Louis D, Verma IM: An alternative approach to somatic cell gene therapy. Proc Natl Acad Sci USA 85: 3150, 1988.

63.   Palmer TD, Thompson AR, Miller AD: Production of human factor IX in animals by genetically modified skin fibroblasts: potential therapy for hemophilia B. Blood 73: 438, 1989.

ANTITHROMBIN III GENETICS, STRUCTURE and FUNCTION

Susan Clark Bock

Temple University School of Medicine
Microbiology & Immunology Department and
The Thrombosis Research Center
3400 N. Broad Street, Philadelphia, PA   19140

INTRODUCTION

Antithrombin III (ATIII) is an important endogenous anticoagulant protein which functions at the level of serine protease inhibition. ATIII inactivates thrombin, factor Xa and other enzymes in the intrinsic coagulation pathway, thereby decreasing fibrin formation. Inhibition occurs when stable, stoichiometric ATIII-enzyme complexes form as a result of interactions between the reactive site of ATIII and the active site of the protease target.[1] The rate of complex formation increases substantially in the presence of heparan sulfate proteoglycans on the surface of the vascular endothelium *in vivo*,[2] or after addition of heparin *in vitro* or pharmaceutically.[1]

A structural description of antithrombin III has emerged from investigations at the protein, carbohydrate and cDNA levels. Biochemical, biophysical, kinetic and genetic studies have also provided insights about structure/function relationships in this important anticoagulant protein. The goal of this chapter is to review how genetic information derived from comparisons of normal antithrombin III with mutant ATIIIs and other members of the serpin gene family has contributed to understanding structure/function relationships in the reactive center and heparin binding regions of ATIII.

BACKGROUND

Human antithrombin III is a 58,000 dalton glycoprotein consisting of a single chain 432 amino acid polypeptide with six cysteines forming three disulfide bridges.[3,4,5] Most circulating ATIII molecules have been posttranslationally modified to contain four glucosamine-based, asparagine-linked oligosaccharides, which account for 15% of the molecular mass.[6,7]   ATIII is synthesized in the liver and has a half life of almost three days in the circulation.[8] The plasma concentration of ATIII is approximately 125ug/ml[9] and does not vary widely among normal people. There is an increased risk of thrombosis when ATIII levels are subnormal.[10]

ATIII works as an anticoagulant by forming stable complexes with thrombin, factor Xa and other serine proteases of the intrinsic pathway. Complex formation involves interactions between the reactive site of antithrombin III and the active sites of its target proteases (see Fig. 1). The anticoagulant activity of antithrombin III is substantially increased by exposure to heparan sulfate proteoglycans on the vascular endothelium[2] and by pharmaceutical application of heparin in clinical settings. Binding of these anionic polysaccharide cofactors to ATIII induces a conformational change in the inhibitor, which is then able to form complexes with target proteases at rates as much as 1000-fold faster than the "progressive" rate.[1] Complexes composed of antithrombin III and inactive target enzyme are finally cleared from the circulation by hepatic receptors that recognize neoepitopes present in the complex, but not its precursor molecules.[11]

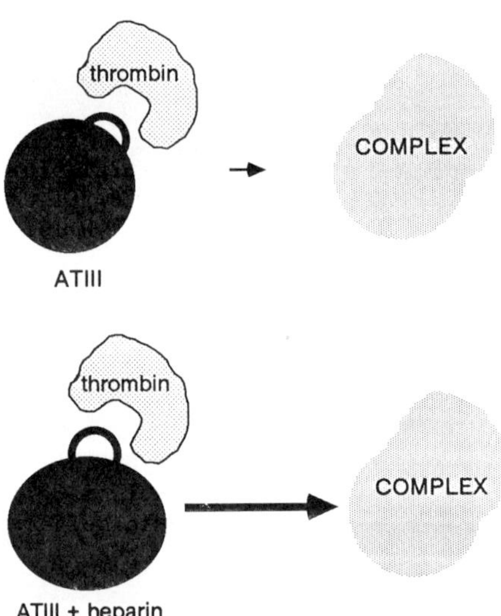

Fig. 1.   *Top:* Antithrombin III inactivates thrombin by presenting a substrate-like sequence to the active site of the enzyme and forming a stable inhibitor-protease complex with it. ATIII also inactivates factors Xa, IXa, XIa and XIIa by a similar mechanism. *Bottom:* Heparin (and heparan sulfate) induce a conformational change in ATIII and increase the rate of complex formation by more than a thousand fold.

ATIII IS A MEMBER OF THE SERPIN GENE FAMILY

Antithrombin III is a member of the serpin (<u>ser</u>ine protease <u>in</u>hibitor) gene family.[12] Inclusion in this group has promoted understanding of ATIII structure/function relationships greatly, particularly with respect to providing a working model of its tertiary structure and a general description of the protease inhibition mechanism.

More than twenty different proteins have been assigned to the serpin gene family by analysis of protein, cDNA and gene sequences. Members of the serpin family began to diverge from a common ancestor more than 500 million years so that representatives of this group are now found widely in nature, and serpins of animal, plant, viral and plasmid origin are known. The most extensively studied serpins come from mammalian blood plasma and include antithrombin III, α1-antitrypsin, C1̄ inhibitor, α2-antiplasmin, PAI-1 (plasminogen activator inhibitor-1), PAI-2, protein C inhibitor, α1-antichymotrypsin and heparin cofactor II. All serpins in this group are functional protease inhibitors in contrast to other members of the serpin family which have no demonstrated protease inhibitor activity (e.g., ovalbumin, angiotensinogen, and thyroxine and corticosteroid binding globulins).

Alignment of serpin protein sequences shows that they consist of a common, homologous region to which nonrelated domains may be attached at the amino and/or carboxyl termini. The nonhomologous domains contribute to the individual character of each serpin. For example, nonhomologous sequences in the amino terminal extension of ATIII participate in heparin binding (discussed below), while nonhomologous sequences in the carboxy-terminal extension of α2-antiplasmin contain a high affinity binding site for plasmin.[13,14]

The homologous region which is common to all serpins extends over about 350 amino acid residues. The overall extent of homology between different members of the serpin family is in the 25-50% range, and conserved amino acids forming the core framework of these molecules are interspersed with elements that vary in sequence and length to form surface structures.

The reactive site region of a serpin is located towards the carboxyl terminal end of the polypeptide sequence. For those serpins which function as protease inhibitors, the reactive site contains a substrate-like sequence that is recognized by its target protease(s). The reactive site is thought to protrude as a loop from the globular, native serpin molecule. An amino acid in the reactive site which is identical or chemically similar to the cleavage site of the serpin's target protease is called its P1 residue. Positions amino-terminal to the P1 residue are numbered P2, P3 and PN, and positions carboxy terminal to it P1', P2', P3', etc. A carboxylic ester bond is thought to form between the P1 residue of a serpin and the serine hydroxyl of its target protease during complex formation.[15,16,17]

α1-ANTITRYPSIN MODEL FOR ATIII TERTIARY STRUCTURE

Understanding of antithrombin III structure/function relationships has benefited greatly from its inclusion in the serpin gene family, particularly with respect to providing a three dimensional working model. Models of serpin tertiary structure are based on the tertiary structure of the prototypical serpin, α1-antitrypsin. The three angstrom crystal structure for a modified form of α1-antitrypsin was solved by Loebermann et al.[18] The ordered portion of this structure corresponds almost exactly to the region of α1-antitrypsin which is homologous to other serpin family members. The α1-antitrypsin molecules which crystallized were cleaved at their reactive sites. Therefore, the P1 and P1' residues, which formed a peptide bond in the native inhibitor, are separated by 69 angstroms in the structure of cleaved, post-complex α1-antitrypsin.

However, many types of evidence indicate that the structure of this cleaved α1-antitrypsin molecule is a valid model for the tertiary structures of other cleaved serpins, including antithrombin III.[19] For example, many amino acids that are conserved or invariant in serpin protein sequences correspond to structurally important hydrophobic, hydrogen bonding, proline and salt bridge residues in the crystal structure, and would be expected to generate the same folded molecule for other members of the family. Furthermore, the positions of mapped disulfide bridges in antithrombin III and PAI-2 are compatible with the α1-antitrypsin crystal structure. Additionally, serpin sites of carbohydrate attachment which have been chemically determined or deduced from sequence analyses map to the surface of the α1-antitrypsin crystal structure.

The validity of using the α1-antitrypsin structure as a model for ATIII structure is also supported by a recent study indicating that crystallographic parameters for bovine antithrombin are similar to those determined for human α1-antitrypsin.[20] It should be noted, however, that although disulfide information[3] indicates linkage of ATIII residues C8 and C128 (α1-antitrypsin D helix) and C21 and C95 (between α1-antitrypsin B and C helices), the α1-antitrypsin structure and presently available information from bovine antithrombin III crystals do not provide further details about this interesting region.

DYSFUNCTIONAL ANTITHROMBIN III MUTANTS

Genetically abnormal antithrombin III variants provided by nature have contributed significantly to understanding of ATIII structure/function relationships. Many of these mutant molecules were identified because their carriers suffered from hereditary thrombosis. Although genetic ATIII deficiencies are rare, and the subclass of patients who make variant ATIIIs is rarer still, episodes of thrombosis in these individuals lead to their recognition, and investigation of the affected proteins or genes. In other cases, dysfunctional ATIII variants that are not usually associated with symptomatic clinical thrombosis in the heterozygous state were identified

through screening programs. Fig. 2 summarizes the locations
and types amino acid substitutions present in dysfunctional
antithrombins from 31 independently ascertained kindred.[21-45,70]

Two important facts emerge from consideration of the data
in Fig 2. The first is that mutations have recurred over and
again at certain "hot spots" in the ATIII molecule. The
second is that groups of mutations cluster in certain areas of
the ATIII molecule. Patients carrying substitutions mapping to
a given cluster exhibit a common phenotype, suggesting that
the areas defined by these clustered, mutationally sensitive
residues represent distinct ATIII structure/function elements.

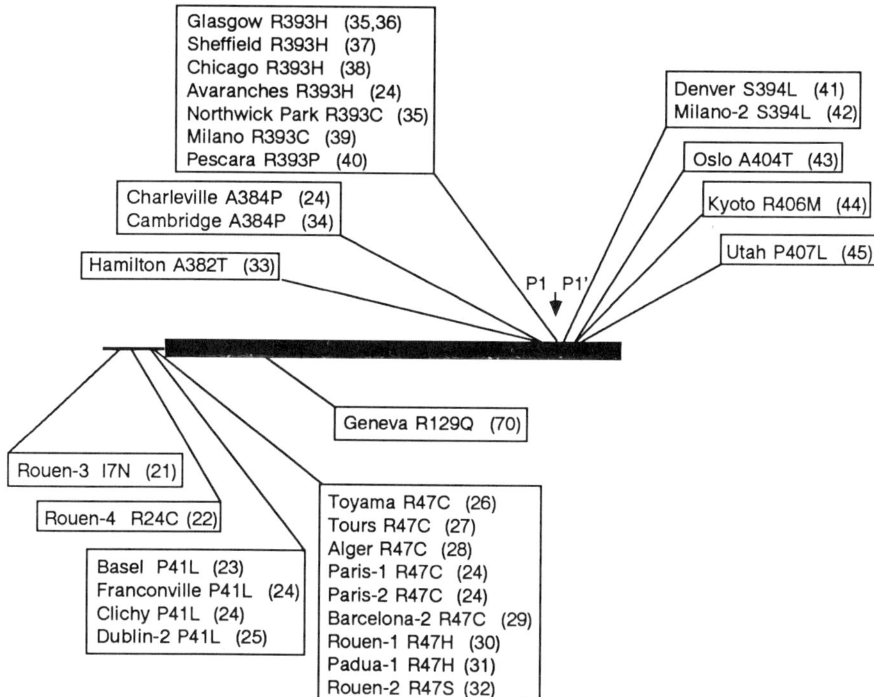

Fig.2   *Dysfunctional antithrombin IIIs.*   The
polypeptide chain of antithrombin III is depicted
in linear fashion with the thick line representing
the domain that is homologous to other members of
the serpin gene family and the thin line indicating
the nonhomologous domain. The arrow points to the
peptide bond between the P1 and P1' reactive site
residues.   The positions of amino acid
substitutions present in dysfunctional molecules
from 31 independently ascertained kindred are
noted. Substitution mutations are indicated by a
three part code in which the first letter and
number signify the identity and position of the
normally occurring residue, and the last letter
indicates the substituted amino acid. Each box
corresponds to an amino acid residue at which one
or several substitutions have occurred. Numbers in
parentheses indicate bibliographic reference
number.

FOUR MUTATIONAL HOTSPOTS IN THE ATIII GENE CONTAIN CpG DINUCLEOTIDES

Fig. 2 shows four mutational hotspots in the ATIII gene. These hotspots are proline-41 (four leucine substitutions), arginine-47 (six cysteine, two histidine and one serine substitution)[46], arg-393 (four histidine, two cysteine and one proline substitution), and serine-394 (two leucine substitutions). Due to the presence of CpG dinucleotides at these sites, it is likely that the observed mutations represent independent events. CpG dinucleotides are substrates for methylation in vertebrate genomes, and the cytosines contained in such sequences are often methylated at the 5 position. Deamination of 5-methyl-cytosine generates the naturally occurring base, thymine, which is not recognized as unusual by repair systems and therefore not excised or corrected.[47] This leads to systematic substitution of CGX-encoded arginines by histidines and cysteines, as observed at residues 47 and 393, and is also consistent with the proline to leucine substitutions and serine to leucine substitutions which have recurred at residues 41 and 394, respectively. The histidine and cysteine substitutions at arginine-393, the P1 residue of ATIII, are also notable in that they reflect the particular sensitivity of CGX-encoded arginines at thrombin cleavage sites to mutational inactivation. A high incidence of similar substitutions occurs at residue 16 in the Aα chain of fibrinogen, where twenty independent histidine and eleven independent cysteine substitutions have also been reported.[48]

CLUSTERS OF MUTATIONS IDENTIFY FUNCTIONALLY IMPORTANT STRUCTURAL ELEMENTS

Fig. 2 also shows that mutationally sensitive amino acids cluster at several sites within the antithrombin III molecule. Variants carrying mutations that map to a given cluster exhibit similar functional deficits, and the nature of these defects can be used to deduce the role of the structural element in question. The mutants shown in Fig. 2 can be divided into three groups. Those in the nonhomologous amino terminal domain (residue 7, 24, 41 and 47 mutants) define a region of ATIII which is involved in heparin binding and conferring heparin cofactor activity. A second group of mutations near the carboxy terminal end of the serpin homology region (residue 382, 384, 393 and 394 mutants) identifies the reactive center; reactive center mutations disrupt the ability of ATIII to form stable complexes with its protease targets. Finally, the third cluster of mutants (residue 404, 406 and 407 mutants) is also near the carboxy terminus and defines an element whose structural integrity is necessary to obtain normal levels of circulating antithrombin III.

GENETIC EVIDENCE FOR A HEPARIN BINDING REGION IN THE NONHOMOLOGOUS AMINO TERMINAL DOMAIN OF ATIII

Variant antithrombins with amino acid substitutions at residues 7, 24, 41 and 47 in the nonhomologous, amino terminal domain exhibit normal progressive ATIII activity, but decreased heparin binding and heparin cofactor activity. Heterozyous carriers of these mutations do not usually experience clinical thrombosis. The normal progressive ATIII activity of these variants is presumably responsible for the recessive inheritance and lower incidence of thrombosis in these families.

Clustering of heparin binding mutants indicates that the amino terminus of ATIII participates in heparin binding and heparin cofactor activation of the anticoagulant reaction. Heparin is a negatively charged, sulfated polysaccharide and the salt dependence of its *in vitro* binding to ATIII suggests a role for ionic interactions between its negatively charged groups and positively charged groups on ATIII. The nature of the substitutions present in ATIII variants with heparin binding defects are consistent with this model. For example, positively charged arginine residues are abolished in the arginine-24 and arginine-47 mutants.[22,24,26-32] Similarly, leucine substitutions at proline-41 may alter the conformation of this region thereby preventing heparin interactions.[23-25] Finally, substitution of an asparagine residue for isoleucine-7 introduces a new glycosylation site. Attachment of a carbohydrate prosthetic group at the newly created N-glycosylation consensus sequence could sterically interfere with heparin binding in ATIII Rouen-3.[21]

Because residues 7, 24, 41 and 47 are contained in the nonhomologous amino terminal extension of antithrombin III, it is not possible to precisely locate them in the α1-antitrypsin tertiary structure model. However, the general location and attachment points of the nonhomologous amino terminal extension to the serpin core of ATIII can be deduced since it immediately precedes helix A in the primary structure and disulfide bonds link ATIII cysteines 8 and 128 and 21 and 95. Cysteine 128 maps to helix D and cysteine 95 maps between the B and C helices of α1-antitrypsin. This information places the nonhomologous heparin binding polypeptide near to tryptophan 49 and helix D, which have been proposed as part of the heparin binding site based on evidence from chemical modification experiments,[49] conservation of helix D basic residues in ATIII and heparin cofactor II,[50] and molecular analysis of ATIII-β, an ATIII isoform which is not glycosylated at asn-135 and which exhibits higher affinity heparin binding and an increased rate acceleration factor in the presence of the cofactor.[51] The helix D region of ATIII contains a high density of positively charged amino acids which could interact with negatively charged sulfate groups on heparin. There is chemical modification evidence that lysines 107, 114, 125 and 136 participate in heparin binding[52-54,] and the recent identification of ATIII Geneva[70] provides genetic evidence that arginine 129 plays an important role in the heparin binding site of ATIII. The contributions of other basic residues in the helix D region (arginine 132, and lysines 133 and 139) have yet to be established.

GENETIC DISSECTION OF STRUCTURAL ELEMENTS IN THE REACTIVE SITE OF ANTITHROMBIN III

Fifteen amino acid substitution mutations map to the reactive site region near the carboxy terminus of ATIII (see Fig. 2). These mutants can be divided into two groups. Variants from one group are present at normal concentrations, but defective in their interactions with thrombin, while variants from the other group are present at severely reduced concentrations in patient plasma. The sites defined by mutations in the first group can be further subdivided into the functionally distinct P1 residue, P1' residue, and the P12-region. The group of mutants causing the quantitative defect map to the P11'-region.

GENETIC EVIDENCE FOR THE IMPORTANCE OF THE P1 RESIDUE

Arginine-393 is the P1 residue of antithrombin III. As discussed previously, the triplet encoding arginine-393 contains a CpG dinucleotide in the first two positions, leading to a high rate of histidine and cysteine substitutions (four and two independently ascertained kindred, respectively, see Fig. 2). A proline substitution at the P1 arginine of ATIII Pescara has also been identified.

Dysfunctional R393H, R393C, and R393P ATIII molecules lack progressive ATIII and heparin cofactor activity. This loss of protease inhibitor activity is in accordance with the loss of an arginine residue in the thrombin recognition sequence of the reactive site loop. Arginine residues are characteristically present as the P1 amino acids of thrombin cleavage sites and play a major role in the recognition of substrate sequences by this enzyme. For example, substitution of an arginine for the P1 methionine residue of α1-antitrypsin-Pittsburg changed it from an elastase inhibitor to a thrombin inhibitor.[55] In addition to loss of their P1 arginines, the R393C mutants ATIII Northwick Park and ATIII Milano may have further steric blockage of their reactive sites regions by disulfide bonding of the substituted cysteine residue to serum albumin.

The affinity of P1 substitution mutants for heparin varies. The R393C and R393P variants bind heparin normally, while R393H variant molecules isolated from patients of four different kindred can be classified on the basis of two different kinds of behavior on heparin-Sepharose chromatography. ATIIIs Sheffield and Avaranches[24,37] coelute with normal antithrombin III, but ATIIIs Glasgow and Chicago[35,36,38] require higher salt concentrations, reflecting increased heparin affinity. These differences in the heparin affinity of the R393H molecules may reflect the presence of additional as yet undetected mutation(s) (full length sequence analysis was not performed for any of the variants), or could result from differences in chromatographic conditions. For instance, pH differences that influence protonation of the substituted histidine residue could lead to different elution profiles. Nevertheless, on the basis of the increased affinity of ATIIIs Glasgow and Chicago for heparin, it has been hypothesized that the status of the reactive center region can influence the status of the heparin binding region.[36]

P1′ RESIDUE REQUIREMENTS

The P1′ residues of serpins are fairly well conserved. Since many serpins have a serine or threonine in this position, it was thought that the side chain hydroxyl group might play a critical mechanistic role in inhibitor function. This hypothesis received initial support from studies on dysfunctional ATIII Denver, a leucine for serine P1′ substitution.[41] However, further investigations using *in vitro* generated mutants with proline, methionine, cysteine, valine, leucine, glycine, alanine or threonine P1′ substitutions showed that the hydroxyl group was not critical for functional activity since thrombin inhibitory activity was retained in the glycine and alanine mutants.[56] Instead, there seems to be a side chain size optimum which is further mediated by hydrophobic effects.

HIGHLY CONSERVED SEQUENCES FLANKING SERPIN REACTIVE SITES SPECIFY THE MUTATIONALLY SENSITIVE AND FUNCTIONALLY IMPORTANT P12- AND P11′-REGIONS

Fig. 3 shows an alignment of serpin sequences in and around the reactive site. Sequences in the reactive site loops of serpins (center portion of diagram) vary greatly in sequence and length, but are flanked by highly conserved regions (indicated by * and + symbols over the alignment) which are functionally important and mutationally sensitive. We will refer to these critical structural elements as the P12- and P11′- regions, based on the positions of key invariant residues in the reference serpin, α1-antitrypsin.

Mutations leading to the production of dysfunctional serpins cluster in the P12- and P11′-regions, and patients carrying dysfunctional mutants from each cluster exhibit *distinct* disease phenotypes. Since patients with mutations at homologous residues of *different* serpins, including ATIII, produce molecules that are defective in similar ways, the P12- and P11′-regions appear to specify general (as opposed to individualized) properties of serpins. Patients with mutations in the P12-region make normal amounts of serpin molecules that act as substrates, rather than inhibitors, of their target proteases. Those with mutations in the P11′-region have only trace amounts of the variant serpin in their blood.

```
          342            350        358     360            370
           Z
            <        sheet 4A       >        <sheet 1C> <      sheet4B       >
          *     *     *  +  +     +                       +    *  *  +  +  +  +
                      P12 P10  P8        P1        P1'          P11'
A1AT:   E K G T E A A G A M FLEAIPM/------SIPPEVK F N K P F V F L M I E Q N
                                                  k         L               d

ATIII:  E E G S E A A A S T AVVIAGR/---SLNPNRVTFK A N R P F L V F I R E V P
                  T   P                           T   M L

C1inh:  E T G V E A A A A S AISVAR/--------TLLVFE V Q Q P F L F M L W D Q Q
                  E   T                                       v

a2AP:   E V G V E A A A A T SIAMSR/-------MSLSSFS V N R P F L F F I F E D T
                  -AAAAA-

pCi:    E S G T R A A A A T GTIFTFR/---SARLNSQRLV F N R P F L M F I V D N N
PAI-1:  E S G T V A S S S T AVIVSAR/------MAPEEII M D R P F L F V V R H N P
PAI-2:  E E G T E A A A G T GGVMTGR/----TGHGGPQFV A D H P F L F L I M H K I

hcII:   E E G T Q A T T V T TVGFMPL/------STQVRFT V D R P F L F L I Y E H R
A1ACT:  E E G T E A S A A T AVKITLL/--SALVETRTIVR F N R P F L M I I V P T D
ctpsn:  E T G T E A A A A T GVIGGIRK/-----AILPAVH F N R P F L F V I Y H T S

bpZ:    E E G T E A G A A T VAMGVAM--SMPLKVDLVDFY A N H P F L F L I R E D I
ova:    E A G R E V V G S A EAGVDAA-------SVSEEFR A D H P F L F C I K H I A
AGTH:   E A D E R E P T E S TQQLNKP--------EVLEVT L N R P F L F A V Y D Q S

TBG:    E K G T E A A A V P EVELSDQ---PENTFLHPIIQ I D R S F M L L I L E R S
CBG:    E E G V D T A G S T GVTLNLT-------SKPIILR F N Q P F I I M I F D H F
```

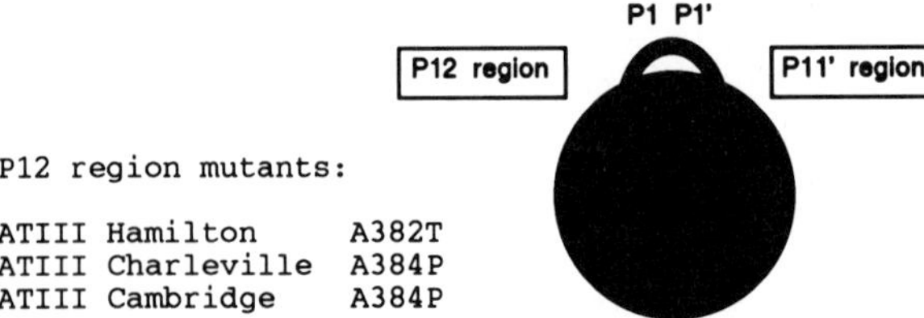

P12 region mutants:

ATIII Hamilton   A382T
ATIII Charleville   A384P
ATIII Cambridge   A384P

C1inh Ma   A434E
C1inh Ca   A436T
C1inh Mo   A436T

a2AP Enschede   insertion of an
                additional alanine
                between A354 and A357

P11' region mutants:

a1AT Heerlen   P369L
ATIII Oslo   A404T
ATIII Kyoto   R406M
ATIII Utah   P407L

Normal variants:

a1AT Christchurch E363K
a1AT M3   E376D
C1inh 458 V or M

FIG. 3

Fig. 3. The P12- and P11'-regions are highly
conserved, mutationally sensitive structural
elements on either side of the reactive site loop.
Top: Aligned amino acid sequences of 15 serpin
family members. Numbers refer to α1-antitrypsin
primary structure and sheet strand designations to
the tertiary structure of cleaved α1-antitrypsin.[18]
(*) appears above positions that are invariant in
α1-antitrypsin and serpins *with inhibitor function*,
and (+) above positions that are highly conserved.
The positions of the P12, P10, P8, P1, P1' and P11'
residues of α1-antitrypsin are marked above the
alignment. Dashes indicate gaps. Slashes mark the
peptide bond between the P1 and P1' residues in
serpins known to function as inhibitors. Positions
where substitution mutations have resulted in
dysfunctional serpins are indicated with upper case
letters below the normal amino acid, and positions
where substitutions generate functionally and
quantitatively normal variants are indicated with
lower case letters. The sequence -AAAAA- indicates
the region of α2-antiplasmin Enschede where
insertion of one alanine residue has occurred.[59] The
P12, P10 and P8 residues proposed to play a
critical role in determining inhibitor/substrate
status are shaded. Serpin abbreviations: a1AT, α1-
antitrypsin; ATIII, antithrombin III; C1inh, C$\overline{1}$
inhibitor; a2AP, α2-antiplasmin; pCi, protein C
inhibitor; PAI-1, endothelial plasminogen activator
inhibitor; PAI-2, placental plasminogen activator
inhibitor; HcII, heparin cofactor II; a1ACT, α1-
antichymotrypsin; ctpsn, mouse contrapsin; bpZ,
barley protein Z; ova, ovalbumin; AGTH,
angiotensinogen; TBG, thyroxine binding globulin;
CBG, cortisol binding globulin. See references 19
and 57 for sources of the original amino acid
sequences.
Center: The relative positions of the P1 and P1'
residues and the P12- and P11'-regions are
schematically indicated on a diagram in which a
native serpin is illustrated as a globular protein
molecule with an exposed reactive center loop.
Bottom: P12-region mutants are listed on the left,
and P11'-region mutants and flanking normal
variants on the right.

GENETIC EVIDENCE THAT THE P12-REGION DETERMINES INHIBITOR/SUBSTRATE STATUS

The P12-region appears to specify a structural determinant which enables a serpin to be an inhibitor, rather than a substrate, of its target protease(s). Substitution mutations at the P12[a] and P10 residues of ATIII and C1̄ inhibitor have produced dysfunctional serpins in six different kindred, and insertion of an extra alanine in this same region lead to α2-antiplasmin Enschede. In those families for whom studies have been performed, the P12-region variant proteins behaved as substrates, rather than inhibitors, of their respective target proteases.

It is notable that the mutationally sensitive amino acids (P12, P10 and P8) in the conserved P12-region alternate (see Fig. 3). In serpins which are functional protease inhibitors, the P12 residue is an invariant alanine. However, the P12 amino acids of ATIII Hamilton[33] and C1̄ inhibitor Ma (K. Skriver, W. Wikoff, A.P. Kaplan and S.C. Bock, unpublished results) have been changed to threonine and glutamic acid, respectively. The P10 alanines of ATIII and C1̄ inhibitor are similarly replaced by prolines in ATIIIs Charleville and Cambridge[24,34] and threonines in C1̄ inhibitors Ca and Mo.[58] Finally, α2-antiplasmin Enschede,[59] which results from insertion of a fifth alanine in the ala-ala-ala-ala sequence normally present between the P12 and P9 positions of α2-antiplasmin, can be viewed as a P8 mutant because an effect of this single amino acid insertion is to displace the threonine usually present in the P8 position.

Alternation of mutationally sensitive residues in the P12-region suggests that a β-pleated sheet structure is involved in determining the inhibitor versus substrate properties of serpins. The P12, P10 and P8 residues are located in sheet strand 4A in the tertiary structure of cleaved α1-antitrypsin, and the side chains of these key amino acids map to the core of the post-complex molecule (Fig. 4). The internal locations of the P12, P10 and P8 residues in the cleaved, post-complex serpin contrast, however, with their disposition in the *native* serpin. In native inhibitors, the P12-, P10- and P8-containing polypeptide chain is the amino terminal half of the reactive site loop which is exposed on the surface of the molecule (see Fig. 1). We propose that transferring the P12, P10 and P8 residues from exposed positions on the reactive site loop to internal positions under sheet A generates a special protein conformation in which the tetrahedral or acyl enzyme intermediate present in complexes between serpins and their target proteases is stabilized.

---

[a] Subsite numbering for all serpins in this discussion is based on the position of corresponding α1-antitrypsin residues in the Fig. 3 alignment.

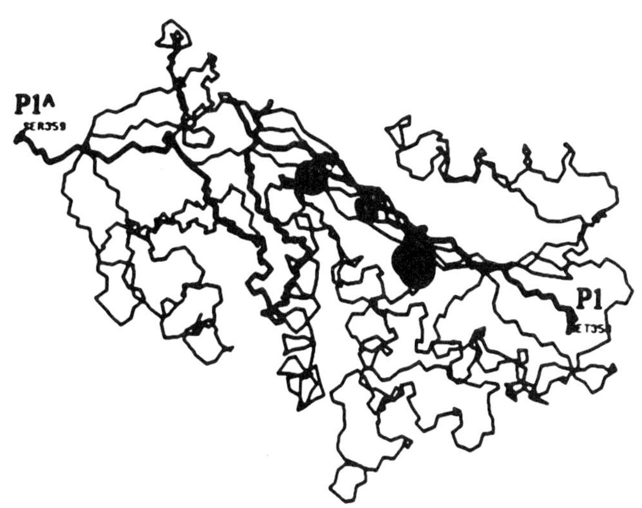

FIG. 4.   Tertiary structure of cleaved α1-antitrypsin.[18] The P1 and P1' residues are separated by 67 angstroms in the crystal structure of post-complex α1-antitrypsin, but form a peptide bond in the intact inhibitor. Sheet A is approximately orthoganol to the page and extends from the top left to the lower right of the illustrated molecule. Polypeptide chains for the (P1'-containing) activation peptide and sheet strand 4A are traced with bold lines.

Genetic, structural, biophysical and immunochemical evidence suggests internalization of the P12 , P10 and P8 sidechains is critical for expressing serpin protease inhibitor activity. Van der waals surfaces for the sidechains of these amino acids are shown; the P12 is on the upper left, the P10 in the middle and the P8 at the lower right.

When the P12, P10 or P8 residues can not be incorporated into the core of a serpin molecule, substrate-like cleavage of the reactive site loop occurs instead. This phenomenon is observed in the six ATIII and C$\overline{1}$ inhibitor variants which have substitutions of larger or less hydrophobic residues (proline, threonine and glutamic acid) for their P12 and P10 alanines and in the α2-antiplasmin Enschede variant which affects the P8 residue (see Fig. 3), and is consistent with predictions about structural requirements for the P12, P10 and P8 residues based on analysis of α1-antitrypsin tertiary structure and patterns of serpin amino acid conservation.

The post-complex tertiary structure of α1-antitrypsin shows invariant and highly conserved amino acids with bulky, hydrophobic sidechains surrounding the P12 alanine of α1-antitrypsin and forming a small, hydrophobic pocket. The small size and very hydrophobic character of this pocket are maintained in all serpins, and explain why an alanine is present at the P12 position of all serpins which function as protease inhibitors (see Fig. 3). The model predicts that substitution of amino acids with larger and less hydrophobic side chains will prevent incorporation of the P12 residue into this little pocket and block stable complex formation. In accordance with this hypothesis, mutants and noninhibitor serpins with non-alanine P12 residues are substrates rather than inhibitors of their target proteases (e.g. ATIII Hamilton (P12 alanine > threonine), $\overline{\text{C1}}$ inhibitor Ma (P12 alanine > glutamic acid), ovalbumin (P12 valine), angiotensinogen (P12 glutamic acid) and cortisol binding globulin (P12 threonine).

The P10 residue also affects the ability of the polypeptide strand to incorporate into the A sheet. Of the ten plasma serpins known to function as protease inhibitors, seven have alanine P10 residues and one each has a glycine, serine or threonine P10 residue. Mutations at the P10 residues of ATIIIs Charleville and Cambridge (P12 alanine > proline) and $\overline{\text{C1}}$ inhibitors Ca and Mo (P12 alanine > threonine) convert these serpins from protease inhibitors to substrates. This argues that the ability of the P10 sidechain to interact with the underlying A sheet is important in the determination of inhibitor/substrate status. However, in contrast to the situation observed for the invariant P12 residue and its very highly conserved pocket, specifics of the interaction between the P10 sidechain and its pocket may differ from serpin to serpin since variability is noted among P10 residues and some amino acids forming the P10 pocket. The individualized nature of P10 - P10 pocket interactions may explain why serine and threonine residues at the P10 positions of PAI-1 and heparin cofactor II are consistent with inhibitor function, while P10 threonine substitutions in the context of a $\overline{\text{C1}}$ inhibitor framework ($\overline{\text{C1}}$ inhibitors Ca and Mo) convert it into a substrate.

The role which the P8 residue plays in determining whether complex formation or cleavage occurs is suggested by the conservation of a hydroxyl bearing serine or threonine residue at this position in most inhibitory serpins (see Fig. 3), and the behavior of α2-antiplasmin Enschede. Conservation of the hydroxyl group and analysis of the α1-antitrypsin crystal structure suggests it may participate in a functionally important hydrogen bond in serpins which act as inhibitors[b]. The substrate-like behavior of α2-antiplasmin Enschede (which corresponds to insertion of a fifth alanine into the P12-P9 ala-ala-ala-ala sequence) could then be explained as the consequence of displacing the functionally important hydroxyl group on the threonine normally present in the P8 position of α2-antiplasmin to the outward facing side of sheet A.

Thus, we believe that the functional protease inhibitor activity of a serpin derives from the overall ability of its

---

[b] The P8 residue of α1-antitrypsin is methionine rather than threonine or serine. Increased hydrophobic interactions between the methionine side chain and the interior of the α1-antitrypsin molecule may compensate for the hydrogen bond interaction proposed to exist in other inhibitory serpins.

P12-region to internalize and shift the molecule into an inhibitor conformation. The dominant interaction appears to involve the P12 residue, but mutant data indicates the P10 and P8 residues are also important. In addition to the genetic and modeling data presented here, a number of biophysical and immunochemical studies also imply that the P12, P10 and P8 residues play a critical role in determining inhibitor/substrate status. Spectroscopic studies of native and cleaved serpins show that a conformational transition[60,61] characterized by an increase in hydrogen bonds[62] occurs in inhibitory, but not non-inhibitory,[63] serpins upon cleavage of the reactive site loop. A monoclonal antibody which recognizes the P12-P8 residues of ATIII has also been reported to switch the antithrombin III-thrombin reaction from the major pathway of complex formation to an alternative pathway in which the inhibitor is cleaved as a substrate.[64,65]

GENETIC EVIDENCE THAT INTEGRITY OF THE P11'-REGION IS NECESSARY FOR OBTAINING NORMAL CIRCULATING SERPIN LEVELS

Mutations in a highly conserved sequence on the carboxy terminal side of the reactive site severely reduce the circulating level of corresponding serpin gene product. This is illustrated for ATIII-Utah in the Fig. 5.

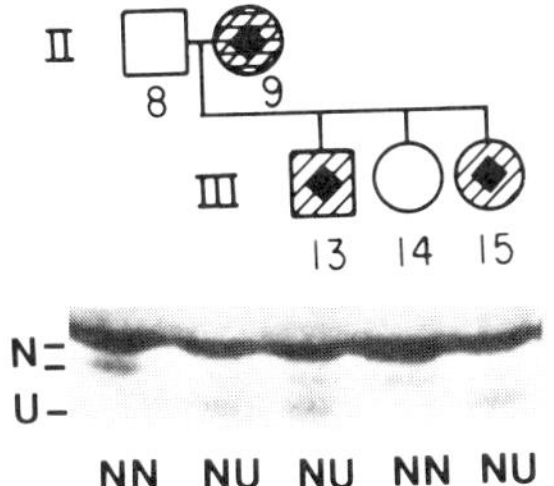

Fig. 5. *Western blot showing low levels of mutant antithrombin in ATIII Utah patients.* The normal man (II-8) who married into the Utah family has normal levels of fully glycosylated antithrombin III (upper N band) and ATIII-β (lower N band), a naturally occurring ATIII isoform which lacks a carbohydrate prosthetic group on asn-135.[51] Plasma samples from his heterozygous ATIII Utah wife (II-9) and affected offspring (III-13 and III-15) contain half normal amounts of normal ATIII and ATIII-β, and a new band (U) corresponding to the ATIII Utah gene product. The circulating level of the ATIII Utah gene product is severely reduced compared to that of the normal ATIII allele. Reprinted from reference 66; see original publication for experimental details.

Similar quantitatively abnormal phenotypes have been observed for three other amino acid substitutions in the P11'-region: ATIII Oslo, ATIII Kyoto and α1-antitrypsin Heerleen.[43,44,67,68] It is interesting to note that ATIII Oslo was actually the first case of genetic ATIII deficiency reported in literature[69] and was thought to be a "classical" ATIII deficiency for many years because the circulating levels of the abnormal protein are so low.

The common phenotype of mutants in the P11' region suggests that structural integrity of this portion[c] of the activation peptide is required for maintaining normal circulating levels of a serpin. The mechanisms by which perturbations in the P11'-region could lead to low circulating variant protein levels include increased intracellular degradation, decreased secretion, or exposure of an epitope recognized by hepatic receptors which clear serpin-protease complexes from the circulation.[11]

RELATIONSHIP OF THE ATIII HEPARIN BINDING AND THROMBIN BINDING SITES

According to the α1-antitrypsin model for the tertiary structure of antithrombin III, the amino terminal heparin binding sequence of ATIII is located on the face of the inhibitor molecule which does not contain the reactive site. In general, the normal progressive ATIII activity of the residue 7, 24, 41 and 47 heparin binding mutants and the normal heparin binding properties of P12, P10, P1', P8', P10', P11' and most P1 ATIII reactive site mutants support the idea that separate regions of the ATIII molecule interact with heparin and thrombin (or other target proteases).

SUMMARY

Genetic evidence from the analysis of serpin gene relationships and dysfunctional antithrombin III variants has contributed greatly to our understanding of ATIII structure/function principles. These analyses have shown that certain residues of ATIII are subject to a high rate of mutational change due to the presence of CG dinucleotides, and have identified functionally important amino acid residues in regions of the molecule which interact with heparin and thrombin, and which govern protease inhibitor/substrate status and circulating protein concentration.

---

[c] The normal circulating levels of α1-antitrypsin Christchurch, α1-antitrypsin M3 and the C1 inhibitor 458 met/val variants suggest that the P5', P15' and P18' residues lie outside of the P11'-region structural element. See Fig. 3.

REFERENCES

1. Rosenberg RD, Damus PS: The purification and mechanism of action of antithrombin-heparin cofactor. J. Biol. Chem. 248:6490-6505, 1973
2. Marcum JA, Rosenberg RD: Anticoagulantly active heparan sulfate proteoglycan and the vascular endothelium. Seminars in Thrombosis and Haemostasis 13:464-474, 1987
3. Petersen TE, Dudek-Wojciechowska G, Sottruup-Jensen L, Magnusson S: Primary structure of antithrombin-III (heparin cofactor). Partial homology between a1-antitrypsin and antithrombin-III, in Collen D, Wiman B, Verstraete M (eds): The physiological inhibitors of coagulation and fibrinolysis. Elsevier-North Holland Biomedical Press, Amsterdam, 1979, pp. 43-54
4. Bock SC, Wion K, Vehar G, Lawn RM: Cloning and expression of the endogenous anticoagulant protein, human antithrombin III. Nucl. Acids. Res. 10:8113-8125, 1982
5. Chandra T, Stackhouse R, Kidd VJ, Woo SLC: Isolation and sequence characterization of a cDNA clone of human antithrombin III. Proc. Natl. Acad. Sci., USA 80:1845-1848, 1983
6. Franzen LE, Svensson S, Larm O: Structural studies on the carbohydrate portion of human antithrombin III. J. Biol. Chem. 255:5090, 1980
7. Mizuochi T, Fujii J, Kurachi K, Kobata A: Structural studies of the carbohydrate moiety of human antithrombin III. Arch. Biochem. Biophys. 203:458, 1980
8. Collen D, Schetz J, DeCock F, Holmer E, Verstraete M: Metabolism of antithrombin III (heparin cofactor) in man: effects of venous thrombosis of heparin administration. Eur. J. Clin. Invest. 7:27-35, 1977
9. Conard J, Brosstad F, Larsen ML, Samama M, Abildgaard U: Molar antithrombin concentration in normal human plasma. Haemostasis 13:363-368, 1983
10. Thaler E, Lechner K: Antithrombin III deficiency and thromboembolism. Clin. in Haematol. 10:369-390, 1981
11. Pizzo SV: Serpin receptor 1: a hepatic receptor that mediates the clearance of antithrombin III-proteinase complexes. Am. J. Med. 87:10-14, 1989
12. Carrell RW, Pemberton PA, Boswell DR: The serpins: evolution and adaptation in a family of protease inhibitors. Cold Spring Harbor Symp. Quant. Biol. LII:527-535, 1987
13. Sugiyama N, Sasaki T, Iwamoto M, Abiko Y: Binding site of a2-plasmin inhibitor to plasminogen. Biochim. Biophys. Acta 952:1-7, 1988
14. Hortin GL, Trimpe BL, Fok KF: Plasmin's peptide binding specificity: characterization of ligand sites in a2-antiplasmin. Thromb. Res. 54:621-632, 1989
15. Owen WG: Evidence for the formation of an ester between thrombin and heparin cofactor. Biochim. Biophys. Acta 405:380-387, 1975
16. Fish WW, Bjork I: Release of a two-chain form of antithrombin from the antithrombin-thrombin complex. Eur. J. Biochem 101:31-38, 1979
17. Jornvall H, Fish WW, Bjork I: The thrombin cleavage site in bovine antithrombin. FEBS Lett. 106:358-362, 1979
18. Loebermann H, Tokuoka R, Deisenhofer J, Huber R: Human a1-proteinase inhbitor. Crystal structure analysis of two

crystal modifications, molecular model and preliminary analysis of the implications for function. J. Mol. Biol. 177:531-556, 1984

19. Huber R, Carrell RW: Implications of the three-dimensional structure of a1-antitrypsin for structure and function of serpins. Biochem. 28:8951-8966, 1989

20. Samama JP, Delarue M, Mourey L, Choay J, Moras D: Crystallization and preliminary crystallographic data for bovine antithrombin III. J. Mol. Biol. 210:877-879, 1989

21. Brennan SO, Borg JY, George PM, Soria C, Soria J, Caen J, Carrell RW: New carbohydrate site in mutant antithrombin (7-ile > asn) with decreased heparin affinity. FEBS Lett. 237:118-122, 1988

22. Borg JY, Brennan SO, Carrell RW, George P, Perry DJ, Shaw J: Antithrombin Rouen-IV 24 Arg > Cys. The amino terminal contribution to heparin binding. FEBS Lett. 266:163-166, 1990

23. Chang JY, Tran TH: Antithrombin III Basel: Identification of a pro-leu substitution in a hereditary abnormal antithrombin with impaired heparin cofactor activity. J. Biol. Chem. 261:1174-1176, 1986

24. Molho-Sabatier P, Aiach M, Gaillard I, Fiessinger JN, Fischer AM, Chadeuf G, Clauser E: Molecular characterization of seven ATIII variants using PCR. Identification of a new mutation: 384 Ala-Pro. J. Clin. Invest. 83:1236-1242, 1989

25. Daly M, Ball R, O'Meara A, Hallinan FM: Identification and characterisation of an antithrombin III mutant (AT Dublin 2) with marginally decreased heparin activity. Thromb. Res. 56:503-513, 1989

26. Koide T, Odani S, Takahashi K, Ono T, Salmagawa N: Antithrombin III Toyama: replacement of arginine-47 by cysteine in hereditary abnormal antithrombin III that lacks heparin binding ability. Proc. Natl. Acad. Sci., USA 81:289, 1984

27. Duchange N, Chasse JF, Cohen GN, Zakin MM: Molecular characterization of the antithrombin III Tours deficiency. Thromb. Res. 45:115-121, 1987

28. Brunel F, Duchange N, Fischer AM, Cohen GN, Zakin MM: Antithrombin III Alger: an new case of arg47>cys mutation. A. J. Hemat. 25:223-224, 1987

29. Owen MC, Shaw GJ, Grau E, Foncuberta J, Carrell RW, Boswell DR: Molecular characterization of antithrombin Barcelona-2: 47 arginine to cysteine. Thromb. Res. 55:451-457, 1989

30. Owen MC, Borg JY, Soria C, Soria J, Caen J, Carrell RW: Heparin binding defect in a new antithrombin III variant: Rouen, 47 arg to his. Blood 69:1275-1279, 1987

31. Caso R, Lane DA, Thomson E, Zangouras D, Panico M, Morris H, Olds RJ, Thein SL, Girolami A: Antithrombin Padua I: Impaired heparin binding caused by an arg-47 to his (CGT to CAT) substitution. Thromb. Res. 58:185-190, 1990

32. Borg JY, Owen MC, Soria C, Soria J, Caen J, Carrell RW: Proposed heparin binding site in antithrombin based on arginine 47: a new variant Rouen-II, 47 arg to ser. J. Clin. Invest. 1292-1296, 1988

33. Devraj-Kizuk R, Chui DHK, Prochownik EV, Carter CJ, Ofosu FA, Blajchman MA: Antithrombin III Hamilton: a gene with a point mutation (guanine to adenine) in codon 382

causing impaired serine protease reactivity. Blood 72:1518-1523, 1988

34. Perry DJ, Harper PL, Fairham S, Daly M, Carrell RW: Antithrombin Cambridge, 384 Ala to Pro: a new variant identified using the polymerase chain reaction. FEBS Lett. 254:174-176, 1989

35. Erdjument H, Lane DA, Panico M, DiMarzo V, Morris HR: Single amino acid substitutions in the reactive site of antithrombin leading to thrombosis. J. Biol. Chem. 263:5589-5593, 1988

36. Owen MC, Beresford CH, Carrell RW: Antithrombin Glasgow, 393 arg to his: a P1 reactive site variant with increased heparin affinity but no thrombin inhibitory activity. FEBS Letts. 231:317-320, 1988

37. Lane DA, Erdjument H, Flynn A, DiMarzo V, Panico M, Morris HR, Greaves M, Dolan G, Preston FE: Antithrombin Sheffield: amino acid substitution at the reactive site (arg 393 to his) causing thrombosis. Brit. J. Heamatol. 71:91-96, 1989

38. Erdjument H, Lane DA, Panico M, DiMarzo V, Morris HR, Bauer K, Rosenberg RD: Antithrombin Chicago, amino acid substitution of arginine 393 to histidine. Thromb. Res. 54:613-619, 1989

39. Erdjument H, Lane DA, Ireland H, DiMarzo V, Panico M, Morris HR, Tripodi A, Mannucci PM: Antithrombin Milano, single amino acid substitution at the reactive site, arg-393 to cys. Thromb. Haemost. 60:471-475, 1988

40. Lane DA, Erdument H, Thompson E, Panico M, DiMarzo V, Morris HR, Leone G, DeStefano V, Thein SL: A novel amino acid substitution in the reactive site of a congenital variant antithrombin. Antithrombin Pescara, arg-393 to pro, caused by a CGT to CCT mutation. J. Biol. Chem. 264:10020-10204, 1989

41. Stephens AW, Thalley BS, Hirs CHW: Antithrombin III Denver, a reactive site variant. J. Biol. Chem. 262:1044-1048, 1987

42. Olds RJ, Lane D, Caso R, Tripodi A, Mannucci PM, Thein SL: Antithrombin III Milano 2: a single base substitution in the thrombin binding domain detected with PCR and direct genomic sequencing. Nucl. Acids Res. 17:10511, 1989

43. Bock SC, Silbermann JA, Wikoff W, Abildgaard U, Hultin MB: Identification of a threonine for alanine substitution at residue 404 of antithrombin III Oslo suggests integrity of the 404-407 region is important for maintaining normal plasma inhibitor levels. Thromb. Haemostas. 62:494A, 1989

44. Nakagawa M: Antithrombin III deficiency and its molecular analysis. XIIth Congress of ISTH, Kyoto Satellite Symposium, August 27-28, 1989

45. Bock SC, Marrinan JA, Radziejewska E: Antithrombin III Utah: Proline-407 to leucine mutation in a highly conserved region near the inhibitor reactive site. Biochem. 27:6171-6178, 1988

46. Perry DJ, Carrell RW: CpG dinucleotides are "hotspots" for mutation in the antithrombin III gene. Twelve variants identified using the polymerase chain reaction. Mol. Biol. Med. 6:239-243, 1989

47. Bird AP: DNA methylation and the frequency of CpG in animal DNA. Nucl. Acids Res. 8:1499-1504, 1980

48. Giddings JC: Molecular Genetics and Immunoanalysis in Blood Coagulation. Ellis Horwood Ltd., Chichester, England. 1988

49. Blackburn MN, Smith RL, Carson J, Sibley CC: The heparin-binding site of antithrombin III: identification of a critical tryptophan in the amino acid sequence. J. Biol. Chem. 259:939-941, 1984

50. Carrell RW, Christey PB, Boswell DR: Serpins: Antithrombin and other inhibitors of coagulation and fibrinolysis, Evidence from amino acid sequences, in Verstraete M, Vermylen J, Lijnen HR, Arnout J (eds): Thrombosis and Haemostasis 1987. Leuven University Press, 1987

51. Brennan SO, George PM, Jordan RE: Physiological variant of antithrombin III lacks carbohydrate side chain at asn-135. FEBS Lett. 219:431-436, 1987

52. Peterson CB, Noyes CM, Pecon JM, Church FC, Blackburn MN: Identification of a lysyl residue in antithrombin which is essential for heparin binding. J. Biol. Chem. 262:8061-8065, 1987

53. Liu CS, Chang JY: The heparin binding site of human antithrombin III. J. Biol. Chem. 262:17356-17361, 1987

54. Chang JY: Binding of heparin to human antithrombin III activates selective chemical modification at lysine 236. Lys-107, lys-125 and lys 136 are situated within the heparin binding site of antithrombin III. J. Biol. Chem. 264:3111-3115, 1989

55. Owen MC, Brennan SO, Lewis JH, Carrell RW: Mutation of antitrypsin to antithrombin. a1-antitrypsin Pittsburgh (358 met > arg), a fatal bleeding disorder. N. Engl. J. Med. 309:694-698, 1983

56. Stephens AW, Siddiqui A, Hirs CHW: Site-directed mutagenesis of the reactive center (serine 394) of antithrombin III. J. Biol. Chem. 263:15849-15852, 1988

57. Bock SC, Skriver K, Nielsen E, Thogersen HC, Wiman B, Donaldson VH, Eddy RL, Marrinan J, Radziejewska E, Huber R, Shows TB, Magnusson S: Human $C\bar{1}$ inhibitor: primary structure, cDNA cloning and chromosomal localization. Biochem. 25:4292-4301, 1986

58. Levy NJ, Ramesh N, Cicardi M, Harrison RA, Davis AE: Type II hereditary angioneurotic edema that may result from a single nucleotide change in the codon for alanine-436 in the $C\bar{1}$ inhibitor gene. Proc. Natl. Acad. Sci., USA 87:265-268, 1990

59. Holmes WE, Lijnen HR, Nelles L, Kluft C, Nieuwenhuis HK, Rijken DC, Collen D: a2-antiplasmin Enschede: alanine insertion and abolition of plasmin inhibitory activity. Science 238:209-211, 1987

60. Bruch M, Weiss V, Engel J: Plasma serine proteinase inhibitors (serpins) exhibit major conformational changes and a large increase in conformational stability upon cleavage at their reactive sites. J. Biol. Chem. 263:16626-16630, 1988

61. Gettins P, Hartens JB: Properties of thrombin- and elastase-modified human antithrombin III. Biochem. 27:3634-3639, 1988

62. Haris PI, Chapman D, Harrison RA, Smith KF, Perkins SJ: Conformational transition between native and reactive center cleaved forms of a1-antitrypsin by Fourier

transform infrared spectroscopy and small-angle neutron scattering. Biochem. 29:1377-1380, 1990

63. Gettins P: Absence of large-scale conformational change upon limited proteolysis of ovalbumin, the prototypic serpin. J. Biol. Chem. 264:3781-3785, 1989

64. Asakura S, Matsuda M, Yoshida N, Terukina S, Kihara H: A monoclonal antibody that triggers deacylation of an intermediate thrombin-antithrombin III complex. J. Biol. Chem. 264:13736-13739, 1989

65. Asakura S, Hirata H, Okazaki H, Hashimoto-Gotoh T, Matsuda M: Hydrophobic residues 382-386 of antithrombin III, ala-ala-ala-ser-thr, serve as an epitope for an antibody which facilitates hydrolysis of the inhibitor by thrombin. J. Biol. Chem. 265:5135-5138, 1990

66. Bock SC, Harris JF, Schwartz CE, Ward JH, Hershgold EJ, Skolnick MH: Hereditary thrombosis in a Utah kindred is caused by a dysfunctional antithrombin III gene. Am. J. Hum. Genet. 37:32-14, 1985

67. Hultin MB, McKay J, Abildgaard U: Antithrombin Oslo: Type Ib classification of the first reported antithrombin-deficient family, with a review of hereditary antithrombin variants. Thromb. Haemost. 59:468-473, 1988

68. Hofker MH, Nukiwa T, van Paassen HMB, Nelen M, Frants RR, Klasen EC, Crystal RG: A Pro > Leu substitution in codon 369 in the alpha-1-antitrypsin variant PIM-Heerlen. Am. J. Hum. Genet. 41:A220, 1987

69. Egeberg O: Inherited antithrombin deficiency causing thrombophilia. Throm. Diath. Haemorrh. 13:516-530, 1965

70. Grandille S, Aiach M, Lane DA, Vidaud D, Molho-Sabatier P, Caso R, de Moerloose P, Fiessinger JN, Clauser E: Important role of Arg-129 in heparin binding site of antithrombin III: identification of a novel mutation arg-129 to gln. J. Biol. Chem. In press.

INTERACTIONS BETWEEN THE FUNCTIONAL DOMAINS OF ANTITHROMBIN III

Paula R. Boerger, Robert M. Wolcott, Morgan Lorio
and Michael N. Blackburn

Department of Macromolecular Sciences, SmithKline
Beecham Pharmaceuticals, King of Prussia PA 19406-
0939 USA and Department of Biochemistry and
Molecular Biology and Department of Immunology
Louisiana State University Medical Center
Shreveport LA 71130 USA

INTRODUCTION

Antithrombin III is the major physiologic inhibitor of
the activated serine proteases of the blood coagulation
cascade[1]. Inactivation of the proteases occurs following
cleavage of the Arg_{393}-Ser_{394} peptide bond and formation of a
stable complex between this reactive site arginine of antithrombin III and the active site serine of the protease[2].
Antithrombin III is a member of a superfamily of protease
inhibitors[3] designated as serine protease inhibitors or
"serpins"[4]. Although all of the members of the serpin family
exhibit a high degree of amino acid sequence homology, particularly in the C-terminal portions of the molecules which contain
the reactive site residues, the specificities of these
inhibitors vary widely[5-8]. Furthermore, with antithrombin III
the relative rates of complex formation with different proteases, for example, thrombin, Factor Xa, Factor IXa, Factor
VIIa, are significantly different despite the fact that each of
these enzymes displays selectivity for cleavage at arginine
side chains. At present, it is not clear what structural features, in addition to the amino acid sequence at the p1 and p1'
positions, determine the kinetics of protease-inhibitor docking
and the stability of the resulting complex. However, a variety
of approaches, including site-directed mutagenesis, antibody
mapping and protein crystallography, are being used to explore
these questions.

The rate of complex formation between antithrombin III
and the coagulation proteases can be accelerated at least one
thousand fold by the mucopolysaccharide, heparin. This enhanced activity is known as heparin cofactor activity.
However, only about one-third of a typical commercial sample of
heparin has anticoagulant activity[9]. These molecules have been
shown to contain a particular pentameric sequence of saccharide
units containing a unique 3-O-sulfate moiety[10]. This pentamer
provides a high affinity binding site for antithrombin III on
the heparin chain, and is capable of activating antithrombin

III towards some coagulation proteases such as Factor Xa. In
contrast, a larger heparin chain with at least 16-18 saccharide
units, including the pentameric high affinity antithrombin
binding site, is required for enhanced thrombin inhibition[11-14]. Presumably, these additional residues are required for
interaction with thrombin to form a ternary complex between
thrombin, antithrombin and the heparin chain, which serves as a
template[15].

Identification of the Heparin Binding Domain

Numerous studies with chemically altered forms of antithrombin III or naturally occurring variants of the protein
have been used to map the heparin binding domain and to identify amino acid residues which are involved in the binding of
heparin to antithrombin III. We have shown that modification
of a single tryptophan with the reagent hydroxynitrobenzyl
sulfonium bromide blocked heparin binding without altering the
intrinsic ability of antithrombin III to react with thrombin or
with Factor Xa[16,17]. These studies implicated Trp_{49} as a critical residue in the heparin binding domain[18]. More importantly,
they provided evidence that the heparin binding domain of antithrombin III resides in the disulfide cross-linked loop at the
amino terminal end of the molecule whereas the protease complexing site is near the carboxyl-terminus (Fig. 1).
Antithrombin III, thus, appears to be organized into two functional domains, a heparin binding domain and a protease complexing domain, which are separated within the amino acid
sequence of the protein[16]. Guanidine hydrochloride denaturation studies, which show that antithrombin unfolds with a
biphasic denaturation pattern, also indicate a two domain
structure for the protein[19,20].

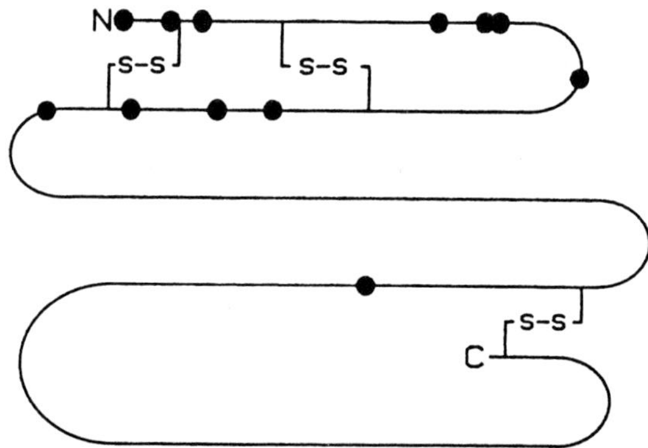

Fig. 1. A Schematic Model of the Antithrombin
Showing Amino Acid Residues Implicated in
Heparin Binding. Beginning at the N-terminus,
residues shown are: His_1, Ile_7, Lys_{11}, Pro_{41},
Arg_{47}, Trp_{49}, His_{65}, Lys_{107}, Lys_{114}, Lys_{125},
Lys_{136}, Lys_{275}

Lysine residues were first shown by Rosenberg and Damus[21]
to be necessary for heparin binding. Pecon and Blackburn[22]
found that phosphopyridoxylation of a single antithrombin III

lysine residue blocked both heparin binding and heparin-catalyzed inhibition of thrombin. In a continuation of this work, Peterson et al. showed that Lys_{125} was essential for heparin binding and furthermore that other lysine residues, including Lys_{11}, Lys_{136}, Lys_{275} and Lys_{287} were protected from phosphopyridoxylation by the bound heparin. Chang and coworkers[23,24] have obtained similar results using trinitrobenzene sulfonate and 4-N-N-dimethylamino-azobenzene-4'-isothiocyano-2' sulfonic acid. They have also identified Lys_{107} and Lys_{114} as essential residues. Of special interest, Chang has shown that heparin binding *increases* the reactivity of Lys_{236} and suggested that this increase in lysine reactivity is a result of the conformational changes occurring within antithrombin III upon binding heparin.

Other amino acid residues have also been implicated. Based on proton NMR studies of heparin binding to antithrombin, Gettins[25] has suggested that several histidine residues, including His_1 and His_{65}, may be involved in interaction of the protein with the mucopolysaccharide. However, it is unlikely that His_1 is required for heparin binding since the amino terminal histidine is not present in rat or chicken antithrombins[26].

Numerous groups have identified patients with coagulation disorders arising from amino acid substitutions within the antithrombin III protein that result in reduced heparin cofactor activity and many of these patients exhibit a marked tendency towards thrombotic episodes. These amino acid substitutions tend to map in the amino terminal portion of the protein molecule. For instance, Antithrombin Toyama[27] and Antithrombin Alger[28] correspond to Arg_{47} to Cys_{47} substitutions, Antithrombin Rouen I[29] and Antithrombin Rouen II[30] represent Arg_{47} to His_{47} and Ser_{47}, respectively. Antithrombin Basel[31] (Pro_{41} to Leu_{41}) represents another mutation which influences heparin binding to antithrombin.

In contrast to these mutations which reduce heparin affinity, a naturally occurring variant, Antithrombin$_\beta$, identified by Peterson and Blackburn[32], lacks glycosylation at Asn_{135}[33]. The non-glycosylated protein displays increased heparin affinity and enhanced rates of thrombin inhibition at subsaturating heparin concentrations compared to the fully glycosylated form of the inhibitor. These results suggest that the carbohydrate side chain attached at Asn_{135} interferes with the binding of the sulfated glycosaminoglycan in the heparin binding domain. Brennan et al.[34] have also shown that glycosylation at Asn_7 in a mutant antithrombin decreases heparin affinity.

Rosenfeld and Danishefsky[35] and Smith and Knauer[36] have reported the isolation of heparin binding fragments of antithrombin III. Cleavage of the protein with cyanogen bromide produced a fragment spanning amino acids 104-251[35] whereas proteolysis of antithrombin with endoproteinase V8 yielded a peptide corresponding to residues 114-156[36]. The results obtained with isolated peptide fragments thus support the conclusions from chemical modification and genetic analysis. As can be seen from Fig. 1, which shows the location of essential amino acid residues within a schematic representation of anti-

thrombin III, these essential residues tend to be clustered
near the amino terminus of the inhibitor.

Mapping of Antithrombin III By Immunochemical Methods

Despite the progress made in identification of the heparin binding and protease complexing domains, little is known
concerning the mechanisms by which structural information is
communicated between these parts of the molecule. Although
crystals of bovine antithrombin III have recently been reported[37], a crystal structure of the inhibitor has not yet been
solved. Structural models, however, have been generated from
the crystal structure of alpha-1-antitrypsin[38]. While these
models may provide substantial insight into the organization of
these homologous serpins, they are subject to several limitations, particularly with respect to the elucidation of heparin
interactions. Alpha-1-antitrypsin, in contrast to antithrombin
III, does not bind specifically with heparin and shows little
sequence homology with the first fifty residues in the antithrombin sequence[8], which includes several residues implicated
in heparin-antithrombin interactions. Additionally, the form
of alpha-1-antitrypsin, for which a structure has been solved,
represents the proteolytically cleaved protein, clipped at the
P1-P1' position. The corresponding form of antithrombin displays only weak affinity for heparin[39]; hence the high affinity
binding site has been lost in the conformational rearrangements
that take place upon nicking of the protein molecule (see
below). The computer generated models do, however, provide a
basic framework which can be used for the interpretation of
antithrombin structure/function relationships, although detailed information must await high-resolution structures of the
native and heparin-complexed inhibitor. In the absence of an
antithrombin crystal structure many alternative methods have
been utilized to analyze antithrombin structure.

Immunochemical methods have been used by several groups
in the study of antithrombin III. Lau and Rosenberg[40],
McDuffie et al[41] and Wallgren et al[42] identified neoantigenic
sites developed by formation of the antithrombin-thrombin complex. Since these sites were not present on the individual
proteins, it is likely that the appearance of at least some of
these epitopes may reflect the conformational changes in the
antithrombin molecule that occur upon cleavage of the reactive
site bond. Herion et al.[43] developed a panel of monoclonal
antibodies to antithrombin to define antigenic and functional
regions of the protein. Although one of these monoclonal
antibodies could be used to differentiate heparin dependent and
non-heparin dependent pathways for reaction of antithrombin
with thrombin, the location of these antibody recognition sites
within the protein amino acid sequence was not determined.
Hence, based on these results, it is not possible to localize
heparin-induced or thrombin-induced conformational changes
within the antithrombin molecule.

In order to achieve this result, Peterson and Blackburn[44]
combined site-specific chemical modification with spectral and
immunochemical techniques. As described above, reaction of
antithrombin with hydroxynitrobenzyl sulfonium bromide (HNB)
specifically labels Trp_{49} within the heparin binding domain
with the HNB chromophore. The HNB moiety was used as an envi-

ronmentally sensitive reporter group to detect local conformational changes in the heparin domain when thrombin binds at the
protease complexing site. Upon reaction of HNB-labeled antithrombin with thrombin, a shift in the spectrum of the HNB
chromophore was observed at 410 nm (Fig. 2, upper panel). As
originally demonstrated by Villanueva and Danishefsky[45] and as
shown in the lower panel of Fig. 2, reaction of antithrombin
with thrombin also induces changes in the ultraviolet spectrum
of the proteins with negative peaks in the difference spectrum
at 284 and 289 nm. Identical spectral changes in the ultraviolet region were observed with labeled and unlabeled antithrombins.

The change in the HNB spectrum, which shows a maximal
increase at 410 nm with a decrease at 320 nm, was, of course,
specific to the labeled protein. This thrombin-induced
perturbation of the HNB chromophore results from an increased
exposure of the HNB group and reflects a shift in the
environment of HNB-Trp$_{49}$.

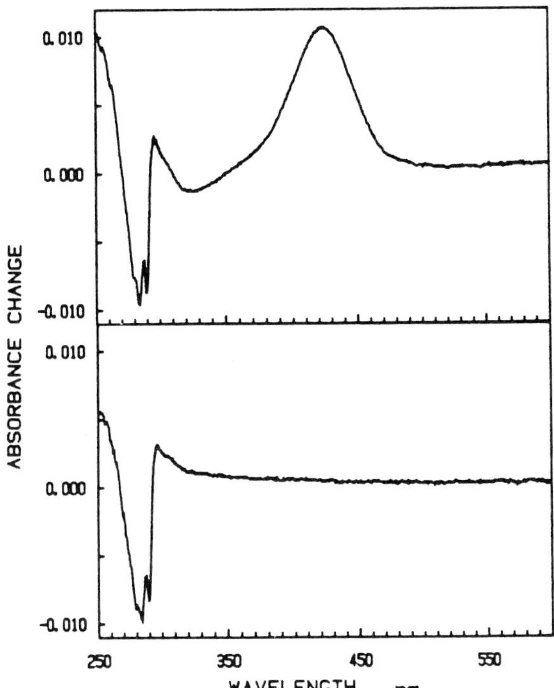

Fig. 2 Spectral Changes that Occur upon
Formation of Complexes Between Thrombin and
Native Antithrombin or HNB-Antithrombin.
Difference spectra were obtained by mixing HNB-
antithrombin (1.44 x 10^{-5} M) with thrombin
(1.15 x 10^{-5} M) in the upper panel and native
antithrombin (1.20 x 10^{-5} M) with thrombin (0.99
x 10^{-5} M) in the lower panel. All proteins
were dissolved in 10 mM sodium phosphate, 0.15
M sodium chloride, pH 7.4, containing 0.02%
sodium azide.

    To confirm this conclusion, antibodies specific to the
HNB moiety were produced using HNB as a hapten (using HNB-
labeled lysozyme for immunization). These antibodies reacted
with HNB-antithrombin or with a labeled tripeptide, Gly-
(HNB)Trp-Gly, but not with native antithrombin nor with the unmodified tripeptide. In competitive ELISA assays (Fig. 3), the
anti-HNB antibodies were shown to exhibit greater affinity for
HNB-antithrombin-thrombin complex than for HNB-antithrombin.

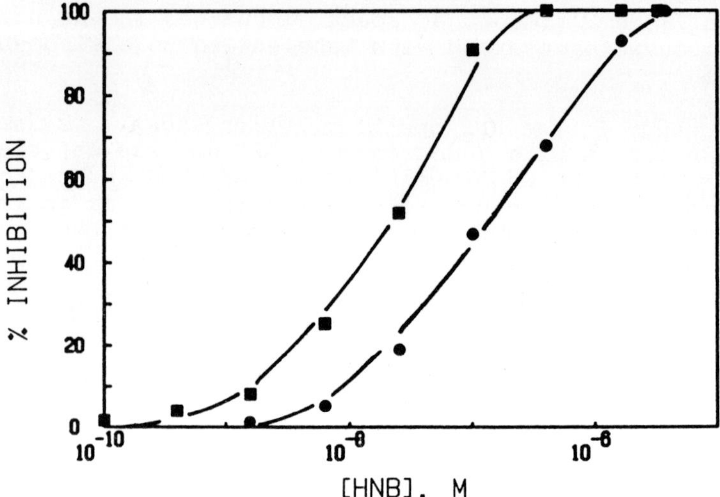

Fig.3. Interaction of Anti-HNB antibodies with
HNB-Antithrombin or with HNB-Antithrombin-
Thrombin Complex. Relative affinities of anti-
HNB antibodies with the HNB labeled proteins
were measured in a competitive ELISA procedure
in which HNB-ovalbumin was adsorbed to the
solid phase. Data are presented as percent inhibition of the ELISA assay vs the concentration of the HNB hapten for HNB-antithrombin
(●) and for HNB-antithrombin-thrombin complex
(■).

    These immunochemical results, like the spectral studies,
show increased solvent exposure of HNB-Trp_{49}. Thus, thrombin
complex formation, which decreases heparin affinity for antithrombin, is closely linked to a structural change involving
Trp_{49} within the heparin binding domain.

    To extend these studies we have examined a battery of
monoclonal antibodies for potential effects on antithrombin activity and have identified antibodies which appear to neutralize heparin dependent antithrombin activity by competition for
binding within the heparin binding domain. Also, we have used
synthetic peptides derived from the antithrombin sequence to
further map the heparin binding site.

    Reduced, denatured and deglycosylated antithrombin was
used to generate monoclonal antibodies which were selected
using an ELISA screen with native antithrombin. This approach
was chosen to enhance the probability of selecting antibodies
which recognize surface features of the native structure and
which depend primarily upon the linear sequence of amino acids

rather than upon the folded three-dimensional conformation of
antithrombin.

   As shown in Fig. 4, a number of hybridoma cell lines were
found to secrete antibodies which when added to a heparin cofactor activity assay were shown to substantially decrease
antithrombin activity. Since interaction of the antibody with
antithrombin completely blocks heparin cofactor activity, we
concluded that this antibody may recognize an epitope at or
near the heparin binding domain and we initiated experiments to
identify this antibody recognition site.

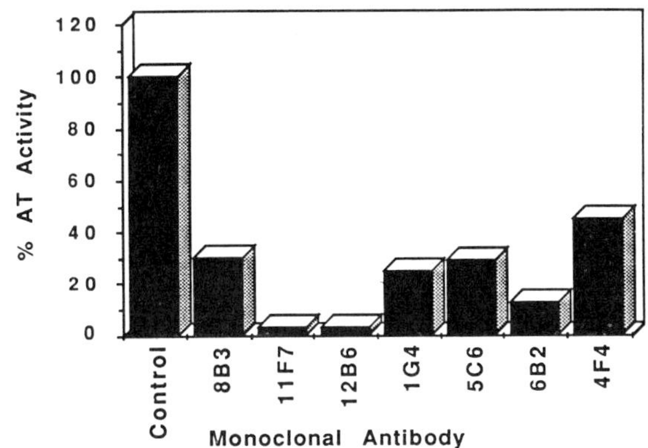

Fig. 4. Effect of anti-Antithrombin III
Monoclonal Antibodies on Antithrombin III
Heparin Cofactor Activity. Inactivation of
thrombin by heparin-activated antithrombin
was monitored at 405 nm in a continuous
chromogenic assay in the presence and absence of 50 ul of supernatant from various
anti-ATIII antibody cell cultures. Pseudo-
first-order rates were normalized to the
activity of 200 nM antithrombin with 2 nM
thrombin, 5 nM heparin, and .19 mM
Chromozym TH substrate.

   As described above, chemical modification experiments and
sequence identification of naturally occurring antithrombin
variants had implicated the sequence adjacent to Trp_{49} in heparin binding. Thus, we synthesized a peptide (designated
HBD_{41-58}) spanning amino acid residues 41 to 58. The sequence
of this peptide is shown below. This basic peptide contains
three Arg and one Lys; an amino terminal Cys residue, which is
not part of the antithrombin sequence, was added to the peptide
to facilitate labeling with spectral probes.

Cys(41)ProGluAlaThrAsnArgArgValTrpGluLeuSerLys

AlaAsnSerArgPhe(58)

This peptide was then used in immunoassays to determine
if the antibodies raised against antithrombin would cross-
react. Many of these monoclonal antibodies were found to recognize this peptide. Indeed, from two independent immuniza-
tions and hybridoma fusion experiments a majority of the cloned
hybridoma cell lines produced antibodies that reacted with
HBD_{41-58}. From these experiments, however, it is not clear
whether this region of the denatured protein is exceptionally
immunogenic or if our ELISA screening procedure preferentially
selects for antibodies to this part of the antithrombin
molecule.

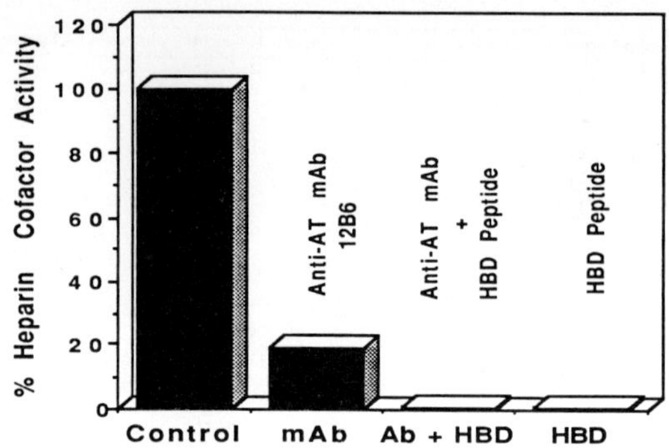

Fig. 5. Effects of anti-Antithrombin III
Antibody and HBD_{41-58} Peptide on Antithrombin
III Activity. Antithrombin III heparin cofactor activity was determined in a chromogenic
assay for thrombin activity In the assay, 50
nM antithrombin III, 2 nM thrombin, 5 nM heparin, and .19 mM Chromozym TH substrate were
reacted while monitoring absorbance at 405 nm.
The rate of this reaction was then compared to
that in the presence of 30 ul of culture supernatant from the 12B6 anti-antithrombin III cell
line. In experiments with added HBD_{41-58}, the
peptide concentration was 43 ug/ml.

Since many of these antibodies interact with both the
HBD_{41-58} peptide and native antithrombin we reasoned that the
peptide might compete with native antithrombin III for binding
to the neutralizing monoclonal antibodies and therefore, addition of the peptide to the assay would restore heparin cofactor
activity which was blocked by the antibody. For this experiment we utilized the antibody, 12B6, which binds tightly to
both antithrombin and to the HBD_{41-58} peptide. As shown in
Fig. 5, monoclonal antibody 12B6 effectively neutralizes antithrombin heparin cofactor activity. However, despite the fact
that 12B6 binds the peptide HBD_{41-58}, addition of the peptide
did not reverse the activity blocking effect of the antibody.
The reason for this apparent discrepancy was discovered when we
determined the effect of the peptide on antithrombin heparin

cofactor activity. When the peptide was added to the assay, in
the absence of added antibody, HBD$_{41-58}$ was found to inhibit
antithrombin mediated heparin cofactor activity.

In order to establish the mechanism by which the peptide
HBD$_{41-58}$ might interfere with the ability of antithrombin to
inactivate thrombin, we determined if the peptide altered
thrombin or antithrombin activities. HBD$_{41-58}$ was found to
have no effect on the rate of thrombin hydrolysis of the
tripeptide substrate, tosyl-Gly-Pro-Arg-p-nitroanilide.
Additionally, the peptide did not decrease the rate of thrombin
inactivation by antithrombin when this activity was measured
without added heparin in a progressive antithrombin activity
assay (Fig. 6).

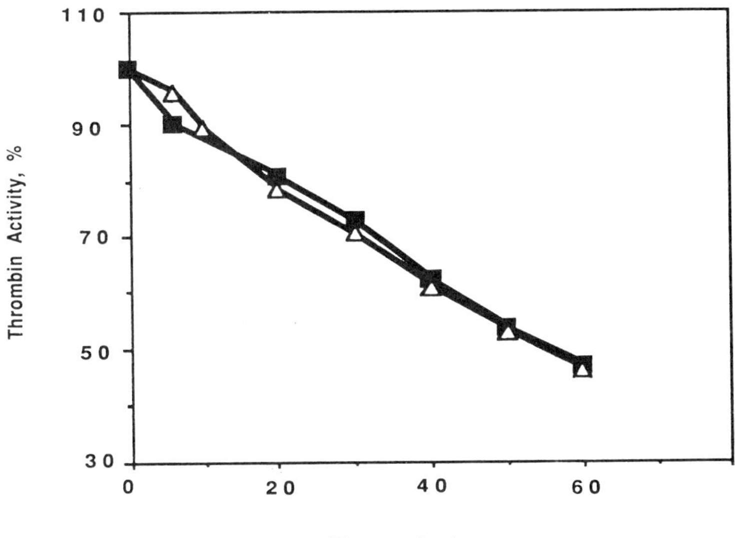

Fig.6. Effects of HBD$_{41-58}$ peptide on progressive antithrombin III activity. Inhibition of
thrombin by antithrombin III was monitored at
405 nm over time in the presence (Δ) and
absence (■) of 100 nM HBD$_{41-58}$ peptide. At
each indicated time, the residual amount of
thrombin activity was determined with the addition of 0.19 mM Chromozym substrate to 2 nM
thrombin and 20 nM antithrombin III at 30 °C.
Reactions were stopped by the addition of 5%
acetic acid.

These data suggest that the HBD$_{41-58}$ peptide must bind
heparin directly and thus must compete with antithrombin for
the binding site on the heparin chain. An estimate of the
affinity of interaction between the peptide and heparin was
obtained by analyzing the ability of the peptide to interfere
with heparin cofactor activity as a function of peptide
concentration. In this assay, the psuedo-first-order rate
constant for the heparin catalyzed inactivation of thrombin by
antithrombin was determined using a chromogenic substrate
assay. This approach can be formally treated like an enzyme

catalyzed reaction in which antithrombin and thrombin are considered as the 'substrates' for the reaction catalyzed by heparin; the HBD$_{41-58}$ peptide, in this formalism, represents an inhibitor of the reaction. The results of this analysis are presented in Fig. 7. Full antithrombin activity (i.e., 100% activity) corresponds to the rate constant for the thrombin-antithrombin reaction in the absence of added peptide. The peptide shows potent inhibition of the heparin catalyzed reaction with 50% inhibition achieved between 50 and 100 nM peptide. The dissociation constant for antithrombin and heparin is approximately 10^{-7} to 10^{-8} M[46]. Thus, the synthetic HBD$_{41-58}$ peptide, representing only eighteen amino acids from within the human antithrombin III sequence, exhibits an affinity towards heparin which is comparable to the affinity of the native intact protein for heparin.

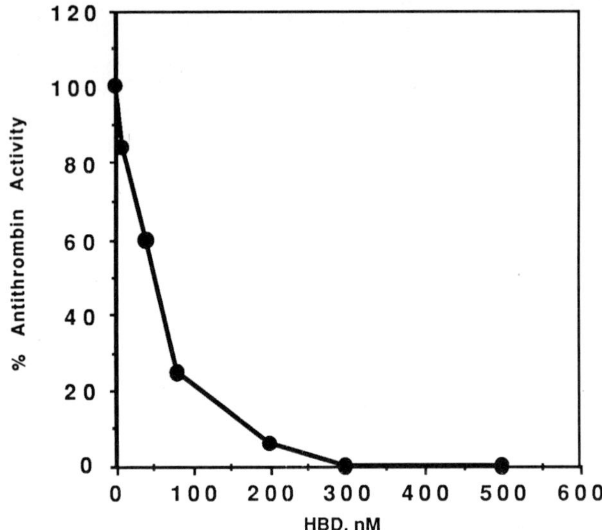

Fig. 7. Inhibition of Antithrombin III Heparin Cofactor Activity by HBD41-58 Peptide. Antithrombin III inhibitory activity was determined via a fixed time point assay Total antithrombin activity is represented by 20 nM antithrombin reacted with 2 nM thrombin, 5 nM heparin, and .19 mM Chromozym TH in PBS pH 7.4 at 30 °C. Reactions were stopped with 5% acetic acid and absorbance was read at 405 nm. Increasing amounts of HPLC purified peptide were used to titrate the antithrombin III activity.

Comparison with the crystal structure of the homologous serpin, alpha-1-antitrypsin, would suggest that the portion of antithrombin III spanning amino acid residues 41 through 58 is largely alpha helical[38]. Fig. 8 presents a helical wheel projection of the amino acid sequence corresponding to HBD$_{41-58}$. From this representation it can be seen that HBD$_{41-58}$ is composed chiefly of hydrophilic residues. The few hydrophobic

side chains (Val$_{48}$, Trp$_{49}$, Leu$_{51}$, and Phe$_{58}$) are located along
one face of the helix. The opposite face contains all of the
charged amino acid residues (Glu$_{42}$, Arg$_{46}$, Arg$_{47}$, Glu$_{50}$, Lys$_{53}$,
and Arg$_{57}$). Which of these residues interact with heparin? In
the isolated peptide the positively charged groups of these
residues may be able to interact with the negatively charged
sulfates of heparin. However, it is not clear that all of
these amino acids also interact with heparin in the native
antithrombin structure, since some may participate in the maintenance of antithrombin structure through formation of hydrogen
bonds and salt bridges. Delineation of the roles of these
amino acids in heparin binding will require additional experimental approaches.

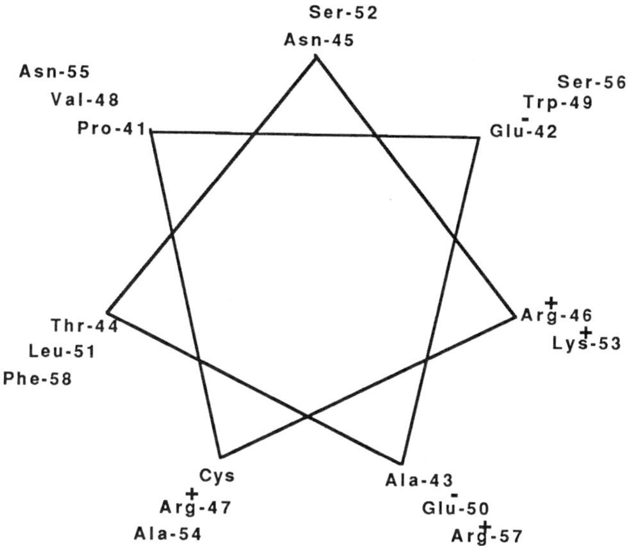

Fig. 8. Helical Wheel Projection of HBD41-58
Peptide

The Heparin Catalytic Cycle

Several lines of evidence indicate that binding of heparin to the inhibitor induces a conformational change in antithrombin III. This induced conformational change, or allosteric effect, has been proposed[21] to be responsible for accelerated activity of antithrombin III in the presence of added
heparin. However, direct evidence for a structural change involving the reactive site residues has not been obtained and
the precise role of this conformational change in enhanced
thrombin inactivation remains controversial[14,47,48].
Proteolytic cleavage of the reactive site bond decreases heparin binding affinity[39] and facilitates heparin dissociation.
These binding events and linked conformational changes form a
catalytic cycle for the interaction of heparin with antithrombin as illustrated in Fig. 9.

Beginning on the left we see the binding of antithrombin
(**A**) to heparin (**H**) to form the heparin-antithrombin complex
(**A•H**). Alterations in tryptophan fluorescence[49], in tyrosine
and tryptophan absorption spectra[50] and in chemical reactivity
of various amino acid side chains[24,51] indicate that antithrombin then isomerizes to yield the species **A*•H**. Olson et
al.[52] have shown that this heparin induced conformational
change also stabilizes the heparin-antithrombin complex. Both
of these heparin-inhibitor complexes can bind thrombin (**T**) to
give a non-covalent ternary complex (**A*•H•T**). Heparin thus
serves as a template surface to bring the inhibitor and the
enzyme together. Thrombin sees the Arg_{393}-Ser_{394} reactive site
as if it were a substrate, forms an initial Michaelis complex
and cleaves the scissile bond to yield a covalent adduct
(**A⁺•T**), linked through the active site Ser of the protease to
the Arg of the inhibitor/substrate. An acyl enzyme complex is
a normal transient intermediate in the mechanism of serine
proteases, however with antithrombin this complex is very
stable. Indeed, one can isolate the complex by SDS
polyacrylamide gel electrophoresis. However, as shown by Bjork
and Fish, in an alternative pathway, not shown on this figure,
the complex can hydrolyze to yield a cleaved inactive form of
antithrombin, designated as AT_m, and to regenerate active
thrombin[39]. At very low ionic strength, conditions that
stabilize the interaction of heparin with these proteins, bond
cleavage and release of AT_m is the major pathway.

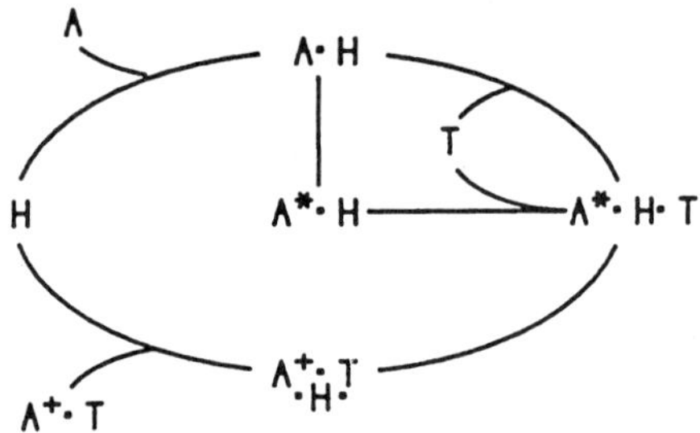

Fig. 9. A Model for the Catalytic Cycle for
Heparin-Promoted Inactivation of Thrombin by
Antithrombin. The cycle illustrates the steps
involved in reaction of antithrombin with
thrombin. **H** represents heparin, **A** and **T** represent antithrombin and thrombin, respectively.
A* represents antithrombin that has undergone
the heparin-promoted conformational change and
A⁺ the cleaved form of antithrombin.

Cleavage of the reactive site bond elicits a conformational change in the heparin domain[39,44] that triggers the
release of heparin through a decrease in the affinity of this
form of the thrombin-antithrombin complex (**A⁺•T**) for heparin.
Heparin, also freed in this final step of the cycle, can once

again bind antithrombin and thrombin to initiate another round
of protease inactivation.

In summary, the physiologic activity of antithrombin to
regulate the blood coagulation cascade requires the communication of conformational changes between the two functional
domains of the protein. The heparin catalytic cycle plays a
key role in the maintenance of hemostasis. It begins with the
initial heparin induced conformational change which stabilizes
the formation of the ternary complex, and is followed by the
reciprocal thrombin promoted conformational change which forces
the heparin chain from the heparin binding domain of antithrombin III. The release of heparin from the inhibitor-
protease complex allows heparin to catalyze additional rounds
of protease inactivation and it seems likely that this series
of conformational changes and altered heparin affinities plays
a major role in the catalytic mechanism by which the mucopolysaccharide functions to accelerate the inactivation of
proteases and modulate the pathways for blood coagulation.

ACKNOWLEGMENTS

This research was supported by National Institutes of
Health Grants HL-24846 and HL-26723, from the National Heart,
Lung and Blood Institutes, by a grant to R. W. from the
American Heart Association-Louisiana Inc., by a Specialized
Center Grant from the Northwest Louisiana Foundation for
Biomedical Research, and by student research grants to M. L.
from the Southern Medical Association and the American Heart
Association-Louisiana, Inc.

REFERENCES

1.   Rosenberg, RD: "Regulation of the Hemostatic Mechanism."
     in: The Molecular Basis of Blood Diseases. G
     Stamatoyannopoulos, AW Nienhuis, P Leder and PW Majerus
     eds. W.B.Saunders Company. Philadelphia, 534-574, 1987.

2.   Jornvall, H, Fish, WW and Bjork, I. "The Cleavage Site in
     Bovine Antithrombin." FEBS Letters. 106: 358-362, 1979.

3.   Hunt, LT and Dayhoff, MO. "A Surprising New Superfamily
     Containing Ovalbumin, Antithrombin III and Alpha-1-
     Proteinase Inhibitor." Biochem. Biophys. Res. Commun. 95:
     864-871, 1980.

4.   Carrell, RW and Travis, J. "Alpha-1-Antitrypsin and the
     SERPINS." TIBS. 10: 20-24, 1985.

5.   Petersen, TE, Dudek-Wojciechowska, G, Sottrup-Jensen, L and
     Magnusson, S: "Primary Structure of Antithrombin-III
     (Heparin Cofactor). Partial Homology Between alpha-1-
     Antitrypsin and Antithrombin-III." in: The Physiological
     Inhibitors of Coagulation and Fibrinolysis. D Collen, B
     Wiman and M Verstraete eds. Elsevier/North Holland
     Biomedical Press. 43-54, 1979.

6.   Bock, SC, Wion, KL, Vehar, GA and Lawn, RM. "Cloning and
     Expression of the cDNA for Human Antithrombin III." Nuc.
     Acid. Res. 10: 8113-8125, 1982.

7. Prochownik, EV, Markham, AF and Orkin, SH. "Isolation of a cDNA Clone for Human Antithrombin III." J. Biol. Chem. 258: 8389-8394, 1983.

8. Kurachi, K, Chandra, T, Friezner Degen, SJ, White, TT, Marchioro, TL, Woo, SLC and Davie, EW. "Cloning and sequence of cDNA coding for alpha-1-antitrypsin." Proc. Natl. Acad. Sci. U.S.A. 78: 6826-6830, 1981.

9. Lam, LH, Silbert, JE and Rosenberg, RD. "The Separation of Active and Inactive Forms of Heparin." Biochem. Biophys. Res. Commun. 69: 570-577, 1976.

10. Lindahl, U, Backstrom, G, Thunberg, L and Leder, IG. "Evidence for a 3-O-sulfated D-glucosamine Residue in the Antithrombin-Binding Sequence of Heparin." Proc. Natl. Acad. Sci. U.S.A. 77: 6551-6555, 1980.

11. Casu, B, Oreste, P, Torri, G, Zopetti, G, Choay, J, Lormeau, J and Petitou, M, Sinay, P. "The Structure of Heparin Oligosaccharide Fragments with High Anti-(Factor Xa) Activity Containing the Minimal Antithrombin III-Binding Sequence." Biochem. J. 197: 599-609, 1981.

12. Oosta, GM, Gardner, WT, Beeler, DL, and Rosenberg, RD. "Multiple Functional Domains of the Heparin Molecule." Proc. Natl. Acad. Sci. U.S.A. 78: 829-833, 1981.

13. Danielsson, A, Raub, E, Lindahl, U and Bjork, I. "Role of Ternary Complexes, in Which Heparin Binds Both Antithrombin and Proteinase, in the Acceleration of the Reactions between Antithrombin and Thrombin or Factor Xa." J. Biol. Chem. 261: 15467-15473, 1986.

14. Griffith, MJ. "Kinetics of the Heparin-enhanced Antithrombin III/Thrombin Reaction. Evidence for a Template Model for the Mechanism of Action of Heparin." J. Biol. Chem. 257: 7360-7365, 1982.

15. Pomerantz, MW, and Owen, WG. "A Catalytic Role for Heparin. Evidence for a Ternary Complex of Heparin Cofactor, Thrombin, and Heparin." Biochim. Biophys. Acta. 535: 66-77, 1978.

16. Blackburn, MN and Sibley, CC. "The Heparin Binding Site of Antithrombin III. Evidence for a Critical Tryptophan Residue." J. Biol. Chem. 255: 824-826, 1980.

17. Blackburn, MN, Smith, RL, Sibley, CC and Johnson, VA. "Heparin Binding and Protease Inhibitor Activity of Chemically Modified Antithrombin III." Ann. NY Acad. Sci.. 370: 700-707, 1981.

18. Blackburn, MN, Smith, RL, Carson, J and Sibley, CC. "The Heparin Binding Site of Antithrombin III. Identification of a Critical Tryptophan in the Amino Acid Sequence." J. Biol. Chem. 259: 939-941, 1984.

19. Villanueva, GB and Allen, N. "Refolding Properties of Antithrombin III. Mechanism of Binding to Heparin." J. Biol. Chem. 258: 14048-14053, 1983.

20. Fish, WW, Danielsson, A, Nordling, K, Miller, SH, Lam, CF
    and Bjork, I. "Denaturation Behavior of Antithrombin in
    Guanidine Hydrochloride. Irreversibility of Unfolding
    Caused by Aggregation." Biochemistry. 24: 1510-1517, 1985.

21. Rosenberg, RD and Damus, PS. "The Purification and
    Mechanism of Action of Human Antithrombin-Heparin
    Cofactor." J. Biol. Chem. 248: 6490-6505, 1973.

22. Pecon, JM and Blackburn, MN. "Pyridoxylation of Essential
    Lysines in the Heparin-Binding Site of Antithrombin III."
    J. Biol. Chem. 259: 935-938, 1984.

23. Liu, CS and Chang, JY. "The Heparin Binding Site of Human
    Antithrombin III. Selective Chemical Modification at Lys
    114, Lys 125, and Lys 287 Impairs Its Heparin Cofactor
    Activity." J. Biol. Chem. 262: 17356-17361, 1987.

24. Chang, JY. "Binding of Heparin to Human Antithrombin III
    Activates Selective Chemical Modification at Lysine 236.
    Lys-107, Lys-125, and Lys-136 are Situated within the
    Heparin-Binding Site of Antithrombin III." J. Biol. Chem.
    264: 3111-3115, 1989.

25. Gettins, P and Wooten, EW. "On the Domain Structure of
    Antithrombin III. Tentative Localization of the Heparin
    Binding Region Using 1-H NMR Spectroscopy." Biochemistry.
    26: 4403-4408, 1987.

26. Koide, T, Ohta, Y and Odani, S. "Chicken Antithrombin.
    Isolation, Characterization, and Comparison with Mammalian
    Antithrombins and Chicken Ovalbumin." J. Biochem. 91:
    1223-1229, 1982.

27. Koide, T, Odani, S, Takahashi, K, Ono, T and Sakuragawa, N.
    "Antithrombin III Toyama: Replacement of Arginine-47 by
    Cysteine in Hereditary Abnormal Antithrombin III that Lacks
    Heparin-Binding Ability." Proc. Natl. Acad. Sci. U.S.A.
    81: 289-293, 1984.

28. Brunel, F, Duchange, N, Fischer, AM, Cohen, GN and Zakin,
    MM. "Antithrombin III Alger: A New Case of Arg 47 - Cys
    Mutation." Am. J. Hematol. 25: 223-224, 1987.

29. Owen, MC, Borg, J, Soria, C, Soria, J, Caen, J and Carrell,
    RW. "Heparin Binding Defect in a New Antithrombin III
    Variant: Rouen, 47 Arg to His." Blood. 69: 1275-1279,
    1987.

30. Borg, J, Owen, MC, Soria, J, Caen, J and Carrell, RW.
    "Proposed Heparin Binding Site in Antithrombin Based on Arg
    47. A New Variant Rouen II, 47 Arg to Ser." J. Clin.
    Invest. 81: 1292-1296, 1988.

31. Chang, JY and Tran, TH. "Antithrombin III Basel.
    Identification of a Pro-Leu Substitution in a Hereditary
    Abnormal Antithrombin with Impaired Heparin Cofactor
    Activity." J. Biol. Chem. 261: 1174-1176, 1986.

32. Peterson, CB and Blackburn, MN. "Isolation and
    Characterization of an Antithrombin III Variant with

Reduced Carbohydrate Content and Enhanced Heparin Binding."
J. Biol. Chem. 260: 610-615, 1985.

33. Brennan, SO, George, PM and Jordan, RE. "Physiological
    Variant of Antithrombin III Lacks Carbohydrate Sidechain at
    Asn 135." FEBS Letters. 219: 431-436, 1987.

34. Brennan, SO, Borg, J and George, PM. "New Carbohydrate site
    in Mutant Antithrombin (7Ile --> Asn) with Decreased
    Heparin Affinity." FEBS Letters. 237: 118-122, 1988.

35. Rosenfeld, L and Danishefsky, I. "A Fragment of
    Antithrombin that Binds both Heparin and Thrombin." Biochem
    J. 237: 639-646, 1986.

36. Smith, JW and Knauer, DJ. " A Heparin Binding Site in
    Antithrombin III. Identification, Purification and Amino
    Acid Sequence." J Biol Chem. 262: 11964-11972, 1987.

37. Samama, JP, Delarue, M, Mourey, L, Choay, J and Moras, D.
    "Crystallization and Preliminary Crystallographic Data for
    Bovine Antithrombin III." J. Mol. Biol. 210: 877-879,
    1989.

38. Huber, R and Carrell, RW. "Implications of the Three-
    dimensional Structure of Alpha-1-Antitrypsin for Structure
    and Function of Serpins." Biochemistry. 28: 8951-8966,
    1989.

39. Bjork, I and Fish, WW. "Production in Vitro and Properties
    of a Modified Form of Bovine Antithrombin, Cleaved at the
    Active Site by Thrombin." J. Biol. Chem. 257: 9487-9493,
    1982.

40. Lau, HK and Rosenberg, RD. "The Isolation and
    Characterization of a Specific Antibody Population Directed
    against the Thrombin-Antithrombin Complex." J. Biol. Chem.
    255: 5885-5893, 1980.

41. McDuffie, FC, Peterson, JM, Clark, G and Mann, KG.
    "Antigenic Changes Produced by Complex Formation between
    Thrombin and Antithrombin-III." J. Immunol. 127: 239-244,
    1981.

42. Wallgren, P, Nordling, K and Bjork, I. "Immunological
    Evidence for a Proteolytic Cleavage at the Active Site of
    Antithrombin in the Mechanism of Inhibition of Coagulation
    Serine Proteases." Eur. J. Biochem. 116: 493-496, 1981.

43. Herion, P, Francotte, M, Siberdt D, Garduno Soto, G,
    Urbain, J and Bollen, A. "Monoclonal Antibodies Against
    Plasma Protease Inhibitors: Production and
    Characterization of 15 Monoclonal Antibodies Against Human
    Antithrombin III. Relation Between Antigenic Determinants
    and Functional Sites of Antithrombin III." Blood. 65:
    1201-1207, 1985.

44. Peterson, CB and Blackburn, MN. "Antithrombin Conformation
    and the Catalytic Role of Heparin. I. Does Cleavage by
    Thrombin Induce Structural Changes in the Heparin-Binding
    Region of Antithrombin?" J. Biol. Chem. 262: 7552-7558,
    1987.

45. Villanueva, G and Danishefsky, I. "Conformational Changes
    Accompanying the Binding of Antithrombin III to Thrombin."
    Biochemistry. 18: 810-817, 1979.

46. Nordenman, B and Bjork, I. "Binding of Low-Affinity and
    High-Affinity Heparin to Antithrombin. Ultraviolet
    Difference Spectroscopy and Circular Dichroism Studies."
    Biochemistry. 17: 3339-3344, 1978.

47. Olson, ST and Shore, JD. "Transient Kinetics of Heparin-
    catalyzed Protease Inactivation by Antithrombin III. The
    Reaction Step Limiting Heparin Turnover in Thrombin
    Neutralization." J. Biol. Chem. 261: 13151-13159, 1986.

48. Peterson, CB and Blackburn, MN. "Antithrombin Conformation
    and the Catalytic Role of Heparin. II. Is the Heparin-
    Induced Conformational Change in Antithrombin Required for
    Rapid Inactivation of Thrombin?" J. Biol. Chem. 262: 7559-
    7566, 1987.

49. Einarsson, R and Andersson, LO. "Binding of Heparin to
    Human Antithrombin III as Studied by Measurements of
    Tryptophan Fluorescence." Biochim. Biophys. Acta. 490:
    104-111, 1977.

50. Villanueva, G and Danishefsky, I. "Evidence for a Heparin-
    Induced Conformational Change on Antithrombin III."
    Biochem. Biophys. Res. Commun. 74: 803-809, 1977.

51. Peterson, CB, Noyes, CM, Pecon, JM, Church, FC and
    Blackburn, MN. "Identification of a Lysine Residue in
    Antithrombin Which is Essential for Heparin Binding." J.
    Biol. Chem. 262: 8061-8065, 1987.

52. Olson, ST, Srinivasan, KR, Bjork, I and Shore, JD. "Binding
    of High Affinity Heparin to Antithrombin III. Stopped Flow
    Kinetic Studies of the Binding Interaction." J. Biol. Chem.
    256: 11073-11079, 1981.

PROTEIN C: GENE STRUCTURE AND PROTEIN SYNTHESIS

George L. Long

Department of Biochemistry
University of Vermont
Burlington, VT 05405

INTRODUCTION

Protein C is a vitamin K-dependent plasma glycoprotein that, after activation by thrombin-thrombomodulin upon the endothelial cell surface, serves as a feedback down-regulator of the coagulation cascade by specifically degrading the protein cofactors VIIIa and Va (Figure 1). The biological role of protein C and the vascular endothelium has been recently reviewed in a succinct fashion[1]. Protein C in a less understood manner also enhances the process of fibrinolysis. Because of its antithrombotic as well as pro-fibrinolytic properties, protein C is considered to be a potential agent for antithrombotic therapy and a candidate for production by recombinant DNA technologies. The scope of this paper is to review: (i) the organization of the human protein C gene, particularly as it relates to the genes for homologous vitamin K-dependent proteins and to genetic protein C deficiency; and (ii) the biosynthesis of recombinant protein C in forms that are properly processed and biologically active.

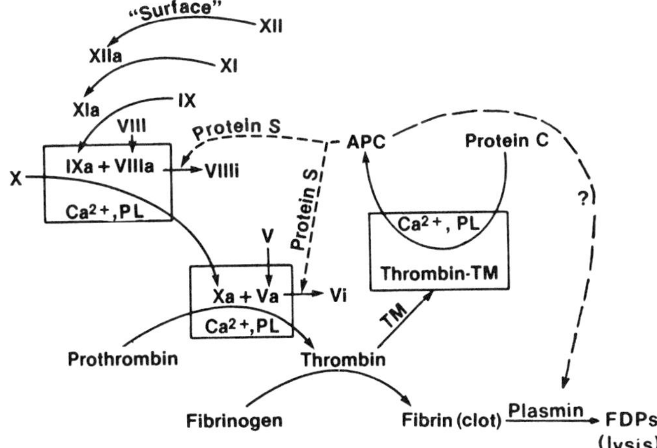

Fig. 1.   Pathway of blood coagulation and clot lysis. Boxes enclose multicomponent enzymatic complexes on the vascular cell surface. Abbreviations: phospholipid (PL), activated protein C (APC), thrombomodulin (TM), fibrin degradation products (FDPs).

Two laboratories have independently isolated and characterized overlapping lambda-human genomic clones for protein C, with essentially identical results[2,3]. The report of Foster and coworkers presents a continuous 11.6 kilobase nucleotide sequence containing the protein C gene[2]. A more complete 5' cDNA sequence reported by Beckmann et al.[4] in concert with primer extension studies upon mRNA template[3] has revealed the existence of an intervening sequence (intron) in the 5' non-coding portion of the initial transcript, not identified by Foster et al.[2]. Analyses of mRNA and cDNA species by several laboratories indicate that there are two alternatively used poly A recognition sites separated by 229 nucleotides in the 3' non-translated region of the gene. The distance from the proposed transcription start site to the more distant polyadenylation site is 10,772 nucleotides.

The protein C gene is composed of eight introns and nine exons, as shown schematically in Figure 2. Intron F, interrupting the codon for amino acid #137 of the mature protein, is the largest (2,669 nucleotides) and contains two "Alu" family sequences. The Alu family consensus sequence (~150-300 nucleotides in length) is distributed throughout the human genome at an average frequency of about once every 5 kilobases[5]. In contrast, introns D and E are only 92 and 102 nucleotides in length, respectively.

Southern hybridization mapping and chromosome location using human-mouse hybrid cell lines indicate that there is one locus for the protein C gene and that it is located on chromosome 2 [6,7].

Relationship of intron positions to structural domains

The vitamin K-dependent plasma proteins, including protein C, are composed of relatively well defined structural domains or modules that have been conserved during their evolution from common ancestral proteins (reviewed recently by Furie and Furie[8]). An analysis of the position of introns relative to the structural domains of these proteins[8,9] presents one of the strongest arguments for the hypothesis put forth several years ago by Gilbert[10] that introns separate protein structural-functional domains and play a role in the genetic interchange of common protein structural modules.

Figure 3 shows the position of introns relative to the nascent protein C sequence. The positions of introns are in good agreement with defined structural domains. The first coding exon (Exon 2) contains the signal peptide. Exon 3 consists of the majority of the propeptide and the γ-carboxyglutamate (Gla) domain. Exon 4 is only nine amino acids in length and corresponds to the "aromatic stack" structure recently described by Soriano-Garcia et al.[11]. Exons 5 and 6 are each domains homologous to epidermal growth factor[12]. Exon 7 contains the peptide cleavage sites for both single to two-chain conversion prior to secretion and activation of the circulating zymogen by thrombin. The trypsin-like serine protease domain is interrupted by only one intron in the case of protein C.

Table 1 shows the position of introns in the coding portions of the protein C gene and those of homologous vitamin K-dependent plasma protein genes. Once maximum alignment of the proteins are made to one another, allowing for small insertions and deletions, the position of introns are identical for the homologous portions of the proteins, reflecting a remarkable conservation of intron position during the course of the genes' evolution. As also shown in Table 1, even the type of intron, based upon position in the amino acid triplet codon, has been conserved for these genes.

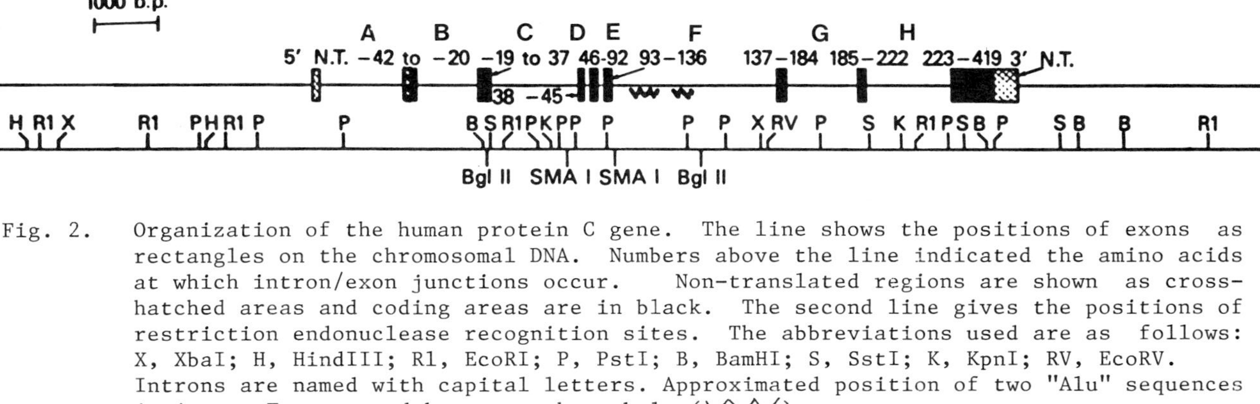

Fig. 2.   Organization of the human protein C gene. The line shows the positions of exons as rectangles on the chromosomal DNA. Numbers above the line indicated the amino acids at which intron/exon junctions occur. Non-translated regions are shown as cross-hatched areas and coding areas are in black. The second line gives the positions of restriction endonuclease recognition sites. The abbreviations used are as follows: X, XbaI; H, HindIII; R1, EcoRI; P, PstI; B, BamHI; S, SstI; K, KpnI; RV, EcoRV. Introns are named with capital letters. Approximated position of two "Alu" sequences in intron F are noted by sawtooth symbols ($\wedge\!\wedge\!\wedge$).

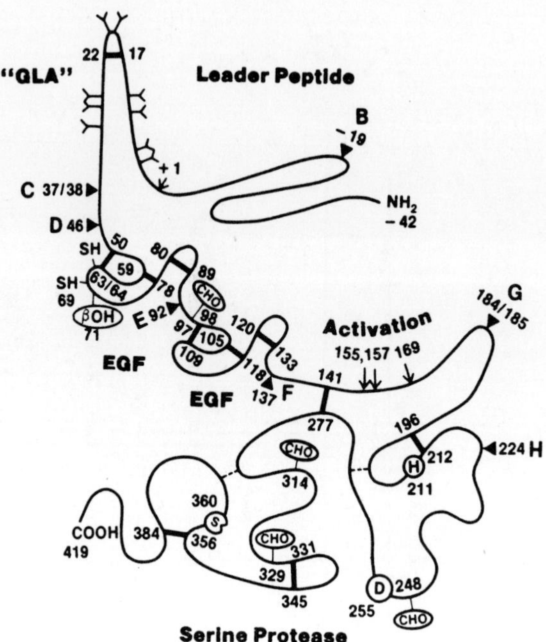

Fig. 3. Schematic representation of protein C structure and position of introns. Single chain precursor is represented by the thin curving line. Numbers refer to amino acid positions in the precursor protein. Proposed positions of the disulfide bridges (thick bars), β-hydroxyaspartate (β-OH), γ-carboxyglutamate (Y), and carbohydrate attachment (CHO) in the mature protein are also shown, and are based upon homology with related proteins. The catalytic triad (Ser, His, Asp) of the serine protease domain are represented by the circled letters S, H and D, respectively. Known proteolytic cleavage sites resulting in two-chain, activated protein C are shown with arrows. The corresponding positions of introns B-H in the gene are portrayed by closed triangles (▲).

Table 1. Comparison of Intron Position and Type for Human Protein C with Homologous Vitamin K-dependent Plasma Proteins

Intron	B	C	D	E	F			G
Protein C	-19	37/38	46	92	137			184/185
Protein S	-16	37/38	46	(75)[b]	116	160	202	242/243
Factor VII	-17	37/38	46	84	131			167/168
Factor IX	-17	38/39	47	85	128			195/196
Factor X	-17	37/38	46	84	128			209/210
Prothrombin	-17	37/38	46					
Intron Type[a]	I	0	I	I	I	I	I	0

[a]Position in coding frame.
[b]The exon between Introns D and E of protein S is non-homologous.

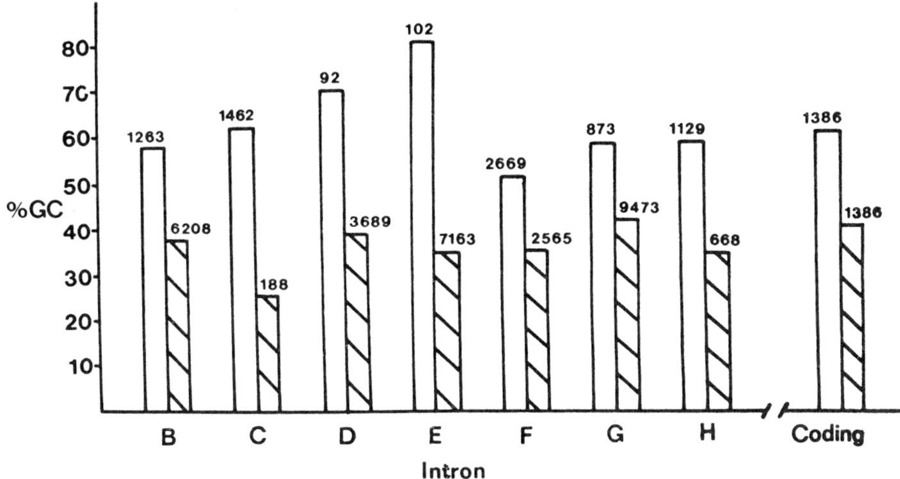

Fig. 4.   Comparison of the guanine + cytosine base content for the human
protein C and factor IX genes. The percent G+C for introns within
the coding portions of human protein C (open bar) and factor IX
(striped bar) genes are shown. Also shown is the G+C content for
the two coding sequences. Numbers above each bar show the number
of nucleotides in each intron. Intron initials are as shown in
Fig. 2. Compositional data are taken from Foster et al.[2] and
Yoshitake et al.[27].

In contrast to the conservation of intron position and type for the
vitamin K-dependent plasma protein genes, there is no apparent conservation
of intron size or base composition. Figure 4 compares the size and base
composition of introns, as well as the coding regions, for the protein C and
Factor IX genes. Comparison of the guanine plus cytosine content for the
sum of positionally common introns shows that protein C is 19% higher in G
+ C content than Factor IX. The coding segments of protein C is similarly
higher in G + C content (21%) compared to Factor IX. The G + C content for
Factors VII and X, as well as codon usage patterns, are similar to those of
protein C[13]. No clear explanation for the compositional divergence of the
Factor IX gene exists, but may be related to the location of this gene on
the X chromosome and consequential spontaneous deamination of methylated
cytosine, resulting in C→T and G→A (complimentary strand) transition[13].

Genetic analysis of protein C deficiency

On the basis of the frequency of genetic protein C deficiency in
patients with thrombosis and the incidence of thrombosis in the general
population, it was estimated that the level of genetic heterozygous protein
C deficiency in the normal population is about 1/16,000[14]. As pointed out
by Miletich and coworkers, this would predict the incidence of homozygous
protein C deficiency to be about $1/10^9$, far below the observed level of
newborns with purpura fulminans attributable to absent or very low levels
of protein C[15]. Additionally, members of purpura fuminans families have an
expected distribution of 35-65% normal protein C levels, predicted for the
heterozygote, but do not have higher levels of thrombosis than the normal
population[15]. This led Miletich and his coworkers to examine genetic protein
C deficiency in the normal population. Based upon measurements of protein

C levels in over 5,000 healthy individuals and detailed family and personal histories on people having $\leq 65\%$ normal plasma levels, they concluded that the incidence of heterozygous protein C deficiency in the normal population is about 1 in 200-300, significantly higher than that suggested by Broekmans et al.[14]. The second important conclusion from their study is that in most cases heterozygous protein C deficiency is not sufficient by itself to result in thrombosis, and that other presently undefined factors are also involved.

Currently, over fifty genetically protein C deficient families have been analyzed for mutations in the protein C gene. Considering the large number of families studied and the apparently high incidence of genetic protein C deficiency, the number of documented instances of protein C mutation is disappointingly low.

Table 2. Genetic mutations associated with protein C deficiency

Reference	% Antigen	% Activity	Mutation	Explanation
16	~50	~50	deletion of one allele	half gene dosage
17	Father 50	50	deletion of one allele	half gene dosage
	Daughter 18	0	deletion of one allele $Arg_{169}(CGG)\rightarrow Trp(TGG)$	half gene dosage no cleavage of activation peptide
	Mother 80	60	$Arg_{169}(CGG)\rightarrow Trp(TGG)$	no cleavage of activation peptide
18	30	30	$Arg_{169}(CGG)\rightarrow Trp(TGG)$	no cleavage of activation peptide
19	50	50	Pvu II RFLP $Arg_{306}(CGA)\rightarrow Stop(TGA)$	truncated protein no active site Ser.
19	50	50	Bam HI RFLP $Trp_{402}(TGG)\rightarrow Cys(TGC)$	change in conserved residue and/or abnormal S-S bonding-unstable?

Table 2 summarizes the results of studies in which a clear genetic mutation has occurred and for which explanations for reduced levels of protein C can be offered. The first case is that involving two siblings each having half the normal band intensity upon Southern hybridization with a protein C cDNA probe[16]. These data were interpreted as resulting from the deletion of one entire protein C allele, leading to half normal gene dosage and antigen levels. The same type of mutation has been reported in a Japanese family in which the propositus is a double heterozygote for protein C[17]. She, as well as her father carry a mutation involving deletion of the entire protein C gene. DNA sequencing of all the exonic segments for the other allele revealed that she and her mother also carry a point mutation

(CGG→TGG) resulting in Arg_{169}→Trp. Normal activation of protein C by thrombin proceeds by cleavage on the carboxyl side of Arg_{169}, probably explaining the absence of any protein C activity. Antigen levels in the mother and daughter were also somewhat lowered, suggesting that the Arg→Trp mutation may also affect the processing, stability, or clearance of the protein. The same C→T mutation in codon 169 has been independently reported by Grundy _et al_. in an unrelated individual with a history of thrombosis[18].

Romeo and coworkers have used restriction fragment length polymorphism (RFLP) to study several protein C deficient families with thrombosis and have reported two independent RFLPs[19]. One variant is a _Bam_ HI RFLP resulting from the disappearance of a _Bam_ HI restriction endonuclease site in one allele. Direct sequencing of this mutant site revealed a missense mutation (TGG→TGC) resulting in Trp_{402}→Cys. As pointed out by the authors, Trp_{402} is highly conserved in serine proteases, and is thought to interact with an adjacent hydrophobic face of the protein. Mutation to Cys might significantly alter the required stability of this hydrophobic interaction. Alternatively, introduction of a Cys residue might lead to altered disulfide bonding and resultant changes in structure and stability. The second variant exhibited a _Pvu_ II RPLP resulting from creation of a new _Pvu_ II restriction site. DNA sequencing revealed a nonsense mutation (CGA→TGA) resulting in Arg_{306}→Stop. This mutation would result in a truncated protein lacking 114 carboxy-terminal amino acids that might alter its processing, stability or clearance and lead to the observed 50% of normal circulating antigen level. Additionally, the active site Ser_{360} would not be present.

Most of the studies aimed at detecting alterations in the gene have involved RFLP analysis and have probably missed many point mutations or small insertions/deletions. With the advent of polymerase chain reaction (PCR) and sophisticated denaturing gel and hybridization techniques it is reasonable to expect that the detection of mutations will be more fruitful. However, it is important to keep in mind, based upon the large range of normal protein C levels, that some individuals believed to be genetically protein C deficient may not actually be so. Also, in some cases where protein C is manifested as an inherited trait, the actual cause for the reduction in circulating levels or enzyme activity may not be due directly to the expression of the gene or the structure of the protein, but rather to genetic changes affecting components that interact with protein C, such as binding proteins or inhibitors.

BIOSYNTHESIS OF PROTEIN C

The second part of this review will address the topic of the biosynthesis of protein C, and will present selected studies relating to three aspects of protein C production: i) the role of the propeptide and γ-carboxylation; ii) cleavage of single-chain to two-chain protein C; and iii) activation of the protein C zymogen. The discussion will focus on studies performed with mammalian cell cultures using a variety of expression vectors. In the case of protein C, requiring diverse and extensive post-translational modification for biological activity, the challenge has not been so much one of increasing expression levels, but rather that of producing properly modified protein. Consequently, little attention in this brief review will be made to the particular expression vectors and attempts to increase expression levels. Obviously, for production purposes and utility as a recombinant therapeutic agent, the issue of amounts must ultimately also be addressed.

Based upon the amino acid sequence similarities between vitamin K-dependent proteins, including the bone protein osteocalcin, it was suggested that the propeptide located between a typical signal peptide and the amino terminus of the mature protein might play an important role in directing γ-carboxylation[20,21]. Subsequently, a large body of data from several laboratories, most notably those of the Furies, has strengthened this hypothesis.

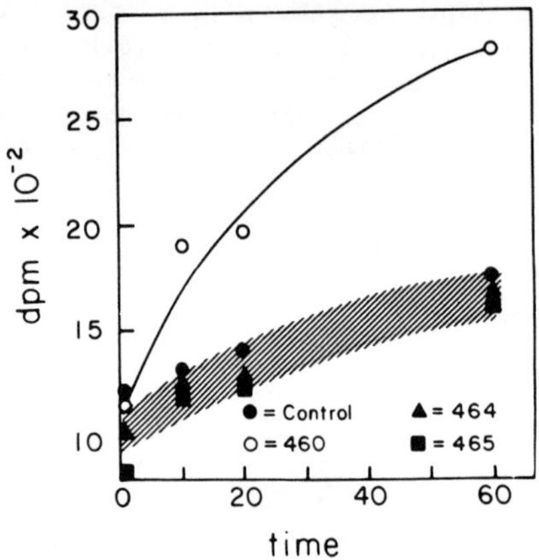

Fig. 5.   Time course of $^{14}CO_2$ incorporation (γ-carboxylation) into protein: Effect of the carboxyl-terminal half of the protein C propeptide. Protein C produced in E. coli was used as a substrate for the rat liver microsomal γ-carboxylase, and TCA-precipitable counts were determined as a function of time (min). Protein C substrate consisted of: nascent protein C amino acids 1-419 plus -1 to -10 of the propeptide (mutant 460); nascent protein C amino acids 1-419 (mutant 464); and nascent protein C lacking the Gla domain, amino acids 42-419 (mutant 465).

An early experiment to test the effect of the propeptide on γ-carboxylation was performed using recombinant protein C produced in E. coli as a substrate for the rat liver microsomal enzyme system[22]. Figure 5 shows the extent of $^{14}CO_2$ incorporation into TCA precipitable counts (γ-carboxylation) as a function of time. Protein C lacking the propeptide or lacking the propeptide plus the Gla domain (amino acids 1-41) showed no vitamin K-dependent incorporation above the normal endogenous (control) levels. In contrast however, protein C containing the carboxyl terminal ten amino acids of the propeptide resulted in a significant and protein-dependent (not shown) increase in γ-carboxylation.

Foster and coworkers have taken a different tact and examined the role of the propeptide in cultured human kidney 293 cells[23]. In this study deletion mutants involving the propeptide region were constructed and the expressed protein from the media analyzed. As one indication of the extent of γ-carboxylation, barium citrate precipitation was determined with the results shown in Figure 6. As indicated in Figure 6, progressive deletions of amino acids from the carboxy terminus of the propeptide (residue #-1) results in progressive reduction in γ-carboxylation. The nine amino acid (Δ9) and seventeen amino acid (Δ17) deletion mutants as well as wild-type recombinant protein C were purified and further characterized. Amino acid N-terminal sequencing of wild-type protein C revealed proper cleavage between amino acids -1 and +1. In contrast, both the Δ9 and Δ17 mutants began with amino acid -24, indicating a lack of propeptide cleavage and strongly suggesting that the signal peptide cleavage site is between residues -25 and -24. The amino acid sequencing also showed significant increases in Glu content for the Δ9 and Δ17 mutants at positions +6 and +7 of the mature protein, indicating reduction in γ-carboxylation. All three species appeared to be cleaved to the two-chain form and activated by the snake venom protease Protac C to catalyze small synthetic substrates but neither the Δ9 nor the Δ17 proteins were effective in a plasma-cephalin-kaolin clotting assay.

The group at Eli Lilly & Co. have also looked in some detail at the effect of recombinant protein C γ-carboxylation on biological activity. Table 3 presents data taken from two recent reports[24,25]. The first is that using human kidney 293 cells and the second using a transformed syrian hamster tumor cell line AV12. The table indicates that γ-carboxylation, as measured by chemical Gla analysis, is complete and independent of expression levels in clonal 293 cell lines even up to secretion levels greater that 20 μg/10^6 cells, 24 hr. Two different species of protein C having different Gla content are produced in the syrian hamster cell line. Biological activity of thrombin-thrombomodulin activated recombinant protein C was determined in a clotting assay and, as shown in Table 3, is proportional to the extent of γ-carboxylation. Extrapolation of the data to zero inhibition of clotting would occur at a Gla content of 6 Gla/mol enzyme. In contrast, amidolytic activity using a small synthetic substrate (S-2238) is relatively independent of Gla content.

Conversion of single-chain to two-chain

Shortly prior to or upon secretion, plasma protein C is cleaved by a trypsin-like and a carboxypeptidase B-like protease to generate the NH_2-terminal light chain, the COOH-terminal heavy chain, and a Lys-Arg connecting dipeptide. Cleavage from the single to the two-chain form has no apparent effect on the ability of protein C to be activated or its biological activity. Studies involving the expression of recombinant protein C in a variety of cell lines has provided useful information on the single to two-chain cleavage process. Table 4 summarizes the extent of conversion for several expression systems commonly used. Not shown in the table is the general observation that within a particular expression system, as the levels of expression increase they reach a point where the extent of cleavage becomes reduced.

Of particular interest to the group at Zymogenetics was the marked reduction of protein C cleavage by BHK cells compared to 293 cells, compared to nearly total cleavage of recombinant Factor X by both cell lines[26]. In an elegant set of mutational experiments portrayed in Figure 7, mutations were made in the region of protein C cleavage; and the extent of expressed recombinant protein C cleavage measured. The results indicate that the protease system in BHK cells strongly prefers the presence of Arg at

<pre>   PRE   PRO   MATURE
-42   -24   -21   -18   -15   -12   -9   -6   -3   +1
  M V Q L T S L L L F V A T W G I S G T P A P L D S V F S S S E R A H Q V L R I R K R A N S F L γ γ L...
                            ↑   ↑
</pre>

Table II: Adsorption and Precipitation of Recombinant Protein C with Barium Citrate[a]

plasmid	% in barium citrate	
	supernatant	pellet
native protein C	30	70
Δ4 mutant	41	59
Δ9 mutant	60	40
Δ12 mutant	47	53
Δ15 mutant	74	26
Δ16 mutant	94	6
Δ17 mutant	94	6

[a] Human kidney cells were transfected with protein C expression plasmids as detailed under Materials and Methods. After 48 h, the culture media were collected, and the protein C was precipitated with barium citrate as described under Materials and Methods. Protein C present in the barium citrate supernatant and the redissolved pellet was measured by ELISA. The results presented are the average of three independent measurements.

Taken from Foster et al. Biochemistry 26 7003 (1987).

Fig. 6.   Effect of Protein C propeptide deletions on γ-carboxylation. Top: Amino terminal sequence of the protein C precursor. Arrows show cleavage sites for generation of the pre or signal peptide, the propeptide, and the mature light chain. Bottom: Table II taken from Foster et al. Biochemistry 26: 7003 (1987), with permission. Mutants in Table II consisted of deletions from position # -1 toward the left in the sequence at the top of the figure. Example: Δ 4 mutant lacks amino acids -1 to -4. The degree of protein C precipitation is assumed to be proportional to the level of γ-carboxylation.

Table 3.   Relationship of γ-carboxylation to functional activity.

Source of Human Protein C	Protein C Produced (μg/10^6 cells, 24 hr)	GLA (mol/mol HPC)	Functional activity (units/mg)	
			anticoagulant	amidolytic
Plasma derived	-	8.5 ± 0.5	250	20 ± 2
293-CC31	1.5		350 ± 25	25 ± 2
293-21-10-3	2.6	9.3 ± 0.6	337 ± 35	24
293-21-10b-10	5.3		340 ± 38	19 ± 2
293-hdA6	21.0		333	19
AV12-664	17	9.1 ± 0.3	283	24 ± 3
		7.0 ± 0.3	75	

Adapted from Walls et al. Gene 81 139 (1989) and Ehrlich et al. J. Biol. Chem. 264 14298 (1989). Used with permission.

position -4 relative to the trypsin-like cleavage site, and that activity is abolished if a negatively charged amino acid (Glu) resides in this position. The location of Lys or Arg in the dipeptide positions seems unimportant but replacement of these positively charged residues by neutral amino acids (Ala) abolishes enzymatic conversion. Sequence analysis of wild-type, PC962, and PC1401 protein indicated that cleavage had occurred carboxy-terminal of Leu_{155} and Arg_{157} in all three cases (positions -3 and -1 in Fig. 7).

Table 4. Percent of two-chain protein C produced by different cell types.

Source	% Two-chain Human Protein C
Human Plasma (liver)	80-90
Human Kidney 293	80-90
Baby Hamster Kidney (BHK)	30
Mouse C127	50
Human Hep G2	50

Mutant	-4 -3 -2 -1 light chain heavy chain	Approximate % Two-chain from BHK
PC - wt	...Ser-His-Leu⌉Lys-Arg⌈Asp-Thr-Glu...	30
FX - wt	Ser-Glu-Arg-Arg-Lys-Arg-Ser-Val-Ala	100
PC962	His-Leu-Arg-Arg-Lys-Arg-Asp-Thr-Glu	100
PC1401	Ser-(Arg)-Leu-Lys-Arg-Asp-Thr-Glu	100
PC1567	Ser-(Leu)-Leu-Lys-Arg-Asp-Thr-Glu	30
PC1713	Ser-(Glu)-Leu-Lys-Arg-Asp-Thr-Glu	0
PC1461	Ser-His-Leu-(Arg)-Arg-Asp-Thr-Glu	30
PC1869	Ser-His-Leu-Lys-(Lys)-Asp-Thr-Glu	30
PC1477	Ser-His-Leu-(Ala)-(Ala)-Asp-Thr-Glu	0

Fig. 7. Influence of amino acid sequence on conversion of single chain to two-chain protein C in baby hamster kidney cells. Modified from Foster et al. Biochemistry 29: 347 (1990), with permission. The left column shows the protein C wild-type (PC-wt), Factor X wild-type (FX-wt) and protein C mutants tested. The middle part of the figure shows the protein amino acid sequences in the region of single to two-chain conversion. Arrows show the normal sites of cleavage to generate the mature two-chain form of protein C and boxes indicate the resulting carboxy and amino terminii of the light and heavy chains, respectively. Circles indicate single amino acid substitutions. The column on the right reports the extent of cleavage, as indicated by the amount of immunoprecipitated metabolically labeled (^{14}C-Cys) protein C migrating as heavy chain upon SDS-polyacrylamide gel electrophoresis.

It is believed that protein C must be in its activated form to be therapeutically useful. Presently, protein C can be activated by the proteolytic action of thrombin-thrombomodulin or the snake venom protease, both of which are problematic in reference to generating a therapeutically satisfactory product. Consequently a major hurdle to be overcome in the development of protein C as an antithrombotic agent is its activation prior to administration.

A recent report from the group at Eli Lilly presents a possible solution to this problem. They have designed mutants in the activation peptide region of protein C in an attempt to generate secreted protein C in the already activated state[25]. Figure 8 shows the amino acid sequence in the area of the activation peptide and the two constructed mutants. The first mutant consists of a deletion of the entire twelve amino acid activation peptide. The second mutant has in place of the activation peptide an octapeptide that corresponds to the carboxy-terminus of the processed human insulin receptor α-chain. All cells are thought to be capable of cleaving the α-chain precursor prior to its expression upon the cell surface by a trypsin-like protease on the carboxy-terminal side of the octapeptide sequence. As indicated in the lower part of Figure 8, secreted recombinant protein C with the insulin receptor segment was found to be fully active in a synthetic substrate assay. Full anticoagulant activity was also seen in an activated partial thromboplastin time clotting assay. In contrast, wild-type protein C required activation by thrombin-thrombomodulin. The activation peptide deletion mutant exhibited no biological activity prior to or following thrombin-thrombomodulin treatment, even though it migrated on SDS-PAGE in a manner similar to wild-type activated protein C. Amino acid sequencing on the insulin receptor α-chain octapeptide mutant demonstrated that the trypsin-like cleavage was at the predicted Arg-Leu site. However, no data were provided relating to whether the octopeptide is cleaved from the generated light chain.

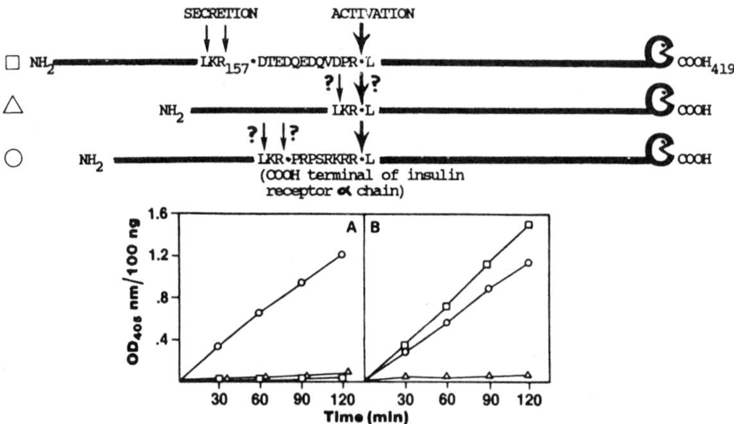

Expression of amidolytic activity by AV12 cells transiently transfected with pLPC(human protein C zymogen) (□), pLPC(ΔAP) (△), and pLPC(IRΔAP) (○). The conditioned serum-free media were mixed directly (A) or after incubation with thrombomodulin-thrombin (B) with S2238, and the increase in $A_{405\,nm}$ was observed over 2 h. $A_{405\,nm}$ readings obtained with serum-free media from MOCK-transfected cells were subtracted from all data.

Fig. 8. Mammalian cell expression of activated protein C. Top: Schematic representation of wild-type and mutant protein C cleavage (arrows) by AV12 mammalian cells. Unconfirmed sites of cleavage are shown with question marks. Bottom: Wild-type and mutant protein C amidolytic activity. Taken from Ehrlich et al. J. Biol. Chem. 264: 14298 (1989), with permission.

ACKNOWLEDGEMENTS

The author wishes to thank the groups at Eli Lilly & Co. and Zymogenetics for agreeing to presentation of their work in table and figure form, and to Lisa McNaney for secretarial assistance.

REFERENCES

1. Esmon NL, Esmon CT: Protein C and the endothelium. Sem in Thromb and Hemostas 14:210, 1988

2. Foster D, Yoshitake S, Davie EW: The nucleotide sequence of the gene for human protein C. Proc Natl Acad Sci USA 82:4673, 1985

3. Plutzky J, Hoskins JA, Long GL, Crabtree GR: Evolution and organization of the human protein C gene. Proc Natl Acad Sci USA 83:546, 1986

4. Beckmann RJ, Schmidt RJ, Santerre RF, Plutzky J, Crabtree GR, Long GL: The structure and evolution of a 461 amino acid human protein C precursor and its messenger RNA, based upon the DNA sequence of cloned human liver cDNAs. Nuc Acids Res 13:5233, 1985

5. Deininger PL, Jolly DJ, Rubin CM, Friedman T, Schmid CW: Base sequence studies of 300 nucleotide renatured repeated human DNA clones. J Mol Biol 151:17, 1981

6. Rocchi M, Roncuzzi L, Santamaria R, Archidiacono N, Dente N, Romeo G: Mapping through somatic cell hybrids and cDNA probes of protein C to chromosome 2, factor X to chromosome 13, and α_1-acid glycoprotein to chromosome 9. Hum Genet 74:30, 1986

7. Long GL, Marshall A, Gardner JC, Naylor SL: Genes for human vitamin K-dependent plasma proteins C and S are located on chromosomes 2 and 3, respectively. Somat Cell and Molec Genet 14:93, 1988

8. Furie B, Furie BC: The molecular basis of blood coagulation. Cell 53:505, 1988

9. Patthy L: Evolution of the proteases of blood coagulation and fibrinolysis by assembly from modules. Cell 41:657, 1985

10. Gilbert W: Why genes in pieces? Nature 271:501, 1978

11. Soriano-Garcia M, Park CH, Tulinsky A, Ravichandran KG, Skrzypczak-Jankun E: Structure of Ca^{2+} prothrombin fragment 1 including the conformation of the Gla domain. Biochemistry 28:6805, 1989

12. Banyai L, Varadi A, Patthy L: Common evolutionary origin of the fibrin-binding structures of fibronectin and tissue-type plasminogen activator. FEBS Lett 163:37, 1983.

13. Long GL: Structure and evolution of the human genes encoding protein C and coagulation factors VII, IX, and X. Cold Spr Harbor Symp 51:525, 1986

14. Broekmans AW, van der Linden IK, Velkamp JJ, Bertina RM: Prevalence of isolated protein C deficiency in patients with venous thrombotic disease and in the population. Thromb Haemost 50:350, 1983

15. Miletich J, Sherman L, Brose G: Absence of thrombosis in subjects with heterozygous protein C deficiency. N Engl J Med 317:991, 1987

16. Crabtree GR, Plutzky J, Marler R, Griffin J, Zarcharski L, Gruppo R, Sela N, Long G: Analysis of the organization of the human protein C gene in normal and protein C deficient individuals. Blood 64:261, 1984

17. Matsuda M, Sugo T, Sakata Y, Murayama H, Mimuro J, Tanabe S, Yoshitake S: A thrombotic state due to an abnormal protein C. N Engl J Med 319:1265, 1988

18. Grundy C, Chitolie A, Talbot S, Bevan D, Kakkar V, Cooper DN: Protein C London 1: recurrent mutation at Arg 169 (CGG→TGG) in the protein C gene causing thrombosis. Nuc Acids Res 17:10513, 1989

19. Romeo G, Hassan HJ, Staempfli S, Roncuzzi L, Cianetti L, Leonardi A, Vincent V, Mannucci PM, Bertina R, Peschle C, Cortease R: Hereditary thrombophilia: Identification of nonsense and missense mutations in the protein C gene. Proc Natl Acad Sci USA 84:2829, 1987

20. Long GL, Belagaje RM, MacGillivray RTA: Cloning and sequencing of liver cDNA coding for bovine protein C. Proc Natl Acad Sci USA 81:5653, 1984

21. Pan LC, Price PA: The propeptide of rat bone γ-carboxyglutamic acid protein shares homology with other vitamin K-dependent protein precursors. Proc Natl Acad Sci USA 82:6109, 1985

22. Suttie JW, Hoskins JA, Engleke J, Hopfgartner A, Ehrlich H, Bang NU, Belagaje RM, Schoner B, Long GL: Vitamin K-dependent carboxylase: Possible role of the substrate "propeptide" as an intracellular recognition site. Proc Natl Acad Sci USA 84:634, 1987

23. Foster DC, Rudinski MS, Schach BG, Berkner KL, Kumar AA, Hagen FS, Sprecher CA, Insley MY, Davie EW: Propeptide of human protein C is necessary for γ-carboxylation. Biochemistry 26:7003, 1987

24. Walls JD, Berg DT, Yan SB, Grinnell BW: Amplification of multicistronic plasmids in the human 293 cell line and secretion of correctly processed recombinant human protein C. Gene 81:139, 1989

25. Ehrlich HJ, Jaskunas SR, Ginnell BW, Yan SB, Bang NU: Direct expression of recombiant activated human protein C, a serine protease. J Biol Chem 264:14298, 1989

26. Foster DC, Sprecher CA, Holly RD, Gambee JE, Walker KM, Kumar AA: Endoproteolytic processing of the dibasic cleavage site in the human protein C precursor in transfected mammalian cells: Effects of sequence alterations on efficiency of cleavage. Biochemistry 29:347, 1990

27. Yoshitake S, Schach BG, Foster DC, Davie EW, Kurachi K: Nucleotide sequence of the gene for human factor IX (Anti-hemophilic factor B). Biochemistry 24:3736, 1985.

STRUCTURAL AND FUNCTIONAL PROPERTIES OF PROTEIN C

Frederick J. Walker

American Red Cross Blood Services and
The Departments of Medicine and Laboratory
Medicine, University of Connecticut
Farmington, CT 06032

INTRODUCTION

The discovery of protein C and its effects upon blood coagulation, fibrinolysis, and inflammation has changed the way we now think about the pathophysiology of thromboembolic disease and related disorders. Interestingly, this discovery was not the result of a planned approach to understanding a disease, but rather from a basic approach of attempting to understand the biochemistry of the vitamin K-dependent family of proteins. In fact all of the clinical information about protein C stems from information obtained in basic research laboratories. The first clue of the existence of protein C can be found in a 1947 paper that reports that factor Va is unstable in serum (1). It was subsequently observed that the instability could be removed if the plasma was adsorbed with aluminum hydroxide, an insoluble support that was known to adsorb vitamin K-dependent proteins. Seegers group observed that treatment of prothrombin with thrombin could elicit an inhibitor of coagulation. This led to the inhibitor's characterization and designation as autoprothrombin IIa (2). Marciniak (3) confirmed the observation and noted that the inhibitor was species specific. Some time later, Stenflo characterized a new vitamin K-dependent protein (4), named protein C (5) which turned out to be the zymogen of autoprothrombin IIa (6). Characterization of protein C and the discovery that activated protein C was a potent inhibitor of blood coagulation lead to the realization that protein C may be one of the keys to understanding the molecular basis of thrombotic disease. This paper describes structural properties that give rise both to its unique functional properties as well as the complex scheme by which it is regulated.

STRUCTURE OF PROTEIN C

Protein C is a 56 kDa vitamin K-dependent protein that can be isolated from plasma by adsorption to barium salts (4). It was initially characterized by its ability to bind to these salts and was arbitrarily given the name protein C. It was found to be a phospholipid-binding zymogen of a serine protease (5). The predominant form found in plasma contains two chains that are disulfide-linked (7). A small percentage of the circulating protein C

contains only one chain and is missing a cleavage at an asp-lys bond that
separates the two chains (8). Protein C has considerable sequence homology
with other vitamin K-dependent proteins including factor X and factor IX
(9,10). An examination of the sequence reveals that the amino terminal region
contains the gamma carboxyglutamic acid (Gla) residues (9). These have been
shown to be essential for the interaction of protein C with membranes in the
presence of calcium ions (5). Following the Gla region, activated protein C
has an epidermal growth factor region which is duplicated. In this region a
second unusual amino acid is found, ß-hydroxy aspartic acid (11). This amino
acid has been shown to be important for the interaction between protein C and
protein S (11). Following the EGF domains is the protease domain which is
homologous with other serine proteases (10).

    Calcium is essential for protein C function. Equilibrium binding
studies have revealed that protein C is able to bind calcium through
approximately 16 high affinity sites (12). Several lines of evidence suggest
that there are at least two types of sites and two functional consequences of
calcium binding; lipid binding and thrombin activation. The first type of
site is associated with the Gla residues (13). Removal of the Gla region by
treatment of protein C with chymotrypsin only partially reduces the calcium
binding and reveals the presence of a high affinity site (13). This
derivative of protein C, known as Gla-domainless protein C, has no
anticoagulant activity and is unable to bind to membranes (13). Ohlin and
coworkers have isolated a tryptic fragment of protein C that contains the
second EGF domain of protein C (14). The fragment from this second type of
site is able to bind calcium with a dissociation constant of 100 uM. It
appears that this binding site requires the presence of ß-hydroxyaspartic acid
for both calcium binding and the interaction between protein C and protein S
(11). The actual function of the calcium ions is unclear. It has been
observed that calcium binding can cause a conformational change in the protein
(13). Studies of the effect of calcium on the intrinsic fluorescence of
protein C have indicated that there is a high affinity site that results in
change in the conformation (15).

    Studies with monoclonal antibodies have also revealed two types of
calcium interactions that may be important for function. Several types of
monoclonal antibodies against protein C have been produced which are ion
dependent in their interaction with the protein. The first type of calcium
dependent antibody is found to be directed towards the amino terminal region
of the protein. Church and co-workers have produced a monoclonal antibody
against protein C that can also recognize factors X, VII and prothrombin
(16,17). Using synthetic peptides they have shown that the consensus sequence
for the epitope is Phe-Leu-Glu-Glu-Xaa-Arg/Lys. Binding of the antibody is
inhibited by metal ions suggesting that ion binding induces a conformational
change in this region of the protein. Orthner and co-workers have reported on
a similar antibody with similar specificity (18). A second type of calcium-
dependent monoclonal antibody has been characterized which binds tighter in
the presence of calcium (11). The binding site of one of these antibodies has
been localized to the first EGF domain of protein C. Manganese has also been
observed to bind to the calcium binding sites and has been used as a
spectroscopic probe for calcium binding (19). In summary, there appear to be
two classes of calcium binding sites in protein C. One type of site is the
Gla-dependent site that is essential for anticoagulant action and lipid
binding. The second type of site is associated with the EGF domain and is
important in causing a conformational change that may be associated either
with activation of protein C or in its interaction with protein S.

    Bovine protein C has also been observed to have a monovalent cation
binding site (20-25). It has been observed in kinetic studies that there are

two monovalent binding sites that are important for the hydrolysis of
synthetic substrates. Kinetic analysis suggests that ion binding enhances
acylation reactions by forming an enzyme-ion complex that reacts more readily
with the substrate. Using spin resonant probes it has been observed that upon
activation of protein C there is a subtle change in the conformation of the
monovalent cation binding site. As yet, there is no evidence that the
occupation of the monovalent cation binding site is important for the
anticoagulant activity of activated protein C.

ACTIVATION OF PROTEIN C

Trypsin (5) and thrombin (7) were the first enzymes observed to activate
protein C. Trypsin was observed to promote the incorporation of DFP into the
heavy chain of protein C (5) and while thrombin acted by catalyzing the
removal of a tetradecapeptide from the amino terminal of the heavy chain and
making no changes in the light chain (7). Thrombin activation was inhibited
by calcium, and even in the absence of calcium proceeded slowly suggesting
that it was not a physiological activator (7). It was subsequently observed
that other enzymes including factor Xa (26), and enzymes from the venom of
Russell's viper and the Southern Copperhead snake (Agkistrodon contortrix
contortrix) (28-30) could activate protein C.

Esmon and Owen observed that if you passed protein C and thrombin
through a heart using the Lagendorf heart perfusion model, protein C would be
activated approximately 20,000 times faster than observed in buffer with
thrombin (31). They observed that diisopropylfluorophosphate inactivated
thrombin could inhibit protein C activation both in the myocardium and by
cultured endothelial cells. These results suggest that endothelial cells
contain a cofactor for thrombin activation of protein C. To test this
hypothesis they successfully purified the cofactor and coined the name,
thrombomodulin (32). This protein has a number of effects on thrombin
function. Specifically it can enhance protein C activation (33), inhibit
cleavage of fibrinogen (34), inhibit the activation of platelets (35) and in
some cases, stimulate antithrombin inhibition of thrombin (36-38).

Structure of Thrombomodulin

Thrombomodulin is a intrinsic membrane glycoprotein that has been
purified from mouse (39), rabbit (32), and bovine lung (40) as well as human
placenta (41,42). Its extraction from lung endothelium requires the presence
of detergents for solubility suggesting that a portion of the protein is
embedded into the membrane bilayer. In detergents the protein exists in a
heterogenous mixture that includes the monomer (Mr 65,000), trimer and higher
oligomers (43).

The sequence of thrombomodulin has been determined through the cloning
of the cDNA (44) which codes for a protein with a molecular weight of 60,300.
The gene which has no introns and is identical to the cDNA has also been
cloned (45). The organization of thrombomodulin is similar to that of the
low density lipoprotein receptor (46). The amino terminal 223 amino acids has
some homology with lectins (47). It is followed by an epidermal growth factor
(EGF) domain that contains 6 EGF regions. Within this domain one residue of
ß-hydroxyaspartic acid is found (46). The gene structure of thrombomodulin is
unusual because the gene structure of proteins with EGF domains usually
contains introns. The EGF regions are homologous to a number of proteins
including factor VII, factor IX, factor XII, protein C, tissue plasminogen
activator and urokinase. The EGF domain is followed by a serine/threonine
rich region, a membrane spanning region and a cytoplasmic region of 38 amino
acids. The gene for thrombomodulin is found on chromosome 20 (45).

<u>Functional Properties of Thrombomodulin</u>

Rabbit thrombomodulin has been characterized by its ability to alter thrombin activity in a number of functional assays that include: protein C activation, inhibition of fibrinogen clotting, inhibition of activation of platelets and acceleration of the interaction between thrombin and antithrombin III. For some time it has been puzzling that thrombomodulin that has been extracted from other animal species does not exhibit all of these activities. For example, only rabbit lung thrombomodulin has been observed to stimulate antithrombin III inactivation of thrombin (36-38). While rabbit (34) and bovine (48) thrombomodulins have been observed to inhibit the clotting of fibrinogen by thrombin, human placenta thrombomodulin has not (50). These results raise the questions of whether the different activities represent actual molecular differences in thrombomodulin or isolation artifacts. In the past few years a body of evidence has begun to accumulate suggesting that structural features of thrombomodulin can explain these functional differences. It has been observed that rabbit lung thrombomodulin actually consists of two distinct fractions that can be separated by ion exchange. One of these fractions is very acidic compared to the other (49). Both fractions can enhance the activation of protein C, but only the acidic fraction can inhibit thrombin clotting of fibrinogen and accelerate the inhibition of thrombin by antithrombin III. Treatment of rabbit thrombomodulin with either chondroitin ABC-lyase (51) or other glycanases (52) abolished the ability of thrombomodulin to both inhibit the fibrinogen clotting of thrombin and to enhance the inhibition of thrombin by antithrombin III. This had no effect on thrombin activation of protein C. The acidic fraction of rabbit thrombomodulin was observed to inhibit effects of heparin on the inactivation of thrombin by antithrombin (53,55) suggesting that this functional property of thrombomodulin was not due to the presence of contaminating heparin. In addition, heparinase had no effect on this property (51). It has been observed that vitronectin is a contaminant of many human thrombomodulin preparations. It has also been observed that vitronectin can inhibit the direct anticoagulant activity of rabbit thrombomodulin (i.e. clotting of fibrinogen) (51). This result suggests that vitronectin contamination may be the reason for the lack of direct anticoagulant activity in human thrombomodulin preparations. In general, these results suggest that thrombomodulin has two thrombin binding sites. The primary site is involved in the enhancement of thrombin activation of protein C. The secondary site, which appears to be associated with glycosaminoglycans that are covalently linked to thrombomodulin, appears to be involved in the direct anticoagulant activity of thrombomodulin as well as in the enhancement of thrombin inactivation by antithrombin III.

<u>Thrombin Binding Domains</u>

Studies using fluorescent energy transfer between the active site of thrombin and the phospholipid membranes in which thrombomodulin has been incorporated have indicated that the distance from the active site to the membrane surface is 45 angstroms (61). This indicates that thrombin must sit on top of thrombomodulin. A number of studies have been carried out to localize the sequence of the thrombin binding site. These studies indicate that only a portion of the thrombomodulin molecule is necessary for the expression of its activity. It was first observed that digestion of thrombomodulin with either elastase or trypsin could produce a soluble form of thrombomodulin that could bind thrombin and accelerate protein C activation (62). Cyanogen bromide fragmentation of thrombomodulin yielded a fragment that could bind thrombin, but had no cofactor activity (63). This fragment was derived from the fifth and sixth epidermal growth factor-like regions. A second CNBr fragment that included residues 310-486, both bound thrombin and accelerated protein C activation (64). This 29 Kda fragment, termed microthrombomodulin, appears to be the smallest size that contains cofactor activity. Using recombinant DNA technology and expression of thrombomodulin mutants in COS-1 cells it has been observed that the cofactor activity is contained within the EGF domains (65). Further mutants were used to isolate the minimum structure for cofactor activity from the fourth, fifth and sixth EGF domains (66).

<u>Protein C Interaction With Thrombin-Thrombomodulin Complex</u>

Effect of ions: Activation of protein C by the thrombin-thrombomodulin complex is dependent upon the presence of divalent cations (32). Of the ions tested, calcium appeared to be the most effective (32). In the absence of thrombomodulin, calcium is an inhibitor of protein C activation by thrombin (13). Titration of the rate of protein C activation utilizing calcium with either thrombin alone or the thrombin-thrombomodulin complex as activators has revealed that there is a high affinity binding site for calcium (Kd= 0.061 mM) that is important for activation (14). This site does not appear to be in the Gla domain of protein C as Gla-domainless protein C is activated equally as well as the native protein by the purified thrombin-thrombomodulin complex (14). However, when associated with endothelial cells, the thrombin-thrombomodulin complex does not activate Gla-domainless protein C. This suggests the possibility of either an additional factor or that in cells thrombomodulin has a different conformation from that of the purified protein (14).

Influence of Membrane structure: Thrombomodulin has been incorporated into lipid vesicles of varying compositions (67,68). Increasing concentrations of phosphatidyl serine lowers the Km for protein C. However, membrane incorporation does not alter activation of Gla-domainless protein C (67,68). Km for protein C in vesicles of pure phosphatidyl choline is most like that of endothelial cells. The data suggests that when thrombomodulin is incorporated into vesicles composed entirely of phosphatidyl choline that a site is exposed that recognizes the Gla domain of protein C (67) since protein C alone does not bind pure phosphatidyl choline vesicles.

Influence of Factor Va on protein C activation: It has been reported that during the coagulation of human blood, protein C is activated (69,70). It was found that the light chain of factor Va can stimulate thrombin catalyzed activation of protein C approximately 50-fold (71). When factor Va was added to endothelial cells it could further stimulate protein C activation (70). However, a comparison of the light chain with thrombomodulin indicated that thrombomodulin was at least a 20-times more effective cofactor (71). At high concentrations the light chain could inhibit protein C activation by the thrombin-thrombomodulin complex (71). The results suggested that factor Va and thrombomodulin might act in concert to modulate protein C activation (72).

<u>Enzymatic Properties of the Thrombin-Thrombomodulin Complex</u>

Are the effects of thrombomodulin due to changes in the catalytic properties of thrombin or to changes in substrate recognition? Answers to these questions are complicated by the recent observation that thrombomodulin has two types of binding sites. However, studies on the effects of thrombomodulin on the enzymatic properties of thrombin are also consistent with two effects on thrombin.

Several hypotheses have been offered to explain the change in catalytic

properties. One possibility is that the active site of thrombin is somehow
different when thrombin is bound to thrombomodulin. A second possibility is
that thrombomodulin alters the substrate recognition site. Thrombomodulin
does not alter the cleavage of small molecule substrates (32,57) suggesting
that thrombomodulin has little effect on the active site. However, spin
resonance studies suggest that thrombomodulin can alter the conformation of
the active site of thrombin (58).

Some of the effects of thrombomodulin can be mimicked by small
molecules. It has been observed that a number of tryptamine analogs including
serotonin, tryptamine, 5-fluorotryptamine and 6-fluorotryptamine could enhance
thrombin activation of protein C as much as 7-fold (59) suggesting the
possibility of some sort of allosteric effect. Studies on the inhibition of
the proteolysis of fibrinogen by thrombin suggests that specific recognition
sites are involved in this effect. In order for thrombomodulin to inhibit
cleavage of the A chain of fibrinogen, amino acids 30-34 of the A chain must
be present. For example the cleavage of the fibrinopeptide from fragment 1-23
or 1-29 from the A chain was not inhibited by thrombomodulin. This suggests
that substrate recognition is modified by thrombomodulin (60). The data
appears to suggest that the primary effect of thrombomodulin is to alter the
substrate recognition site since cleavage of fibrinogen is unaltered if one
uses a small enough substrate.

Thrombomodulin Activity in Cell Culture

Thrombomodulin is found on the surface of endothelial cells (31,33)
including those cultured cells derived from brain, fat and aorta (73). It
appears to be an excellent immunohistological marker for endothelial cells as
it has been observed in both the large and small vessels (74,75). Lung, heart
and intestinal vessels have been observed to have the highest concentrations
of thrombomodulin antigen (76). Kidney glomerular capillary loops appeared to
have the least (74). Thrombomodulin has been found in thromboplastin
preparations from placenta and lung but not from brain (77). In studies that
have screened various tissues for thrombomodulin antigen it has been found in
spleen, pancreas, liver, kidney, skin, heart and aorta (76). Functionally
active thrombomodulin has been found in human platelets, in plasma and in
urine (78).

A number of factors have been observed to alter the amount of
thrombomodulin expression in endothelial cells. Thrombin has been observed to
slow down the rate of internalization of thrombomodulin (79), and protein C
has been observed to inhibit the internalization of thrombin-thrombomodulin
complexes (80). Treatment of cells in culture with endotoxin (81), Il-1 (82),
tumour necrosis factor (TNF) (81) and phorbol myristate acetate (84,85)
decrease the amount of thrombomodulin antigen or activity. TNF has been
observed to enhance internalization of thrombomodulin (83). TNF has been
observed both to have no effect (85) and to decrease (86) thrombomodulin mRNA
levels.

In Vivo Activation of Protein C

Thrombin dependent activation of protein C has been observed **in vivo**.
Infusion of thrombin into a dog led to an increase in systemic anticoagulant
activity. The formation of this activity was dependent upon the presence of
protein C in the plasma and the ability of thrombin to form a cell surface
complex (87).

Several studies have been carried out to determine the effects of

infusion of thrombomodulin into animals. Rabbit thrombomodulin has been
injected into rabbits and the half-life in circulation determined (88). The
half life of detergent solubilized thrombomodulin was approximately 7-8 hours.
In a mouse model it was observed that injection of thrombin into mice caused
death from thromboembolism (89,90). The addition of antibodies against
thrombomodulin reduced the amount of thrombin necessary to cause death, while
addition of soluble thrombomodulin protected the mice from death (89). The
results suggest that soluble thrombomodulin might have some use as an
antithrombotic agent.

Other Activators of Protein C

Factor Xa has been observed to activate protein C in the presence of
phospholipids (26); a high proportion of negatively charged phospholipids was
most effective (26). It has also been observed that the rate of factor Xa
catalyzed activation could also be stimulated by increasing the density of
negative charge with sulphated polysaccharides like heparin or pentosan
polysulfate (91). Thrombomodulin has also been observed to be a cofactor for
the factor Xa-catalyzed reaction (26). The factor Xa-thrombomodulin catalyzed
reaction compared favorably with the rate of the thrombin-thrombomodulin
catalyzed reaction (26). It has been reported that factors IXa and XIa do not
activate protein C (92).

ANTICOAGULANT ACTIVITY OF ACTIVATED PROTEIN C

Inactivation of factor V

Kisiel was the first to observe that activated protein C could
inactivate factor Va (93). Activated protein C was observed to cleave both the
heavy and light chains of factor Va (27). The rate of cleavage was found to
be dependent upon the presence of calcium ions and phospholipid vesicles (27).
Factor V activity has been observed to be stable in the presence of activated
protein C, calcium and phospholipids (27). Va activity is stable in the
presence of activated protein C alone (27). The active form of factor V,
factor Va is a preferred substrate over the procofactor (27). It appears
that cleavage of the heavy chain is sufficient for loss of factor Va activity
(94). Cleavage of the heavy chain reduces the affinity of factor Va for both
factor Xa and prothrombin (95,96). Factor Xa has been observed to protect
factor Va from inactivation by activated protein C (27,97). The degree of
protection can be modulated by protein S suggesting that one possible role for
protein S would be to enhance that activity of activated protein C at the site
of thrombin generation (98). Most of the results suggest that the disruption
of the integrity of the binding sites for prothrombin and factor Xa are the
primary defect caused by activated protein C. In addition, it appears that
factor Xa occupies a site on factor Va that either alters the conformation of
factor Va such that it cannot be cleaved or prevents binding of activated
protein C to an important substrate recognition site.

The light chain of factor Va has been observed to contain a distinct
binding site for activated protein C as well as a site for factor Xa (99). It
is unclear whether occupation of this site by activated protein C is required
for proteolysis of the heavy chain.

Inactivation of factor VIII

Factor VIII has a great deal of sequence homology with factor V and like
factor V, acts as a cofactor for the vitamin K-dependent protein, factor IXa.
Similar to factor Va, it can be inactivated by activated protein C (100-102).

The active form, factor VIIIa, appears to be inactivated more rapidly than the procofactor. This reaction like the inactivation of factor Va, requires the presence of calcium and phospholipid (102). Examination of the pattern by which factor VIII is cleaved has revealed that unlike factor Va, only the heavy chain is cleaved (102). Examination of the sequence reveals that the light chain is missing the homologous cleavage site that is found in the factor Va light chain (103). There are at least four cleavage sites in the factor VIII heavy chain (102). Any cleavage of the heavy chain is sufficient for the loss of factor VIII cofactor activity (102) though it appears that there is an order to the cleavages. Studies on the kinetics of factor VIII inactivation have suggested that there is a preferred order in which the cleavages have been made. The addition of protein S enhanced the rate of all of the cleavages (102). Inactivation of factor VIII can be inhibited by the presence of factor IX (102) and von Willebrand factor (104).

Cleavage of the heavy chain of factor VIII requires the presence of the light chain (105). Isolated factor VIII heavy chain was not observed to be a substrate for activated protein C (105). Excess light chain, but not the heavy chain, was observed to be an inhibitor of factor VIII inactivation (105). Like factor Va, the light chain of factor VIII has been observed to contain a binding site for activated protein C (105). Recombinant fragments of the light chain have been used to determine the area in the light chain to which activated protein C binds. Two overlapping mutants of the light chain were found that could bind activated protein C. An inspection of the sequence in this region revealed a small region of sequence homology.   By using synthetic peptides it was determined that a region in the factor Va light chain that runs from amino acid 1861 to 1874 and in the factor VIII light chain from 2009 to 2018 are a portion of the activated protein C binding site (103).

Factor V and VIII are members of a copper ion-binding family of proteins that include ceruloplasmin. Each contains a triplication of part of its sequence that has been designated the A domain (103). In factor V and VIII, the activated protein C binding site is found in the A3 domain. Homologous sequences are found in the A1 and A2 domains of factor VIII and only in the A1 domain of factor V (103). Only the region in the A1 domain of factor VIII has potential to bind protein C. Ceruloplasmin also inhibits the inactivation of factor Va by activated protein C. An inspection of the sequence of the A3 domain indicats that ceruloplasmin has a region of sequence that is homologous to the protein C binding site found in factors V and VIII. Synthetic peptides from this region in ceruloplasmin are able to interact with activated protein C (107). Peptides from the A1 and A2 domains of ceruloplasmin are unable to bind to activated protein C (107).

<u>Role of Protein S in the Anticoagulant Activity of Activated Protein C</u>

DiScipio and Davie were the first to observe the presence of protein S among other vitamin K-dependent proteins. It was named after Seattle, the city of its discovery for want of a better identity (108). A simpler purification was reported by Stenflo (109). It was subsequently discovered that activated protein C required a cofactor for optimum inactivation of factors Va (110) and VIIIa (102). This discovery was based upon the observation that plasma could enhance the rate of activated protein C catalyzed inactivation of factor Va (110). Subsequent purification of the enhancement factor revealed that the cofactor activity was associated with protein S (110). In addition to being found in plasma it is found in platelets (111) and endothelial cells (112). The human genome contains two protein S genes (113). One contains all of the information for the plasma

protein and the second is a pseudogene which is probably not expressed (113).
It is a single chain protein that has been sequenced and the cDNA has been
isolated (114). The amino acid sequence of protein S has revealed that it is
a unique member of the family of vitamin K-dependent proteins (114,115). Like
other vitamin K-dependent proteins, the amino terminal region of the protein
contains the Gla residues, formed by post-translational modification by a
vitamin K-dependent carboxylase. This is followed by a disulfide-linked loop
that contains a thrombin cleavage site. This is often referred to as the
thrombin sensitive region. The sequence next contains four EGF regions. In
this region is found another unusual amino acid, beta-hydroxyaspartic acid
(114). In the sequence of factors X and IX, the next domain would be the
trypsin or protease domain. In protein S, the protease domain is missing and
in its place is a region that is homologous to rat androgen binding protein
(116) This region makes up approximately 2/3 of the protein S sequence.

In order for protein S to function as a cofactor for activated protein
C, it must be able to bind to membranes. Protein S and activated protein C
form a 1:1 stoichiometric complex on the surface of phospholipid containing
membranes (117). Composition of the membranes is important. Using
phospholipid vesicles as models for natural membranes, it has been observed
that vesicles that contain only phosphatidyl choline will not support factor
Va inactivation (118). Phosphatidyl serine is essential for optimum activity.
The activity of the protein S/activated protein C complex can also be modified
by altering the composition of the fatty acid side chains. Vesicles that are
composed of fatty acyl side chains that are unsaturated or are shorter in
length tend to enhance the rate of inactivation to a greater extent (118).

The primary determinant of protein S binding to lipid vesicles resides
in the Gla region of the protein. When the 10 Gla residues were chemically
modified to gamma-methylene glutamic acid, all cofactor activity and lipid
binding was lost (119). When as few as four of the 10 residues were modified,
greater than 75% of the cofactor activity and lipid binding was lost. However
when the protein was modified in the presence of calcium, four residues were
observed to be modified without appreciable loss of cofactor activity. These
experiments indicate that only four of the Gla residues are absolutely
essential for lipid binding and that lipid binding is essential for expression
of cofactor activity (119).

Cleavage of protein S by thrombin results in the conversion of protein S
from a single chain protein to a disulfide-linked two-chain protein (120,121).
The two-chain protein is unable to bind to lipid vesicles, has reduced calcium
binding and, is unable to function as a cofactor (120). Cleaved protein S is
unusual in the respect that, though it retains its Gla region, it is unable to
bind to membranes (120). This observation suggests that there are addition
structural features to the protein that are essential for lipid binding.

Protein S is unusual in an additional respect. Unlike other plasma
factors, it circulates in an active form. No proform has been identified in
the circulation and no proteolytic cleavage of protein S has been observed to
enhance its cofactor activity.

Dahlback was the first to observe that, in plasma, protein S could be
found in two forms: free, and bound to the complement component C4b-binding
protein (122). Though the ratio of bound and free varies somewhat from
individual to individual, approximately one-half of the protein S is bound and
one-half is free. When bound to C4b-binding protein, protein S is not able to
act as a cofactor for activated protein C (123). Hence, the activity of
protein S, and therefore activated protein C, can be modulated by the acute
phase reactant protein C4b-binding protein (124).

C4b-binding protein is found in plasma in two forms. One can bind protein S and one cannot (125). The difference between the two forms is the absence of a 45 kDalton subunit, known as the b chain, which is believed to be the binding domain for protein S (125). A region on protein S that is responsible for binding to C4b-binding protein has been identified near the carboxy terminus of the protein (126). A synthetic peptide from the human protein S sequence, GVQLDLDEAI, has been observed to inhibit the interaction between protein S and C4b-binding protein. When added to plasma, this peptide enhances the anticoagulant activity of activated protein C (126).

<u>Interaction of activated protein C with platelets</u>

Platelets are able to support prothrombin activation by providing factor Va and/or a phospholipid surface. It was first determined by Comp and Esmon that activated protein C could inhibit platelet supported prothrombin activation and inhibit factor Xa binding to platelets (127). Unactivated platelets can support factor Va inactivation by activated protein C only in the presence of protein S (128,129). However, activated platelets do not require protein S to support the activity of activated protein C.

FIBRINOLYTIC PROPERTIES OF ACTIVATED PROTEIN C

In a number of systems activated protein C has been shown to exhibit profibrinolytic activity, though it does not appear to be a direct activator of plasminogen (130). This was first reported by Seegers, who observed that infusion of activated protein C into a dog caused the shortening of the euglobulin clot lysis time (131). Subsequently, it was observed that activated protein C could cause a rise in plasma levels of plasminogen activator (130). It has also been observed that the infusion of thrombin, a potent activator of protein C, can cause a rise in plasma levels of plasminogen activator (87). When human activated protein C was infused into cats (132) it caused a shortening of the whole blood clot lysis time though it had no effect in squirrel monkeys (133). Effects of activated protein C have also been observed in whole blood clot lysis systems. Bertina has observed that in whole blood, activated protein C will simulate clot lysis and that this effect depends on the presence of protein S (134). We have recently confirmed this finding with the observation that the effect of activated protein C on fibrinolysis in whole blood is species specific but its inhibition of plasminogen activator inhibitor is not. Comp and Esmon have observed that euglobulin clot lysis times from blood treated with activated protein C are shortened (135). Finally, a number of studies on the effects of activated protein C have been carried out with systems that use only purified proteins. In these systems, no protein S requirement has been observed (136) and it appears that the effects of activated protein C can be shown to be mediated through the inactivation of plasminogen activator inhibitors. Sakata et al. have shown that activated protein C does appear to interact with a protein known as plasminogen activator inhibitor that is found in endothelial cells and platelets (137). However, kinetic studies have indicated that the interaction between activated protein C and PAI-1 is probably not important in the fibrinolytic action of activated protein C (138). These findings have a number of implications. First, it appears that protein S is important in whole blood clot lysis, but not for the interaction between activated protein C and plasminogen activator inhibitors. Second, these results suggest the possibility of an interaction between activated protein C and a cellular element since the activated protein C effect appeared greater in whole blood than in plasma. Though studies in purified systems indicate that activated

protein C can inhibit plasminogen activator inhibitors, the lack of
involvement of protein S in this interaction suggests that this may not be the
physiological mode for enhanced fibrinolysis.

INHIBITORS OF ACTIVATED PROTEIN C

Activated protein C has a short half-life, both in circulation and in
plasma. Its inhibition in plasma lead to the observation that there are at
least two plasma inhibitors of activated protein C. One is a heparin-
dependent inhibitor known as protein C inhibitor. This inhibitor has a high
association rate constant (6.5×10^3 $M^{-1}s^{-1}$) and a low concentration in plasma
(88 nM) (139). It can be stimulated by heparin (139) and dextran sulfate
(140). This inhibitor is identical to urinary protein C inhibitor and
plasminogen activator inhibitor-3 (141). This inhibitor loses activity if the
plasma is freeze-thawed (142). The protein C inhibitor has been purified by
several groups (139,142,143).

A second inhibitor, heparin-independent, has also been observed. This
inhibitor is rather non-specific and has a low association rate constant (1.1
$\times$ 10 $M^{-1}s^{-1}$), but a high concentration in plasma (40 uM) (144). This inhibitor
has been found to be identical to alpha-1-antitrypsin (145,146). Activated
protein C is not inhibited by alpha-2-macroglobulin (147) or antithrombin
(139).

IN VIVO PROPERTIES OF ACTIVATED PROTEIN C

Protein C levels have been determined in a number of clinical and
experimental situations. Plasma levels of protein C are sensitive to liver
function (148) and are rapidly decreased in liver disease or upon the
administration of oral anticoagulants (149). There is some evidence that
protein C levels are decreased in inflammatory disease (150). In experimental
situations protein C levels have been observed to decrease following the
administration of endotoxin (151).

Taylor and coworkers have given baboons lethal doses of endotoxin
followed by large doses of human activated protein C (152). This treatment
has been observed to spare the animals from death or liver damage. Protein C
itself had no effect and antibodies against protein C have been observed to
enhance the effects of endotoxin. Gruber and coworkers have also infused
human protein C into baboons, observing that it will cause a prolongation of
the APTT and will inhibit the formation of a platelet-dependent thrombus
(153). As discussed above, a number of investigators have observed effects on
fibrinolysis. In most of the experiments, large doses have been administered
in order to overcome the effects due to species specificity.

SUMMARY

All available evidence indicates that activated protein C is an
important regulator of hemostasis. Though the relationship between its
anticoagulant, profibrinolytic, and anti-inflammatory activities are not well
understood, it is clear from the above studies that its effects are important.
The activity of protein C is under one of the most complex regulatory schemes
yet studied in the field of blood coagulation. Its activation is, to some
extent, modulated by immune effectors as well as thrombin, the central
procoagulant factor. Its activity is controlled by protein S, the activity of
which is modulated by C4b-binding protein, an acute-phase reactant. It has
also been suggested that protein C activity may be modulated by ceruloplasmin,
another acute-phase reactant. Lastly, it can be inactivated by two plasma

inhibitors that have differing modes of regulation. The results suggest that activated protein C may be a mediator of both hemostasis and immunologic reactions. This suggests that its ultimate importance has yet to be understood.

REFERENCES

1.   Owren PA: The coagulation of blood. **Acta Med. Scand. Suppl.** 194: 6-56, 1947
2.   Mammen EF, Thomas WR, Seegers WH: Activation of purified prothrombin to autoprothrombin I or autoprothrombin II (platelet cofactor II) or autoprothrombin IIa. **Thrombos. Diath. Haemorrh.** 5: 218-249, 1960
3.   Marcinak E.: Coagulation inhibitor elicited by thrombin. **Science** 170: 452-453, 1970
4.   Stenflo J: A new vitamin K-dependent protein. J. **Biol. Chem.** 251: 255-263, 1976
5.   Esmon CT, Stenflo J, Suttie JW, Jackson CM: A new vitamin K-dependent protein A phospholipid-binding zymogen of a serine protease. J. **Biol. Chem.** 251: 3052-3056, 1976
6   Seegers WH, Novoa E, Henry RL, Hassouna HI: Relationship of "new" vitamin K-dependent protein C and "old" autoprothrombin IIA. **Thromb. Res** 8: 543-546, 1976
7.   Kisiel W, Ericsson LH, Davie EW: Proteolytic activation of protein C from bovine plasma. **Biochemistry** 15: 4893-4900, 1976
8.   Fair DS, Marlar RA: Biosynthesis and secretion of factor VII, protein C, protein S, and the protein C inhibitor from a human hepatoma cell line. **Blood** 67: 64-70, 1986
9.   Fernlund P, Stenflo J: Amino acid sequence of the light chain of bovine protein C. J. **Biol. Chem.** 257: 12170-12179, 1982
10.   Fernlund P, Stenflo, J: Amino acid sequence of the heavy chain of bovine protein C. J. **Biol. Chem.**. 257: 12180-12190, 1982
11.   Ohlin A, Landes G, Bourdon P, Oppenheimer C, Wydro R, Stenflo J: beta hydroxyaspartic acid in the first epidermal growth factor-like domain of protein C. Its role in calcium binding and biological activity. J. **Biol. Chem.**. 263: 19240-19248, 1988
12.   Amphlett GF, Kisiel W, Castellino, FJ: Interaction of calcium with bovine plasma protein C. **Biochemistry** 20: 2156-2162, 1981
13.   Esmon N, DeBault L, Esmon C: Proteolytic formation and properties of gamma-carboxyglutamic acid domainless protein C. J. **Biol. Chem.** 258: 5548-5553, 1983
14.   Johnson AE, Esmon N, Laue T, Esmon CT: Structural changes required for activation of protein C are induced by calcium binding to a high affinity site that does not contain gamma carboxyglutamic acid. J. **Biol. Chem.** 258: 5554-5560, 1983
15.   Ohlin A, Linse S, Stenflo J: Calcium binding to the epidermal growth factor homology region of bovine protein C. J. **Biol. Chem.** 263: 7411-7417, 1988
16.   Church WR, Bhushan FH, Mann KG, Bovill EG: Discrimination of normal and abnormal prothrombin and protein C in plasma using calcium ion inhibited monoclonal antibody to a common epitope on several vitamin K-dependent proteins. **Blood** 74: 2418-2425, 1989
17.   Church WR, Messier T, Horare PJ, Amiral, D. Meyer, Mann KG: A conserved epitope on several human vitamin k-dependent proteins. J. **Biol. Chem.** 263: 6259-6267, 1988
18.   Orthner CL, Madurawe RD, Velander WH, Drohan WN, Battey FD, Strickland DK: Conformational changes in an epitope localized in the NH_2-terminal region of protein C. Evidence for interaction of protein C domains. J. **Biol. Chem.** 264: 18781-18788, 1989

19. Hill KA, Castellino FJ: The binding of manganese to bovine plasma protein C des(1-41)-light chain protein C and activated des (1-41)-light chain activated protein C. **Arch. Biochem. Biophys.** 254: 196-202, 1987

20. Steiner SA, Castellino FJ: Kinetic Mechanism for stimulation by monovalent cations of the amidase activity of the plasma protein bovine activated protein C. **Biochemistry** 24: 609-617, 1985

21. Steiner SA, Castellino FJ: Kinetic studies of the role of monovalent cations in the amidolytic activity of activated bovine plasma protein C. **Biochemistry** 21: 4609-4614, 1982

22. Hill KAW, Castellino FJ: The stimulation by monovalent cations of the amidase activity of bovine des 1-41 light chain activated protein C. **J. Biol. Chem.** 261: 14991-14996, 1986

23. Hill KAW, Castellino FJ: The effect of monovalent cations on the pre-steady state reaction kinetics of bovine activated protein C and des 1-41 light chain activated protein C. J. Biol. Chem. 262: 140-146, 1987

24. Hill KAW, Castellino FJ, Tl^{3+} as a spectroscopic probe of the monovalent cation binding site of bovine plasma activated protein C and des 1-41 light chain activated protein C. J. Biol. Chem. 262: 7098-7104, 1987

25. Hill KAW, Castellino FJ: Topographical relationships amoung the monovalent cation binding sites of bovine plasma activated protein C and des(1-41) light chain activated protein C and a nitroxide spin label bound to their active site serine residues. J. **Protein Chemistry** 6: 489-496, 1987

26. Haley PE, Doyle MF, Mann KG: The activation of bovine protein C by factor Xa. J. **Biol. Chem.** 264: 16303-16310, 1989

27. Walker FJ, Sexton PW, Esmon CT: The inhibition of blood coagulation by activated protein C through the selective inactivation of activated factor V. Biochim. Biophys. **Acta** 571: 333-342, 1979

28. Klein JD, Walker FJ: Purification of a protein C activator from the venom of the southern coopperhead snake (Agkistrodon Contortrix Contortrix). **Biochemistry** 25: 4175-4179, 1986

29. Orthner CL, Bhattacharya P, Strickland DK: Characterization of a protein C activator from the vencm of Agkistrodon contortrix contortrix. **Biochemistry** 27: 2558-2564, 1988

30. Francis RB, Seyfert U: Rapid amidolytic assay of protein C in whole plasma using an activator from the venom of Agkistrodon Contortrix. **A.J.C.P.** 87: 619-625, 1987

31. Esmon CT, Owen WG: Identification of an endothelial cell cofactor for thrombin-catalyzed activation of protein C. **Proc. Natl. Acad. Sci. U.S.A.** 78: 2249-2252, 1981

32. Esmon NL, Owen WG, Esmon CT: Isolation of membrane-bound cofactor for thrombin-catalyzed activation of protein C. J. **Biol. Chem.** 257: 859-864, 1982

33. Owen WG, Esmon CT: Functional properties of an endothelial cell cofactor for thrombin-catalyzed activation of protein C. J. **Biol. Chem.** 256: 5532-5537, 1981

34. Esmon CT, Esmon NL, Harris KW: Complex formation between thrombin and thrombomodulin inhibits both thrombin-catalyzed fibrin formation and factor V activation. J. Biol. Chem. 257: 7944-7950, 1982

35. Esmon NL, Carrol RC, Esmon CT: Thrombomodulin blocks the ability of thrombin to activate platelets. J. Biol. Chem. 258: 12238-12242, 1983

36. Bourin MC, Ohlin AK, Lane DA, Stenflo J, Lindahl U: Relationship between anticoagulant activities and polyanionic properties of rabbit thrombomodulin. J. **Biol. Chem.** 263: 8044-8052, 1988

37. Hofsteenge J, Taguchi H, Stone SR: Effect of thrombomodulin on the kinetics of the interaction of thrombin with substrates and inhibitors. **Biochem J.** 237: 243-251, 1986

38. Preissner KT, Delvos U, Muller-Berghaus G: Binding of thrombin to thrombomodulin accelerates inhibition of the enzyme by antithrombin III. Evidence for a heparin-independent mechanism. **Biochemistry** 26: 2521-2528, 1987

39. Dittman WA, Kumada T, Sadler JE, Majerus PW: The structure and function of mouse thrombomodulin. Phorbol myristate acetate stimulates degradation and synthesis of thrombomodulin without affecting mRNA levels in hemangioma cells. J. Biol. Chem. 263: 15815-15822, 1988

40. Suzuki K, Kusumoto H, Hashimoto S: Isolation and characterization of thrombomodulin from bovine lung. **Biochim Biophys Acta** 882: 343-352, 1986

41. Salem HH, Maruyama I, Ishii H, Majerus PW: Isolation and characterization of thrombomodulin from human placenta. **J. Biol. Chem.** 259: 12246-12251, 1984

42. Kukrosawa S, Aoki N: Preparation of thrombomodulin from human placenta. **Thromb. Res.** 37: 353-364, 1985

43. Winnard PT, Esmon CT, Laue TM: The molecular weight and oligomerization of rabbit thrombomodulin as assessed by sedimentation equilibrium. **Arch. Biochem. Biophys.** 269: 339-344, 1989

44. Jackman RW, Beeler DL, VanDeWater L, Rosenberg RD: Characterization of thrombomodulin cDNA reveals structural similarity to the low density lipoprotein receptor. Proc. Natl. Acad. Sci U.S.A. 83: 8834-8838, 1986

45. Jackman RW, Beeler DL, Fritze L, Soff G, Rosenberg RD: Human thrombomodulin gene is intron depleted: Nucleic acid sequences of the cDNA and gene predict protein structure and suggest sites of regulatory control. **Proc. Natl. Acad. Sci. U.S.A.** 84: 6425-6429, 1987

46. Stenflo J, Ohlin AK, Owen WG, Schneider WJ: beta-hydroxyaspartic acid or beta-hydroxyasparagine in bovine low density lipoprotein receptor and in bovine thrombomodulin. J. Biol. Chem. 263: 21-24, 1988

47. Petersen TE: The amino terminal domain of thrombomodulin and pancreatic stone protein are homologous with lectins. FEBS Lett. 231: 51-53, 1988

48. Jakubowski HV, Kline MD, Owen WG: The effect of bovine thrombomodulin on the specificity of bovine thrombin. J. Biol. Chem. 261: 3876-3882, 1986

49. Bourin, MC: Effect of rabbit thrombomodulin on thrombin inhibition by antithrombin in the presence of heparin. Thromb. Res. 54: 27-39, 1989

50. Hirahara K, Koyama M, Matsuishi T, Kukrata M: The effect of human thrombomodulin on the inactivation of thrombin by human antithrombin III. **Thromb. Res.** 57: 117-126, 1990

51. Maruyama I, Salem HH, Ishii H, Majerus PW: Human thrombomodulin is not an efficient inhibitor of the procoagulant activity of thrombin. J. **Clin. Invest.** 75: 987-991, 1984

52. Preissner KT, Koyama T, Muller D, Tschopp J, Muller-Berghaus G: Domain structure of the endothelial cell receptor thrombomodulin as deduced from modulation of its anticoagulant functions. Evidence for a glycosaminoglycan-dependent secondary binding site for thrombin. J. **Biol. Chem.** 265: 4915-4922, 1990

53. Bourin MC, Ohlin AK, Lane DA, Stenflo J, Lindahl U: Relationship between anticoagulant activities and polyanionic properties of rabbit thrombomodulin. J. **Biol.** Chem. 263: 8044-8052, 1988

54. Bourin MC, Boffa MC, Bjork I, Lindahl U: Functional Domains of rabbit thrombomodulin. **Proc. Natl. Acad. Sci.U.S.A.** 83: 5924-5928, 1986.

55. Preissner KT, Delvos U, Muller-Berghaus G: Binding of thrombin to thrombomodulin accelerates inhibition of the enzyme by antithrombin III. Evidence for a heparin-independent mechanism. **Biochemistry** 26: 2521-2528, 1987

56. Bourin, MC: Effect of rabbit thrombomodulin on thrombin inhibition by antithrombin in the presence of heparin. **Thromb. Res.** 54: 27-39, 1989

57. Jakubowski HV, Owen WG: Macromolecular specificity determinants on thrombin for fibrinogen and thrombomodulin. **J. Biol. Chem.** 264: 11117-11121, 1989

58. Misci G, Berliner LJ, Esmon CT: Evidence for multiple conformational changes in the active center of thrombin induced by complex formation with thrombomodulin analysis employing nitroxide spin-labels. **Biochemistry** 27: 769-773, 1988

59. Musci G, Berliner LJ: Ligands which effect humam protein C activation by thrombin. **J. Biol. Chem.** 262: 13889-13891, 1987

60. Hofsteenge J., Stone SR: The effect of thrombomodulin on the cleavage of fibrinogen and fibrinogen fragments by thrombin. **Eur. J. Biochem.** 168: 49-56, 1987

61. Lu RL, Esmon NL, Esmon CT, Johnson AE: The active site of the thrombin-thrombomodulin complex. A fluorescence energy transfer measurement of its distance above the membrane surface. **J. Biol. Chem.** 264: 12956-12962, 1989

62. Kurosawa S, Galvin JB, Esmon NL Esmon CT: Proteolytic formation and properties of functional domains of thrombomodulin. **J. Biol. Chem.** 262: 2206-2212, 1987

63. Kurosawa S, Stearns D, Jackson K, Esmon CT: A 10-kDa cyanogen bromide fragment from the epidermal growth factor homology domain of rabbit thrombomodulin contains the primary thrombin binding site. **J. Biol. Chem.** 263: 5993-5996, 1988

64. Stearns D, Kurosawa S, Esmon CT: Microthrombomodulin, Residues 310-486 from the epidermal growth factor precursor homology domain of thrombomodulin will accelerate protein C activation. **J. Biol. Chem.** 264: 3352-3356, 1989

65. Suzuki K, Hayashi T, Nishioka J, Kosaka Y, Zushi M, Honda G, Yamanoto S: A domain composed of epidermal growth factor-like structures of human thrombomodulin is essential for thrombin binding and for protein C activation. **J. Biol. Chem.** 264: 4872-4876, 1989

66. Zushi M, Gomi K, Yamamoto S, Maruyama I, Hayashi T, Suzuki K: The last three consecutive epidermal growth factor-like structures of human thrombomodulin comprise the minimum functional domain for protein C-activating cofactor activity and anticoagulant activity. **J. Biol. Chem.**. 264: 10351-10353, 1989

67. Galvin JB, Kurosawa S, Moore K, Esmon CT, Esmon NL: Reconstitution of rabbit thrombomodulin into phospholipid vesicles. **J. Biol. Chem.** 262: 2199-2205, 1987

68. Freyssinet JM, Gauchy J, Cazenave JP: The effect of phospholipids on the activation of protein C by the human thrombin-thrombomodulin complex. **Biochem J.** 238: 151-157, 1986

69. Salem H, Broze G, Miletich J, Majerus P: Human coagulation factor Va is a cofactor for the activation of protein C. **Proc. Natl. Acad Sci USA** 80: 1584-1588, 1983

70. Maruykama I, Salem HH, Majerus PW: Coagulation factor Va binds to human umbilical vein endothelial cells and accelerates protein C activation. **J. Clin. Invest.** 74: 224-230, 1984

71. Salem HH, Broze GJ, Miletich JP, Majerus PW: The light chain of factor Va contains the activity of factor Va that accelerates protein C activation by thrombin. **J. Biol. Chem.** 258: 8531-8534, 1983

72. Salem HH, Esmon NL, Esmon CT, Majerus PW: Effects of thrombomodulin and coagulation factor Va-light chain on protein C activation in vitro. **J. Clin Invest.** 73: 968-972, 1984

73. DeBault LE, Esmon NL, Smith GP, Esmon CT: Localization of thrombomodulin antigen in rabbit endothelial cells in culture. An immunofluorescence and immunoelectron microscope study. **Lab. Invest.** 54: 179-187, 1986

74.   DeBault LE, Esmon NL, Olson JR, Esmon CT: Distribution of the thrombomodulin antigen in the rabbit vasculature. **Lab Invest**. 54: 172-178, 1986

75.   Maruyama I, Bell CE, Majerus PW: Thrombomodulin is found on endothelium of arteries, veins capillaries and lymphatics, and on syncytiotrophoblast of human placenta. **J. Cell. Biol.** 101: 363-371, 1985

76.   Ishii H, Salem HH, Bell CE, Laposata EA, Majerus PW: Thrombomodulin, an endothelial anticoagulant protein, is absent from the human brain. **Blood** 67: 362-365, 1986

77.   Freyssinet JM, Gauchy J, Spaethe R, Cazenave JP: Thrombomodulin activity is found in tissue thromboplastin preparations from placenta and from lung but not from brain. Haemostasis, 17: 114-120, 1987

78.   Ishii H, Majerus PW: Thrombomodulin is present in human plasma and urine. **J. Clin Invest**. 76: 2178-2181, 1985

79.   Beretz A, Freyssinet JM, Gauchy J, Schmitt DA, Klein-Soyer C, Edgell CJ, Cazenave JP: Stability of the thrombin-thrombomodulin complex on the surface of endothelial cells from human saphenous vein or from the cell line EA.HY 926. **Biochem. J.** 259: 35-40, 1989

80.   Maruyama I, Majerus PW: Protein C inhibitors endocytosis of thrombin-thrombomodulin complexes in A549 lung cancer cells and human umbilical vein endothelial cells. **Blood** 69: 1481-1484, 1987

81.   Moore KL, Andreoli SP, Esmon NL, Esmon CT, Bang NU: Endotoxin enhances tissue factor and suppresses thrombomodulin expression of human vascular endothelium in vitro. **J. Clin Invest** 79: 124-130, 1987

82.   Cozzolino F, Torcia M, Miliani A, Carossino AM, Giordani R, Cinotti S, Filimberti E, Saccardi R, Bernabei P, Guidi G et al: Potential role of interleukin-1 as the trigger for diffuse intrasvacular coagulation in acute nonlymphoblastic leukemia. **Am. J. Med**. 84: 240-250, 1988

83.   Moore KL, Esmon CT, Esmon NL: Tumor necrosis factor leads to the internalization and degradation of thrombomodulin from the surface of bovine aortic endothelial cells in culture. **Blood** 73: 159-165, 1989

84.   Dittman WA, Kumada T, Sadler JE, Majerus PW: The structure and function of mouse thrombomodulin. Phorbol myristate acetate stimulate degradation and synthesis of thrombomodulin without affecting mRNA levels in hemangioma cells. **J. Biol. Chem**. 263: 15815-15822, 1988

85.   Scarpati EM, Sadler JE: Regulation of endothelial cell coagulant properties. Modulation of tissue factor, Plasminogen activator inhibitors, and thrombomodulin by phorbol 12-myristate 13 acetate and tumor necrosis factor. **J. Biol. Chem**. 264: 20705-20711, 1989

86.   Conway EM, Rosenberg RD: Tumor necrosis factor suppresses transcription of the thrombomodulin gene in endothelial cells. **Mol. Cell Biol**. 5588-5592, 1989

87.   Comp PC, Jacocks RM, Ferrell Gl, Esmon CT: Activation of protein C in vivo. **J. Clin Invest**. 70: 127-134, 1982

88.   Ehrlich HJ, Esmon Nl, Bang NU: In Vivo behavior of detergent solubilized purified rabbit thrombomodulin on intravenous injection into rabbits. **J. Lab. Clin. Med**. 115: 182-189, 1990

89.   Kumada T, Dittman WA, Majerus PW: A role for thrombomodulin in the pathogenesis of thrombin-induced thromboembolism in mice. **Blood** 71: 728-733, 1988

90.   Gomi K, Aushi M, Honda G, Kawahara S, Matsuzaki O, Kanabayashi T, Yamamoto S, Maruyama I, Suzuki K: Antithrombotic effect of recombinant human thrombomodulin on thrombin-induced thromboembolism in mice. **Blood** 75: 1396-1399, 1990

91. Freyssinet JM, Wiesel ML, Grunbaum L, Pereillo JM, Gauchy J, Schuhler S, Freund, G, Cazenave JP: Activation of human protein C by blood coagulation factor Xa in the presence of anionic phospholipids. Enhancement by sulphonated polysaccharides. **Biochem. J.** 261: 341-348, 1989

92. Thompson EA, Salem HH: Factor IXa, Xa, XIa and activated protein C do not have protein C activating ability in the presence of thrombomodulin **Thromb. Haemost** 59: 339-440, 1988

93. Kisiel W, Canfield WM, Ericsson LH, Davie EW: Anticoagulant properties of bovine plasma Protein C following activation by thrombin. **Biochemistry** 16: 5824-5830, 1977

94. DeWaart PK, Van de Bruls H, Hemker H, Lindhout T: Functional properties of factor Va subunits after proteolytic alterations by activated protein C. **Biochim Biophys Acta** 799: 38-44, 1984

95. Guinto ER, Esmon CT: Loss of prothrombin and of factor Xa-factor Va interactions upon inactivation of factor Va by activated protein C. J. **Biol. Chem.** 259: 13986-13992, 1984

96. Lucklow EA, Lyons DA, Ridgeway TM, Esmon CT, Laue TM: Interaction of clotting factor V heavy chain with prothrombin and prethrombin 1 and role of activated protein C in regulating this interaction Analysis by analytical ultracentrifugation. **Biochemistry,** 28: 2348-2354, 1989

97. Nesheim ME, Canfield W, Kisiel W, Mann K: Studies of the capacity of factor Xa to protect factor Va from inactivation by activated protein C. J. **Biol. Chem.** 257: 1443-1447, 1982

98. Solymoss S, Tucker MM, Tracy PB: Kinetics of inactivation of membrane-bound factor Va by activated protein C. Protein S modulates factor Xa protection, J. **Biol. Chem.** 263: 14884-14890, 1988

99. Krishnaswamy S, Williams EB, Mann KG: The binding of activated protein C to factors V and Va. J. **Biol. Chem.** 261: 9684-9693, 1986

100. Marlar R, Kleiss A, Griffin J: Mechanism of action of human activated protein C. A thrombin-dependent anticoagulant protein. **Blood** 59: 1067-1072, 1982

101. Vehar GA, Davie EW: Preparation and properties of bovine factor VIII (antihemophilic factor). **Biochemistry** 19: 401-410, 1980

102. Walker FJ, Chavin SI, Fay PJ: Inactivation of factor VIII by activated protein C and protein S. **Arch. Biochem. Biophys.** 252: 322-328, 1987

103. Walker FJ, Scandella D, Fay PJ: Identification of the binding site for activated protein C on the light chain of factors V and VIII. J. **Biol. Chem.** 265: 1484-1489, 1990

104. Koedam JA, Meijers JCM, Sixma JJ, Bouma BN: Inactivation of human factor VIII by activated protein C --cofactor activity of protein S and protective effect of von Willebrand factor. J. **Clin Invest,** 2: 1236-1243, 1988

105. Fay PJ, Walker FJ: Inactivation of human factor VIII by activated protein C evidence that factor VIII light chain contains the activated protein C binding site. **Biochim. Biophys. Acta.** 994: 142-148, 1989

106. Fass DN, Hewick RM, Knutson GJ, Nesheim ME, Mann KG: Internal duplication and sequence homology in factors V and VIII. **Proc. Natl. Acad. Sci. USA** 82: 1688-1691, 1985.

107. Walker FJ, Fay PW: Characterization of an interaction between protein C and ceruloplasmin. J. **Biol. Chem.** 265: 1834-1836, 1990

108. DiScipio RG, Davie EW: Characterization of protein S, a gamma-carboxyglutamic acid containing protein from bovine and human plasma. **Biochemistry** 18: 899-904, 1979

109. Stenflo J, Jonsson M: Protein S, a new vitamin K-dependent protein from human plasma. **F.E.B.S. Lett.** 101: 377-381, 1979

110. Walker, FJ: The regulation of activated protein C by a new protein A possible function for bovine protein S. J. Biol. Chem. 255: 5521-5524, 1980

111. Schwarz HP, Heeb MJ, Wencel-Drake JD, Griffin JH: Identification and quantitation of protein S in human platelets. Blood 66: 1452-1455, 1985

112. Stern D, Brett J, Harris K, Nawroth P: Participation of endothelial cells in the protein C-protein S anticoagulant pathway the synthesis and release of protein S. J. Cell. Biol. 102: 1971-1978, 1986

113. Ploos van Amstel HK, Reitsma PH, Hamulyak K, de Die-Smulders CEM, Mannucci PM, Bertina RM: A mutation in the protein S pseudogene is linked to protein S deficiency in a thrombophilic family. Thromb. Haemost, 62: 897-901, 1989

114. Lundwall A, Dackowski W, Cohen E, Shaffer M, Mahr A, Dahlback B, Stenflo J, Wydro R: Isolation and the sequence of the cDNA for human protein S, a regulator of blood coagulation. Proc. Natl. Acad. Sci. U.S.A. 83: 6716-6720, 1986

115. Dahlback B, Lundwall A, Stenflo J: Localization of thrombin cleavage sites in the amino-terminal region of bovine protein S. J. Biol. Chem. 261: 5111-5115, 1986

116. Baker ME, French FS, Joseph DR: Vitamin K-dependent protein S is similar to rat androgen binding protein. Biochem. J. 243: 293-296, 1987

117. Walker FJ: Regulation of activated protein C by protein S The role of phospholipid in factor Va inactivation. J. Biol. Chem. 256: 11128-11131, 1981

118. Walker FJ: Interactions of protein S with membranes. Seminars in thrombosis and hemostasis. 14: 216-222, 1988

119. Walker, FJ: Properties of chemically modified protein S effect of the conversion of gamma-carboxy glutamic acid to gamma-methyleneglutamic acid on functional properties. Biochemistry 25: 6305-6311, 1986

120. Walker FJ: Regulation of vitamin K-dependent protein S Inactivation by thrombin. J. Biol. Chem. 259: 10335-10339, 1984

121. Dahlback B: Purification of human vitamin K-dependent protein S and its limited proteolysis by thrombin. Biochem. J. 209: 837-846, 1983

122. Dahlback B, Stenflo J: High molecular weight complex in human plasma between vitamin K-dependent protein S and complement component C4b-binding protein. Proc. Natl, Acad. Sci U.S.A., 78: 2512-2516, 1981

123. Dahlback B: Inhibition of protein Ca cofactor function of human and bovine protein S by C4b-binding protein. J. Biol. Chem. 261: 12022-12027, 1986

124. Comp PC, Doray D, Patton D, Esmon CT: An abnormal plasma distribution of protein S occurs in functional protein S deficiency. Blood 67: 504-508, 1987

125. Hillarp A, Dahlback B: The protein S binding site localized to the central core of C4b-binding protein. J. Biol. Chem. 262: 11300-11307, 1987

126. Walker FJ: Characterization of a synthetic peptide that inhibits the interaction between protein S and C4b-binding protein. J. Biol. Chem. 264: 17645-17649, 1989

127. Comp PC, CT Esmon: Activated protein C inhibits platelet prothrombin converting activity. Blood 54: 1272-1281, 1979

128. Harris KW, Esmon CT: Protein S is required for bovine platelets to support activated protein C binding and activity. J. Biol. Chem. 260: 2007-2010, 1985

129. Suzuki K, Nishioka J, Matsuda M, Murayama H, Hashimoto S, Protein S is essential for the activated protein C-catalyzed inactivation of platelet-associated factor Va. J. Biochem. 96: 455-460, 1984

130. Comp PC, Esmon CT: Generation of fibrinolytic activity by infusion of activated protein C into Dogs. J. Clin. Invest. 68: 1221-1228, 1981

131. Seegers W, McCoy LE, Groben HD, Sakuragawa N, Agrawal BBL: Purification and properties of autoprothrombin IIa An anticoagulant perhaps also related to fibrinolysis. **Thrombosis Res.** 1: 443-448, 1972

132. Burdick MD, Schaub RG: Human protein C induces anticoagulation and increased fibrinolytic activity in the cat. **Thrombosis. Res.** 45: 413-419, 1987

133. Colucci M, Sassen JM, Collen D: Influence of protein C on blood coagulation and fibrinolysis in squirrel monkeys. **J. Clin. Invest. 74**: 200-202, 1984

134. De Fouw NJ, Haverkate F, Bertina RM, Koopman J, Wijngaarden A, van Hinsbergh VWM: The cofactor role of protein S in the acceleration of whole blood clot lysis by activated protein C in vitro. **Blood 67**: 1189-1195, 1986

135. Comp PC, Esmon CT: in "The Regulation of Coagulation" (eds Mann, K.G and Taylor, F.B.) Elsevier North Holland, Inc pp 583-588, 1980

136. De Fouw NJ, de Jong YF, Haverfkate F, Bertina RM: Activated protein C increases fibrin clot lysis by neutralization of plasminogen activator inhibitor--no evidence for a cofactor role of protein S. **Thrombosis and Haemostasis** 60: 328-333, 1988

137. Sakata Y, Curriden S, Lawrence D, Griffin JH, Loskutoff DJ: Activated protein C stimulates the fibrinolytic activity of cultured endothelial cells and decreases anti-activator activity. **Proc Natl Acad Sci USA** 82: 1121-1128 1985

138. Fay WP, and Owen WG: Platelet plasminogen activator inhibitor. Purification and characterization of interactions with plasminogen activators and activated protein C. **Biochemistry** 28: 5773-5778, 1989

139. Suzuki K, Nishioka J, Hashimoto S: Protein C inhibitor purification from human plasma and characterization. **J. Biol. Chem.** 258: 163-168, 1983

140. Kazama Y, Koide T, Saguraga, N: Further characterization of dextran sulfate as a cofactor of protein C inhibition. **Thromb. Res.** 54: 499-504, 1989

141. Heeb MJ, Espana F, Geiger M, Collen D, Stump DK, Griffin JH: Immunological identity of heparin-dependent plasma and urinary protein C inhibitor and plasminogen activator inhibitor-3 **J. Biol. Chem.** 262: 15813-15816, 1987

142. Espana F, Berrettini M, Griffin JH: Purification and characterization of plasma protein C inhibitor. **Thromb. Res.** 55: 369-384, 1989

143. Pratt, CW, Macik BG, Church, FC: Protein C inhibitor: Purification and proteinase reactivity. **Thromb. Res.** 53: 595-602, 1989

144. Heeb MJ, Bischoll R, Courtney M, Griffin JH: Inhibition of activated protein C by recombinant alpha-1-antitrypsin variants with substitution of arginine for leucine for methionine. **J. Biol. Chem.** 265: 2365-2369, 1990

145. Heeb MJ, Griffin JH: Physiologic inhibition of human activated protein C by alpha-1-antitrypsin. **J. Biol. Chem.** 263: 11613-11616, 1988

146. Heeb MJ, Espana F, Griffin JH: Inhibition and complexation of activated protein C by two major inhibitors in plasma. **Blood**, 73: 446-454, 1989

147. Marlar RA and Kressin DC: Activated protein C is not regulated by alpha-2-macroglobulin. **Thrombosis Res.** 54: 177-186, 1989

148. Walker FJ: Protein C in Liver Disease. **Annals of Clinical and Laboratory Science** 20, 106-112, 1990

149. Vigano S, Mannucci PM, Solinas S. Bottasso B, Mariani G: Decrease in protein C antigen and formation of an abnormal protein soon after starting oral anticoagulant therapy. **Br. J. Hematol.** 57: 213-220, 1984

150. Rodeghiero F, Mannucci PM, Vigano S, Barbui T, Gugliotta L, Cortellaro M, Dini E: Liver dysfunction rather than intravascular coagulation as the main cause of low protein C and antithrombin III in acute leukemia. **Blood**, 63: 965-969, 1984

151. Madden RM, Ward M, Marlar RA: Protein C activity levels in endotoxin-induced disseminated intravascular coagulation in a dog model. **Thrombosis Res.** 55: 297-308, 1989

152. Taylor FB, Chang A, Esmon CT, D'Angelo S, D'Angelo V, Blick KE: Protein C prevents the coagulopathic and lethal effects of Escherichia coli infusion in the baboon. **J. Clin. Invest.** 79: 918-925 1987

153. Gruber A, Griffin JH, Harker LA, Hanson SR: Inhibition of platelet dependent thrombus formation by human activated protein C in a primate model. **Blood** 73: 639-642, 1989

MOLECULAR DEFECTS AFFECTING HEMOSTASIS

MOLECULAR DEFECTS IN HEMOPHILIA A

Leon W. Hoyer

Jerome H. Holland Laboratory
American Red Cross Blood Services
Rockville, MD 20855

INTRODUCTION

Hemophilia A, classic hemophilia, is an X chromosome-linked inherited coagulation disorder characterized by deficient factor VIII activity. Clinically, and when characterized by procoagulant or immunologic assays, it is a heterogeneous disorder. However, disease severity usually corresponds to the extent of factor VIII deficiency.[1] Severe hemophilia with recurrent hemarthroses usually means that there is no detectable plasma factor VIII (< 1% of normal factor VIII activity). Individuals with 1-5% of normal factor VIII levels tend to have moderately severe disease, with most bleeding episodes due to trauma. Mild hemophilia (6-25% of normal factor VIII activity) may only be detected when excessive bleeding occurs during surgery or after severe injuries. Nonfunctional, immunoreactive factor VIII-like protein, at levels comparable to those in normal plasma, is detected in 10% of plasmas from patients with mild or moderate hemophilia A.[2,3] In addition, low levels of factor VIII antigen (1-10% of those in normal plasma) can be detected by sensitive immunoassays in the plasma of 20% of patients with severe hemophilia A.[4] The basis for this heterogeneity has been clarified by recent studies that have examined the molecular basis for the factor VIII deficiency in families with hemophilia.

APPROACHES TO THE MOLECULAR CHARACTERIZATION OF HEMOPHILIA

Several excellent reviews of the molecular biology of the hemophilias have been published recently.[5-8] They provide a detailed summary of the extensive and productive research that has been carried out during the past 5 years. With the cloning of the human factor VIII gene has come the necessary information for undertaking this research.[9,10] The work has been greatly facilitated by the general availability of the necessary probes so that an increasing number of research groups have been able to examine these questions.

From study of other inherited diseases, it was recognized that many different types of mutation could affect the factor VIII gene in a way that might cause hemophilia. For example, deletions, missense mutations, and nonsense mutations have been recognized in many conditions. To date, efforts to characterize molecular defects in hemophilia A have been

somewhat less successful than those carried out for other disorders. The
difficulty has been a result of the size of the factor VIII genes, 186 kb,
approximately 0.1% of the X chromosome. However, it is the X chromosome
location and the large size of the factor VIII gene that have led to the
high incidence of hemophilia A, for new mutations are common. In fact,
several studies have provided support for Haldane's suggestion that new
mutations account for one-third of hemophilia cases.[11]

Most of the data now available have been obtained by laboratories that
have screened DNA from hemophilic patients using factor VIII cDNA probes to
screen Southern blots of genomic DNA that has been digested with several
restriction enzymes. Much of the work has been done with the enzyme *Taq I*,
a particularly useful enzyme for detecting differences in the DNA of
individual hemophiliacs. This is due to the fact that the *Taq I*
recognition sequence, TCGA, contains the CpG dimer sequence. As
Youssoufian and others have emphasized, this dimer is a "hot spot" for
mutation, since cytosine can be methylated at the 5' position and
subsequently deaminated spontaneously to thymine.[12] This accounts for the
CG to TG mutations that are frequently seen, and the same mutation in the
antisense strand leads to a CG to CA mutation in the coding DNA.

The screening studies have identified deletions, single nucleotide
changes, and insertions in the factor VIII gene. The best way to
synthesize this data is to examine the frequency of different gene defects
in relatively large series. Table 1 indicates the relatively low incidence
of identifiable mutations using standard techniques. Only 7.6% of the
defects have been characterized in these studies.

Table 1 Molecular Defects Identified in Large Groups of Hemophilia A
Patients

| | | Mutations Identified | | |
Author	Number of Patients Studied	Deletions	Point Mutations	Insertions
Antonarakis and Kazazian[13]	240	14	12	2
Higuchi et al[14]	160	6	7	-
Casula et al[15]	100	2	3	-
Bardoni et al[16]	49	1	-	-
Bernardi et al[17]	70	1	2	-
Mikami et al[18]	70	1	1	-
Gitschier et al[19]	92	2	2	-
Levinson et al[20]	83	4	6	-
Total	864	31 (3.6%)	33 (3.8%)	2 (0.2%)

FACTOR VIII GENE DELETIONS IN HEMOPHILIA A

Many different deletions have been identified in hemophilia A patients. Those reported through May, 1990, are summarized in Table 2. They have been identified by screening with several probes for factor VIII gene coding regions after restriction enzyme digestion of patient DNA. A missing fragment or an altered fragment length identified a deletion of a part of the factor VIII gene, and the extent of the deletion was then verified by mapping the region with other restriction enzymes and subgenic probes. The available data identify no discrete hot spots that have generated frequent deletions, and the boundaries of all characterized deletions are unique. With a single exception, the deletion has caused severe hemophilia A. The exception is the small deletion in exon 22 in which there is no shift in the message reading frame.[21,22] This 5.5 kb deletion causes moderately severe hemophilia.

As noted in Table 2, a number of these patients have formed an antibody to factor VIII (inhibitor formation). This outcome does not appear to correlate with the extent or position of the deletion, however. The data clearly establish that lack of factor VIII due to a complete gene deletion does not necessarily lead to inhibitor formation.[23] This is in contrast to data for patients with total deletions of the von Willebrand gene, in whom there does seem to be a correlation with anti-von Willebrand factor antibody formation.[34] The number of patients that have been evaluated is rather small, however, and it is not certain if this correlation will persist as additional patients are characterized.

POINT MUTATIONS RESPONSIBLE FOR HEMOPHILIA A

While several restriction enzymes have been used to screen hemophilia A DNA samples, most of the useful information has come with *Taq I*. The *Taq I* recognition site occurs 7 times in the factor VIII coding sequence. In 5 of these, it includes an arginine codon.

Table 3 summarizes the reported factor VIII gene point mutations. Each of the 5 arginine codons included in *Taq 1* sites have been converted to stop codons (TTG) in one or more hemophilic patients. These nonsense mutations cause severe hemophilia, and the immunoassays that have been done for these patients do not detect any factor VIII protein. Interestingly, this is even the case for patients in whom the stop codon is introduced into the extreme 3' end of exon 26.[14]

Missense mutations are also recognized in hemophilia A samples. They are associated with a range of clinical conditions, from severe hemophilia with no detectable plasma factor VIII protein to moderately severe factor VIII deficiency.

INSERTION OF L1 REPETITIVE ELEMENTS IN THE FACTOR VIII GENE

Insertions of "L1" repetitive elements in exon 14 have been identified by Kazazian and coworkers in two unrelated patients with severe hemophilia.[51] L1 sequences are human-specific, long, interspersed, repetitive elements present in about 10^5 copies throughout the genome. While the full length of L1 sequence is 6.1 kb, most elements are truncated. They include an A-rich region and two long open reading frames, the second of which potentially encodes a polypeptide with homology to reverse transcriptases. Because the pathogenic role of the L1 elements within hemophilia genes is not known with certainty, Kazazian and co-workers also evaluated the parents of these two hemophiliacs. In both cases, no abnormalities were detected in the factor VIII genes of the

Table 2   Factor VIII Gene Deletions in Hemophilia A

Exons Deleted	Size (kb)	Clinical Severity*	Reference	Family Identification**
1-26	>210	S+I	Casula et al[15]	H1
1-26	>186	S	Casarino et al[23]	-
1-22	>127	S+I	Lillicrap et al[24]	-
1-5	>35	S	Higuchi et al[14]	484
1	>2	S	Youssoufian et al[22]	JH13
Intron 1	~7	S	Levinson et al[20]	1067
2-3	9-12	S	Youssoufian et al[22]	JH21
3-13	60	S+I	Youssoufain et al[22]	JH22
3	~2	S	Higuchi et al[14]	656
4-25	133-145	S	Youssoufian et al[22]	JH23
4-6, 11-12	10-12, 3-29	?	Kariya et al25	-
5	~2	S	Brocker-Uriends et al[26]	-
5-6	2-10	S+I	Levinson et al[20]	2253
6	3-6	S	Levinson et al[20]	2213
6	~7	S	Youssoufian et al[21,22]	A,JH6
6	~10	S	Levinson et al[20]	1059
7-14	40-56	S	Youssoufian et al[22]	JH24
7-9	15-20	S+I	Higuchi et al[14]	505
8 partial	2bp	S	Higuchi et al[27]	-
7-22	65-110	S+I	Casula et al[15]	H2
8	4bp	S	Kogan & Gitschier[28]	H23
11-18	~55	S+I	Antonarakis et al[29]	A
			Youssoufian et al[22]	JH1
14 partial	~6	S+I	Mikami et al[30]	-
14 partial	2.5	?	Woods-Samuel et al[31]	JH37
14	2-3	S	Higuchi et al[14]	580
14	12-16	S+I	Higuchi et al[14]	194/513
14	3	S	Youssoufian et al[21,22]	B,JH7
14-22	~55	S+I	Tuddenham[32]	5
15-18	~13	S+I	Bardoni et al[16]	-
15-22	>14	S+I	Gitschier[6]	-
15	?	S	Youssoufian et al[22]	JH29
22	5.5	M	Youssoufian et al[21,22]	E,JH10
23-25	~39	S+I	Gitschier et al[19]	H96
23-26	?	S+I	Din et al[33]	-
23-25	16-25	S	Youssoufian et al[21,22]	D,JH9
24-25	~7	S	Youssoufian et al[21,22]	C,JH8
26	~14	S	Youssoufian et al[22]	JH26
26	>20	S	Youssoufian et al[22]	JH30
26	~22	S	Gitschier et al[19]	H51
26	~2	S	Higuchi et al[14]	277
26	>1	S	Bernardi et al[17]	H8

*   Severity: S-Severe, M-Moderate, ?-not specified; I-inhibitor detected
**   Family identification as given in cited reference

Table 3   Factor VIII Gene Point Mutations in Hemophilia A

Exon	Amino Acid Codon	Amino Acid Change	Clinical Severity*	VIII:C/VIII:Ag**	Reference	Family Indentification***
4	170	Ser → Leu	Mo	.03/.09	Chan et al[35]	-
7	272	Glu → Gly	Mo	.02/.03	Youssoufian et al[37]	JH20
8	326	Val → Len	S	0/?	Kogan & Gitschier[28]	H44
	329	Cys → Arg	S	0/?	Kogan et al[38]	V107
	336	Arg → STOP	S	0/?	Gitschier et al[39]	H4
	372	Arg → Cys	Mo	.03/.80	Shima et al[40]	FVIII-Okayami
			Mo	.04/1.10	O'Brien et al[54]	H254
	372	Arg → His	Mi	.05/3.25	Arai et al[41]	ARC-1;FVIII Kumamoto
14	1680	Tyr → Phe	Mi	?/?	Higuchi et al[27]	-
	1686	Gln → STOP	S	0/?	Higuchi et al[27]	
	1689	Arg → Cys	S	0/.96	Gitschier et al[39]	H10
			Mi	.02/.87	Arai et al[42]	ARC-5;FVIII East Hartford
			S	.05/2.11	Arai et al[42]	ARC-10
	1709	Tyr → Cys	Mo	0/.18	Traystman et al[43]	JH-41
18	1922	Asn → Asp	S+I	0/?	Traystman et al[43]	JH-51
18	1941	Arg → STOP	S+I	0/?	Antonarakis et al[44]	B
			S	0/?	Youssoufian et al[12]	K
			S+I	0/?	Higuchi et al[14]	273
			S	0/?	Casula et al[15]	H4
	1941	Arg → Gln	Mi	?/?	Levinson et al[20]	1286
22	2116	Arg → STOP	S	0/?	Youssoufian et al[12]	S
			S	0/?	Youssoufian et al[12]	T
			S	0/?	Higuchi et al[14]	450
			S	0/?	Higuchi et al[14]	509
	2116	Arg → Pro	S	0/?	Levinson et al[45]	V99
23	2147	Arg → STOP	S+I	0/0	Youssoufian et al[46]	JH-14
			S	0/?	Levinson et al[47]	1281
			S+I	0/0	Mikami et al[48]	-

Table 3 (Continued)

Exon	Amino Acid Codon	Amino Acid Change	Clinical Severity*	VIII:C/VIII:Ag**	Reference	Family Indentification***
	2166	Leu → Ser	S	0/?	Levinson et al[20]	1308
24	2209	Arg → STOP	S+I	0/?	Gitschier et al[19]	H2
			S+I	0/0	Youssoufian et al[46]	JH15
			S	0/0	Youssoufian et al[46]	JH16
			S+I	0/?	Higuchi et al[14]	418
			S+I	0/?	Higuchi et al[14]	568
	2209	Arg → Gln	S	0/0	Youssoufian et al[46]	JH18
			S	0/?	Youssoufian et al[46]	JH19
			Mo	.02/?	Bernardi et al[17]	H1
			Mo	.02/?	Bernardi et al[17]	H2
			Mo	?/?	Levinson et al[20]	785
			Mo	?/?	Levinson et al[20]	818
			S	0/?	Traystman et al[43]	JH-50
			S	0/?	Casula et al[15]	H5
26	2307	Arg → STOP	S	0/?	Gitschier et al[19]	H22
			S	0/3	Higuchi et al[14]	550
			S+I	0/?	Higuchi et al[14]	693
			S	0/?	Levinson et al[20]	1060
	2307	Arg → Gln	Mi	.09/.06	Gitschier et al[49]	H104
			Mo	.02/?	Casula et al[15]	H6
	2307	Arg → Leu	Mi	.02/.04	Inaba et al[50]	JH32
Intron 4		(CG → CA) Alternative Splice Site	Mi	.07/.41	Youssoufian et al[36]	JH17

*   Severity: S-Severe, Mo-Moderate, Mi-Mild, I-inhibitor detected
**   units/ml (? indicates data not reported)
*** Family identification as given in cited reference

parents, suggesting that the insertional mutation had arisen de novo.[51]
This is a fundamentally different mechanism by which mutation can produce
human disease. The mechanism of the insertion remains unknown, and there
is no information about the proportion of such insertions that are
inheritable.

THROMBIN CLEAVAGE SITE MUTATIONS

As thrombin cleavage at amino acids 372 and 1689 is crucial for factor
VIII procoagulant function,[52] it is not surprising that mutations at these
sites cause hemophilia A. The essential role of thrombin activation in
factor VIII function has been recognized for some time,[53] and was directly
demonstrated in the elegant studies of Pittman and Kaufman in which site-
directed mutagenesis was used to establish the impact of amino acid
substitutions for the arginine.[52]

Two strategies have been used to identify thrombin cleavage site
mutations responsible for hemophilia A. One approach, that of Gitschier
and coworkers,[39] has been to amplify selected factor VIII regions of genomic
DNA by repeated cycles of primer-directed DNA synthesis. The amplified
DNAs are then screened for mutations by discriminate hybridization using
oligonucleotide probes. This process identified a nonsense mutation (stop
codon) at amino acid 336, the activated protein C cleavage site, that
caused severe hemophilia A. The second mutation they identified was a
substitution of cysteine for arginine at amino acid residue 1689, the
critical thrombin cleavage site on the factor VIII light chain. This
patient had no detectable factor VIII activity, but the factor VIII antigen
level was normal, indicating the synthesis of a nonfunctional protein.[39]

A different approach was taken by Arai and coworkers who screened CRM-
positive hemophilia A plasmas for evidence of cleavage site mutations. In
this analysis the nonfunctional protein was immunoadsorbed, treated with
thrombin, and then tested to determine if thrombin cleavage had occurred.
Three of the first 12 CRM-positive plasmas examined in this way had
missense mutations preventing thrombin cleavage.[41,42] In one patient (ARC-1
in Figure 1), the molecular defect was identified as a substitution of
histidine for arginine 372. This protein, factor VIII-Kumamoto, lacked
procoagulant function because of impaired thrombin cleavage of the factor
VIII heavy chain. In two other patients, unrelated individuals with the
identical missense mutation, arginine 1689 to cysteine, the coagulation
defect was a result of impaired factor VIII light chain cleavage. The
factor VIII from one of these patients (ARC-5) is also shown in Figure 1.
It is of interest that the factor VIII activity levels in these 2 patients
were higher than that of a patient with the same molecular defect
identified by Gitschier and coworkers.[39] All 3 patients had normal or
elevated levels of factor VIII protein by immunoassay. In one patient
there was no detectable factor VIII activity and severe hemophilia with
recurrent hemarthroses,[39] while the other 2 had low, but easily detectable,
factor VIII activity (2-5% of normal). In one case this caused mild
hemophilia with no arthropathy and very infrequent factor VIII treatment;
the bleeding disorder in the other patient caused joint destruction and
required frequent factor VIII replacement therapy.[54]

OTHER FACTOR VIII INSERTIONS CAUSING HEMOPHILIA A

It is of interest that very few factor VIII gene insertions have been
identified as a cause of hemophilia A. While they should be just as easy
to detect as deletions on Southern blot analysis, there are as yet no
unequivocal instances in which an insertion has generated a structural
change--other than loss of factor VIII production in the two patients with
the L1 line insertion.[51]

The recent report of a patient with mild hemophilia associated with a duplication of exon 13 does support that another kind of insertion can reduce factor VIII activity.[15] In this case, it is not known if that patient has nonfunctional factor VIII in his plasma, or if the protein has unusual structural properties.[3]

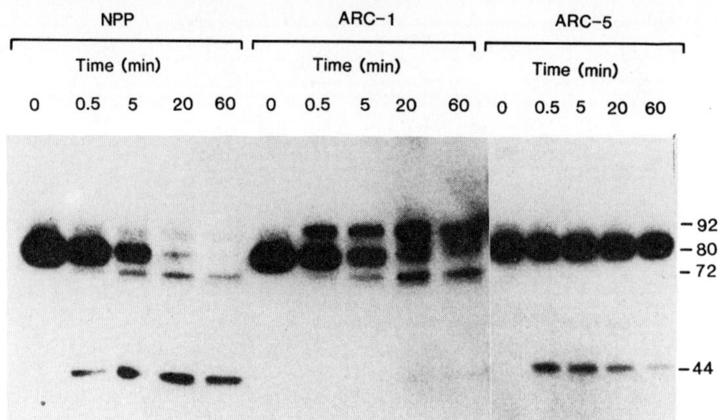

Figure 1. Identification of thrombin cleavage site mutations. Immunoadsorbed normal (NPP) and CRM-positive (ARC-1 and ARC-5) factor VIII detected by immunoblotting using antibodies that react with the factor VIII heavy chain (92 kDa and its 44 kDa fragment) and the factor VIII light chain (80 kDa and its 72 kDa fragment). Factor VIII eluted from the immunoadsorbent beads (time 0) and immunopurified factor VIII incubated with thrombin for 0.5-60 minutes was identified by autoradiography after incubation with the monoclonal anti-factor VIII and I^{125}-labeled anti-mouse immunoglobulin. The ARC-1 factor VIII heavy chain (92 kDa) and the ARC-5 factor VIII light chain (80 kDa) are not cleaved.

In screening CRM-positive hemophilic plasmas for thrombin cleavage site mutations, we have also determined the size of the heavy and light chains for a number of different nonfunctional factor VIII molecules. In the first 20 CRM-positive plasmas that we have examined, there are two instances in which the heavy or light chain fragments have an unusual migration on sodium dodecyl sulfate-polyacrylamide gel electrophoresis (as carried out by the method illustrated in Figure 1). In neither instance has it yet been possible to identify the specific molecular defect responsible for the increased chain size. Both patients have mild hemophilia A with low factor VIII procoagulant activity and normal factor VIII protein by immunoassay.[3] Electrophoresis in the presence of reducing

agents has established that the mobility shift is not due to disulfide bond
formation that affects the protein's electrophoretic mobility.

FUTURE DIRECTIONS

The characterization of factor VIII gene mutations is much more
difficult than that for factor IX. This is due to the size of the factor
VIII gene and the difficulty of screening for point mutations that do not
affect restriction enzyme sites, for very small deletions that are below
the resolution of Southern blot analysis, and for mutations within the
introns. In order to identify a larger proportion of the hemophilia gene
defects, other techniques are beginning to be used.

One of these approaches is denaturing gradient gel electrophoresis
(DGGE)[28,43]. With this technique, single nucleotide changes can be detected
when a relatively large DNA segment is examined. DGGE can identify a
suspicious fragment and DNA sequencing then identifies the specific
nucleotide change. Traystman et al have shown that the sensitivity of DGGE
screening can be improved by the attachment of GC-rich domains to genomic
DNA fragments, so that all sequence changes within a fragment can be
detected.[43] The attachment of such high "GC clamps" to genomic DNA
amplified by the polymerase chain reaction permits detection of some
mutations that could not be detected by DGGE alone.

DGGE has been successfully applied by Kogan and Gitschier to the
acidic region of the factor VIII heavy-chain (the portion of the intron 7
region necessary for proper splicing, and the portion of exon 8 that codes
for amino acids 318-385).[28] Two hemophilia-causing mutations were detected
when DNA was analyzed from 228 unselected hemophilic patients. These
mutations were subsequently characterized by direct sequencing. In one
patient, a T to C transition changed cysteine 329 to arginine. In the
other, a small 4 base pair deletion was detected as well as a
(? irrelevant) polymorphism 5' to exon 8.

Alternative techniques are also likely to be useful, including RNAase
cleavage[55], and chemical modifications.[56] These techniques appear to be
sensitive enough to discern single base mismatches between the normal probe
and the mutant gene.

SUMMARY

The isolation of the human gene for factor VIII has led to many new
insights about factor VIII structure and has provided an early picture of
the molecular pathology responsible for hemophilia A. While the specific
factor VIII abnormality has been identified in less than 8% of patients
studied to date, improved technology will undoubtedly improve the yield.
Approximately 3% of the hemophilic genes that have been examined have
partial or complete deletions, and a similar number have single nucleotide
substitutions causing missense mutations or premature truncation of the
protein due to a stop codon. In addition, insertions have been identified
in a small number of patients and thrombin cleavage site mutations have
been identified as relatively frequent cause of CRM-positive hemophilia A.
The relationship between specific mutations and factor VIII inhibitor
formation has not been clarified by these studies.

The application of advances in molecular biology has led to
improvement in genetic counseling as well, with much improved ability to
detect female carriers and to provide prenatal diagnosis. In the future,
one looks toward the application of this information to produce an even
more dramatic improvement of hemophilic care, the possibility of gene

therapy so that a normal factor VIII gene can be introduced in order to establish factor VIII production.

REFERENCES

1. Hoyer LW: Factor VIII: New perspectives. Trans Med Rev 1:113, 1987
2. Hoyer LW, Breckenridge RT: Immunologic studies of antihemophilic factor (AHF, factor VIII): Cross-reacting material in a genetic variant of hemophilia A. Blood 32:962, 1968
3. Lazarchick J, Hoyer LW: Immunoradiometric measurement of the factor VIII procoagulant antigen. J Clin Invest 62:1048, 1978
4. McMillan CW, Shapiro SS, Whitehurst D, Hoyer LW, Rao AV, Lazerson J, Hemophilia Study Group: The natural history of factor VIII:C inhibitors in patients with hemophilia A: A national cooperative study. II. Observations on the initial development of factor VIII:C inhibitors. Blood 71:344, 1988
5. Antonarakis SE: The molecular genetics of hemophilia A and B in man. Factor VIII and factor IX deficiency. Adv in Hum Gene 17:27, 1988
6. Gitschier J: The molecular genetics of hemophilia A, in Zimmerman TS, Ruggeri ZM: (eds): Coagulation and bleeding disorders: the role of factor VIII and von Willebrand factor, New York, Dekker,M., 1989, p 23
7. White GC, Shoemaker CB: Factor VIII gene and hemophilia A. Blood 73:1, 1989
8. Thompson AR: Molecular biology of the hemophilias, in Coller BS (ed): Progress in Hemostasis and Thrombosis, Volume 10, Philadelphia, WB Saunders, 1990, (in press)
9. Gitschier J, Wood WI, Goralka TM, Wion KL, Chen EY, Eaton DH, Vehar GA, Capon DJ, Lawn RM: Characterization of the human factor VIII gene. Nature 312:326, 1984
10. Toole JJ, Knopf JL, Wozney JM, Sultzman LA, Buecker JL, Pittman DD, Kaufman RJ, Brown E, Shoemaker C, Orr EC, Amphlett GW, Foster WB, Coe ML, Knutson GJ, Fass DN, Hewick RM: Molecular cloning of a cDNA encoding human antihaemophilic factor. Nature 312:342, 1984
11. Haldane JBS: The rate of spontaneous mutation of the human gene. J Genet 31:317, 1935
12. Youssoufian H, Kazazian HH Jr, Phillips DG, Aronis S, Tsiftis G, Brown VA, Antonarakis SE: Recurrent mutations in haemophilia A give evidence for CpG mutation hotspots. Nature 324:380, 1986
13. Antonarakis SE, Kazazian HH Jr: The molecular basis of hemophilia A in man. Trends in Genetics 4:233, 1988
14. Higuchi M, Kochhan L, Schwaab R, Egli H, Brackman H-H, Horst J, Olek K: Molecular defects in hemophilia A: Identification and Characterization of mutations in the factor VIII gene and family analysis. Blood 74:1045, 1989
15. Casula L, Murru S, Pecorara M, Ristaldi MS, Restagno G, Mancuso G, Morfini M, DeBiasi R, Baudo F, Carbonara A, Mori PG, Cao A, Pirastu M: Recurrent mutations and three novel rearrangements in the factor VIII gene of hemophilia A patients of Italian descent. Blood 75:662, 1990
16. Bardoni B, Sampietro M, Romano M, Crapanzano M, Mannucci PM, Camerino G: Characterization of a partial deletion of the factor VIII gene in a haemophiliac with inhibitor. Hum Genet 79:86, 1988
17. Bernardi F, Volinia S, Patracchini P, Gemmati D, Boninsegna S, Schwienbacher C, Marchetti G: A recurrent missense mutation (Arg - Gln) and a partial deletion in factor VIII gene causing severe haemophilia A. Br J Haematol 71:271, 1989
18. Mikami S: Gene analysis in haemophilia A - restriction fragment length polymorphism and molecular defects in the factor VIII gene. Acta Haem Japonica 51:1680, 1988

19. Gitschier J, Wood WI, Tuddenham EGD, Shuman MA, Goralka TM, Chen EY, Lawn RM: Detection and sequence of mutations in the factor VIII gene of haemophiliacs. Nature 315:427, 1985(abstr)

20. Levinson B, Lehesjoki A-E, De La Chapelle A, Gitschier J: Molecular analysis of hemophilia A mutations in the Finnish population. Am J Hum Genet 46:53, 1990

21. Youssoufian H, Antonarakis SE, Aronis S, Tsiftis G, Phillips DG, Kazazian HH Jr: Characterization of five partial deletions of the factor VIII gene. Proc Natl Acad Sci USA 84:3772, 1987

22. Youssoufian H, Kasper CK, Phillips DG, Kazazian HH Jr, Antonarakis SE: Restriction endonuclease mapping of six novel deletions of the factor VIII gene in hemophilia A. Hum Genet 80:143, 1988

23. Casarino L, Pecorara M, Mori PG, Morfini M, Mancuso G, Scrivano L, Molinati AC, Giavarella LT, Giavarella G, Loi A, Perseu L, Cao A, Pirastu M: Molecular basis for hemophilia A in Italians. Ric Clin Lab 16:227, 1986(abstr)

24. Lillicrap DP, Taylor SAM, Grover H, Teitel J, Giles AR, Holden JJA, White BN: Genetic analysis in hemophilia A: Identification of a large F.VIII gene deletion in a patient with high titre antibodies to human and porcine F.VIII. Blood 68:337a, 1986

25. Kariya K, Silverman L, Friedman K, Perry T, Highsmith W, Cooper H: Major rearrangement of the factor VIII gene in severe hemophilia A. Thromb Haem 62:201, 1989(abstr)

26. Dreesen JCFM, Bakker E, Briet E: A somatic mutation that caused inherited hemophilia A. Thromb Haem 62:200, 1989(abstr)

27. Higuchi M, Wong C, Kochran L, Olek K, Aronis S, Kasper CK, Kazazian HH Jr, Antonarakis SE: Characterization of mutations in the factor VIII gene by direct sequencing of amplified genomic DNA. Genomics 6:65, 1989

28. Kogan S, Gitschier J: Mutations and a polymorphism in the factor VIII gene discovered by denaturing gradient gel electrophoresis. Proc Natl Acad Sci USA 87:2092, 1990

29. Antonarakis SE, Waber PG, Smita MS, Kittur SD, Patel AS, Kazazian HH Jr, Mellis MA, Counts RB, Stamatoyannopoulos G, Bowie EJW, Fass DN, Pittman DD, Wozney JM, Toole JJ: Hemophilia A. Detection of molecular defects and of carriers by DNA analysis. N Engl J Med 313:842, 1985

30. Mikami S, Nishimura T, Naka H, Kuze K, Fukui H: A deletion involving intron 13 and exon 14 of factor VIII gene in a haemophiliac with anti-factor VIII antibody. Jpn J Human Genet 33:401, 1988

31. Woods-Samuels AP, Kazazian HH Jr, Antonarakis SE: Molecular characterization of four deletions in the human factor VIII gene in patients with hemophilia A. Amer J Hum Gen 45:A230, 1989(abstr)

32. Reisner HM, Price WA, Blatt PM, Barrow ES, Graham JB: Factor VIII coagulant antigen in hemophilic plasma: A comparison of five alloantibodies. Blood 56:615, 1980

33. Din N, Schwartz M, Kruse T, Vestergaard SR, Ahrens P, Scheibel E, Nordfang O, Ezban M: Factor VIII gene specific probes used to study heritage and molecular defects in hemophilia A. Ric Clin Lab 16:182, 1986(abstr)

34. Shelton-Inloes BB, Chehab FF, Mannucci PM, Federici AB, Sadler JE: Gene deletions correlate with the development of alloantibodies in von Willebrand disease. J Clin Invest 79:1459, 1987

35. Chan V, Chan TK, Tong TMF, Todd D: A novel missense mutation in Exon 4 of the factor VIII:C gene resulting in moderately severe hemophilia A. Blood 74:2688, 1989

36. Youssoufian H, Kazazian HH Jr, Patel A, Aronis S, Tsiftis G, Hoyer LW, Antonarakis SE: Mild hemophilia A associated with a cryptic donor splice site mutation in intron4 of the factor VIII gene. Genomics 2:32, 1988

37. Youssoufian H, Wong C, Aronis S, Platokoukis H, Kazazian HH Jr, Antonarakis SE: Moderately Severe Hemophilia A Resulting from Glu-Gly Substitution in Exon 7 of the Factor VIII Gene. Am J Hum Genet 42:867, 1988

38. Kogan SC, Gitschier J: Detection of hemophilia A mutations near the acidic region of factor VIII by DNA amplifications and denaturing gradient gel electrophoresis. Blood 72:1110, 1988(abstr)

39. Gitschier J, Kogan S, Levinson B, Tuddenham EGD: Mutations of Factor VIII Cleavage Sites in Hemophilia A. Blood 72:1022, 1988

40. Shima M, Ware J, Yoshioka A, Fukui H, Fulcher CA: An arginine to cysteine amino acid substitution at a critical thrombin cleavage site in a dysfunctional factor VIII molecule. Blood 74:1612, 1989

41. Arai M, Inaba H, Higuchi M, Antonarakis SE, Kazazian HH Jr, Fujimaki M, Hoyer LW: Direct characterization of factor VIII in plasma - Detection of a mutation altering a thrombin cleavage site (Arginine-372 - Histidine). Proc Natl Acad Sci USA 86:4277, 1989

42. Arai M, Higuchi M, Antonarakis SE, Kazazian HH Jr, Phillips JA, Janco RL, Hoyer LW: Characterization of a thrombin cleavage site mutation (Arg 1689 to Cys) in the factor VIII gene of 2 unrelated patients with cross reacting material positive hemophilia A. Blood 75:384, 1990

43. Traystman MD, Higuchi M, Kasper CK, Antonarakis SE, Kazazian HH Jr: Use of denaturing gradient gel electrophoresis to detect point mutations in the factor VIII gene. Genomics 6:293, 1990

44. Lillicrap D, Holden JJA, Giles AR, White BN, The Ontario Haemophilia Study Group: Carrier detection strategy in haemophilia A: the benefits of combined DNA marker analysis and coagulation testing in sporadic haemophiic families. Br J Haematol 70:321, 1988

45. Levinson B, Janco R, Phillips III JA, Gitschier J: A novel missense mutation in the factor VIII gene identified by analysis of amplified hemophilia DNA sequences. Nucl Acids Res 15:9797, 1987

46. Youssoufian H, Antonarakis SE, Bell W, Griffin AM, Kazazian HH Jr: Nonsense and missense mutations in hemophilia A: Estimate of the relative mutation rate at CG dinucleotides. Am J Hum Genet 42:718, 1988

47. Hellman L, Smedsrod B, Sandberg H, Pettersson U: Secretion of coagulant factor VIII activity and antigen by in vitro cultivated rat liver sinusoidal endothelial cells. Br J Haematol 73:348, 1989

48. Mikami S, Nishimura T, Naka H, Kuze K, Fukui H, Tone M, Hashimoto-Gotoh T: Nonsense mutation in factor VIII gene of a severe haemophiliac patient with anti-factor VIII antibody. Jpn J Human Genet 33:409, 1988

49. Gitschier J, Wood WE, Shuman MA, Lawn RM: Identification of a missense mutation in the factor VIII gene of a mild hemophilia. Science 232:1415, 1986

50. Inaba H, Fujimaki M, Kazazian HH Jr, Antonarakis SE: Mild hemophilia A resulting from Arg-to-Leu substitution in exon 26 of the factor VIII gene. Hum Genet 81:335, 1989(abstr)

51. Kazazian HH Jr, Wong C, Youssoufian H, Scott AF, Phillips DG, Antonarakis SE: Haemophilia A resulting from de novo insertion of L1 sequences represents a novel mechanism for mutation in man. Nature 332:164, 1988

52. Pittman DD, Kaufman RJ: Proteolytic requirements for thrombin activation of anti-hemophilic factor (factor VIII). Proc Natl Acad Sci USA 85:2429, 1988

53. Rapaport SI, Schiffman S, Patch MJ, Ames SB: The importance of activation of antihemophilic globulin and proaccelerin by traces of thrombin in the generation of intrinsic prothrombinase activity. Blood 21:221, 1963

54. O'Brien DP, Pattinson JK, Tuddenham EGD: Purification and characterization of factor VIII 372-Cys: A hypofunctional cofactor from a patient with moderately severe hemophilia A. Blood 75:1664, 1990

55. Myers RM, Larin Z, Maniatis T: Detection of single base substitutions by ribonuclease cleavage at mismatches in RNA:DNA duplexes. Science 230:1242, 1990
56. Novack DF, Casma NJ, Fischer SG, Ford JP: Detection of single base-pair mismatches in DNA by chemical modification followed by electrophoresis in 15% polyacrylamide gel. Proc Natl Acad Sci USA 83:586, 1986

MOLECULAR DEFECTS IN HEMOPHILIA B

Arthur R. Thompson

Puget Sound Blood Center
921 Terry Avenue
Seattle, WA 98104

INTRODUCTION

Heterogeneity among hemophilic patients was inferred from observations that plasma samples mixed from different patients would occasionally show correction of the prolonged clotting time.[1] In 1952, a second or "B" type was distinguished from the more common "A" type by Aggeler et al[2] and Schulman and Smith.[3] Like hemophilia A, hemophilia B was inherited as an X-linked recessive disorder.[4] About 20 to 25% of patients with hemophilia have the B type, but they are otherwise clinically indistinguishable from patients with hemophilia A. The protein which lacks activity in hemophilia B patients, factor IX, is more stable than factor VIII; when plasma is clotted to serum, the majority of the factor IX molecules remain in the zymogen (precursor) form whereas factor VIII is consumed. Marked genetic heterogeneity among different families with hemophilia B was suggested from comparison of factor IX antigen levels with clotting activities. In patients from different families, specific activities were usually distinctly different.[5]

The cDNA and the entire 34 kb factor IX gene have been sequenced.[6] It is localized within Xp 27.3, centomeric to the fragile X-mental retardation site. Extensive genetic heterogeneity among hemophilia B families is now confirmed after identification of specific mutations by analysis of patients' DNAs. Defects include gross gene alterations, micro-deletions (usually with a frameshift leading to premature termination of translation) and point mutations as single base substitutions. In the non-coding regions, changes in a 5' regulatory region and in splice junctions have been observed. Within the coding sequence, nonsense (new Stop codon) or missense (amino acid substitution) mutations occur. Although many families have their own "private" defect, certain mutations tend to be recurrent, especially transitions within CG dinucleotides.

GROSS GENE ALTERATIONS

Complete Gene Deletions. The entire factor IX gene has been deleted in patients from 15 families (Table 1, upper half). As one would expect, all of these patients have a clinically severe bleeding tendency. By analysis of regions adjacent to either the 5' or 3' ends, the extent of the deletion can be estimated. In two defects (B-1 and B-2) more than 273 kb are missing[8] whereas another (B-7) has at least one breakpoint close to the gene (within 8 kb of the 3' end of exon 8).[13] Ten of these patients have developed alloantibody inhibitors, whereas overall only about 1% of patients with hemophilia B have inhibitors.

    <u>Partial Gene Alterations</u>. Ten patients have gross, partial gene deletions (Table 1, lower half) ranging from 2 kb (B-26)[27] to >40 kb (B-16)[19]. There is no apparent clustering of breakpoints. Sequence through the breakpoints of a 23 kb deletion (B-28)[28] did not reveal homologous sequences at the ends. Sequence through another deletion site (B-24) revealed that it occurred between two 14 bp homologous sequences (with 13 bases identical) that were 10.0 kb apart, suggesting that this deletion occurred by a "looping out" mechanism.[25] Trace amounts of a truncated antigen with amino-terminal epitopes were present in the plasmas and urines of patients from this family.[24] In another patient with a partial deletion (B-21), exon 4 was absent.[26] Since the donor splice junction from exon 3 is in phase with the acceptor splice junction of exon 5, this patient circulates factor IX antigen at about one-third the normal level; it represents a smaller species on polyacrylamide gels.[31] Patient B-23 has a complex factor IX gene alteration in which two segments are deleted, flanking an inverted 3 kb sequence; the breakpoints also revealed insertions of 3 and 5 nucleotides.[22] Patient B-26 also appears to have a deletion and an insertion;[27] since it is within a single intron, however, it may represent an incidental variant, unrelated to his hemophilia. Patients B-22 and B-25 have insertions at the 3' end of exon 4,[21] and an Alu sequence within exon 5,[26] respectively. Patients from each of these families have severe hemophilia B and five have inhibitors.

Table 1. GROSS GENE ALTERATIONS IN HEMOPHILIA B PATIENTS

Defect	Size (kb)	Alteration Type	Location (kb) 5' Extent	Location (kb) 3' Extent	Other Sites Deleted (-), Present (+)	Number of Patients (Inhibitor)	Reference (# of Families)
Complete Gene Deletions							
B-1,2	>273	D	>60	>180	-Prb; -mcf. 2; -HTF	2(2)	7,8(2)
B-3	>115	D	?1	>80	-mcf. 2; +DXS99	1	9
B-4	>115	D	?	~85	-mcf. 2	1(1)	10
B-5	>145	D	?	115	-mcf. 2; +HTF	1(1)	10,11
B-6	>70	D	>40	<40	-Prb; +mcf.2	1	12
B-7	>36	D	>0.5	<8	-BamHI; +HhaI	1	13
B-8	>35	D	?	?	+DXS98, 99, 100	1	14
B-9,10,11	>35	D	?	?	-	3(3)	15,16(2)
B-12,13,14	>35	D	?	?	+HPRT	3(3)	17(3)
B-15	>35	D	?	?	-	1	18
Partial Gene Alterations							
B-16	>40	D	5' Exon 1	Exon 8	-	1(1)	19
B-17	>11-35	D	5' Exon 1	Intron 1	-	1	12
B-18	~9	D	5' Exon 1	Intron 3	-	1(1)	12
B-19	27-42	D	Intron 1	3' Exon 8	+HhaI	1	13
B-20	~8	D	Intron 3	Intron 5	-	1	12
B-21	~5	D	Intron 3	Intron 4	-	1	20
B-22	~6	I	Exon 4 (3' end)		-	1	21
B-23	5.5 and 10.4	2D/R	Intron 4 / Intron 6	Intron 5 and 3' Exon 8	-	1(1)	10,22
B-24	10.0	D	Intron 4	Intron 6	-	1	23,24,25
B-25	?	I	Within Exon 5(Alu)		-	1	26
B-26	~2	D/I	Within Intron 6		-	[1(1)]	27
B-27	~1.5	D	Intron 6	Intron 7	-	1	12
B-28	23.0	D	Intron 5	3' Exon 8	-	1(1)	7,28
B-29	?	D	? Exon 7	3' Exon 8	-	1	29

Other sites:[30] Prb, a locus 35 kb 5' to IX; BamHI and HhaI polymorphic sites, 0.5 kb 5'; 8 kb 3', respectively; mcf.2, a transforming gene-like sequence ~30 kb 3'; HTF, a gene-like sequence ~130 kb 3'; DXS series of loci, flanking Xq 27.3; HPRT (hypoxanthine-guanidine phosphoribosyl transferase), in Xq 26.

Alterations: D, deletion; I, insertion; R, rearrangement.

SPLICE JUNCTION AND TRANSLATION MUTATIONS

Splice Junction Defects. Eight distinct defects have been found including four each involving the 5' acceptor splice sequences and 3' donor splice regions (Table 2). In patient B-37's DNA, there is a 4 bp deletion at an acceptor site[18] whereas the three other acceptor site changes represent single base substitutions (B-36, B-38, B-48).[13,39] The first two of the point mutations are moderately severe, clinically. The four defects occurring at donor splice junctions include a 4 bp deletion (B-35)[13] and two with nucleotide substitutions (B-33 and B-46);[42] these three are within one base 3' to the exon involved. A third substitution (B-41)[39] is 13 bp 3' to exon 5 and, along with the 4 bp deletion near the acceptor splice junction of exon 4 (B-37),[18] is associated with mild hemophilia.

Frameshift Defects. Within the coding regions, "microdeletions" have been described in eight patients (Table 2). These range from one nucleotide (B-30 and B-39)[32,40] to 13 (B-42).[18] In patient B-57, a dinucleotide deletion is accompanied by an insertion of 10 bases at the same site.[51] The latter is the only frameshift mutation which does not result in severe hemophilia B; it occurs 13 codons from the normal termination codon 416, and predicts coding of an additional 22 amino acids.

Table 2. SPLICING AND TRANSLATION MUTATIONS IN HEMOPHILIA B

Defect	Location	Codon[6]	Change	Number (+ Inhibitor)	Factor IX Level IX:C	IX:Ag	Reference (# of Families)
B-29.5	Exon 2	5.3	A→G (? new donor splice)	1	7	-	39
B-30	Exon 2	6.2	ΔT-frameshift	1(1)	<1	1	32
B-31	Exon 2	11.1	C→T, Glu→STOP	1	<1	1	33
B-32	Exon 2	29.1	*C→T, Arg→STOP	4(1)	<1	<1(7)	32,34,35,36
B-33	Exon 3	39.3-40.1	ΔTG, frameshift	1	<1	<1	37
B-34	Intron 3	46.3+3	T→G-donor splice	1	<1	<1	38
B-35	Intron 3	38.3+1 or 2	ΔTGAG-donor splice	1	<1	<1	13
B-36	Intron 3	47.2-1	G→A, acceptor splice	1	3	-	18
B-37	Intron 4	85.2-6 to -9	ΔTTCT, acceptor splice	1	20	-	18
B-38	Intron 4	85.2-2	A→G, acceptor splice	1	3	3	39
B-39	Exon 5	85.2	ΔA, frameshift	1	<1	<1	40
B-40	Exon 5	116.1	*C→T, Arg→STOP	1	<1	<.1	41
B-41	Intron 5	127.3+13	A→G, donor splice	2	10	-	39(2)
B-42	Exon 6	162.3+13bp	Δ13 (AACCATTTGGAT)	1	<1	-	18
B-43	Exon 6	173.1	C→T, Trp→STOP	1	<1	-	18
B-44	Exon 6	191.1	C→T, Glu→STOP	1(1)	<1	-	37
B-45	Exon 6	194.2	G→A, Trp→STOP	1	<1	<1	32
B-46	Intron 6	195.3+1	G→T, donor splice	1	<1	<1	42
B-47	Exon 7	215.2	G→A, Trp→STOP	1	<1	<1	43
B-48	Intron 7	234.2-1	G→A, acceptor splice	2(1)	<1	<1	37,43
B-49	Exon 8	248.1	*C→T, Arg→STOP	7(1)	<1	<1(4)	32,34(4),44,45
B-50	Exon 8	252.1	*C→T, Arg→STOP	4	<1(3)	<1(6)	34,46-48
B-51	Exon 8	277.1+8bp	ΔGAACCCT, frameshift	1	<1	<1	32
B-52	Exon 8	310.2	G→A, Trp→STOP	1	<1	1	44
B-53	Exon 8	313.2-.3	ΔTC, frameshift	1	<1	<2	32
B-54	Exon 8	333.1	*C→T, Arg→STOP	4	<1(6)	<1-1	34,43(3)
B-55	Exon 8	338.1	*C→T, Arg→STOP	6	<1(2)	<1	12,34(2),49(2),50
B-56	Exon 8	346.1 or .2	Insert AA, frameshift	1	<1	<1	13
B-57	Exon 8	402.3-403.1	ΔCC & Ins 10, frameshift (Ins AAGGTACCAA)	1	3	9	51
B-58	Exon 8	408.3	Ins GATT, frameshift	1	<1	-	52
B-59	Exon 8	411.1	A→T, Lys→STOP	1	<1	<1	53

Codon decimals indicate triplet base; IX:C, clotting activity and IX:Ag, antigen level in U/dl (or % of normal pool) with values in parentheses ( ) for "outliers" from the group with that defect. Asterisk (*) is for a CG transition mutation; Δ, microdeletion; Ins, insertion.

    <u>Nonsense Mutations</u>. Premature Stop codons are introduced by single base substitutions at 13 sites ranging from codon 11 (B-31)[33] through codon 411 (B-59).[53] In all, 33 different families with a new Stop codon had been found. All of the nonsense mutations have been associated with severe hemophilia B with trace to undetectable factor IX antigen levels. Multiple occurrences are all as C to T transitions within five CG dinucleotides; three of these have had inhibitors.

REGULATOR REGION DEFECTS

    Mutations at three sites near the first nucleotide (the putative initiation site for transcription) have been described (Table 3, upper portion). All are associated with a "Leyden-like" defect in which the hemophilia becomes more mild or even disappears after puberty.[99] At -20 bp, a T to A transversion was described[54] whereas at +13, families have been found with an A to G transition and one with a single nucleotide deletion.[57] At -6 bp, G to A or G to C mutations occur but have been associated with a less severe bleeding tendency in pre-pubertal years.[55,56] The alteration may involve an androgen sensitive binding protein.[100] Where tested, factor IX antigen levels are equivalent to factor IX clotting activities at different stages of development. Of interest, occasional women within these families may have sufficiently low levels to have mild hemophilia and this persists into adulthood.[99]

Table 3. REGULATOR & MISSENSE MUTATIONS IN HEMOPHILIA B PATIENTS

Defect	Location	Position or Codon	Substitution Nucleotide	Amino Acid	Number Reported	Severity IX:C	IX:Ag	Reference (# of Families)
Regulator Defects								
B-60	5' Exon 1	-20 bp	T→A	-	1	Sev-Mild	-	54
B-61	5' Exon 1	-6 bp	*G→A	-	1	Mod-Mild	-	55
B-62	5' Exon 1	-6 bp	G→C	-	2	Mod-Mild	-	56,104
B-63	5' Exon 1	+13 bp	A→G	-	3	Sev-Mild	-	18,57,58
B-64	5' Exon 1	+13 bp	△ A	-	1	Sev-Mild	-	57
Missense Mutations								
B-65	Exon 2	-4.1	*C→T	Arg→Trp	4	<1-4	26-40	32,34(3)
B-66	Exon 2	-4.2	*G→A	Arg→Gln	8	<1(4)	40-75	32(2),39,45,59-62
B-67	Exon 2	-4.2	G→T	Arg→Leu	1	<1	27	39
B-68	Exon 2	-1.3	G→T/C	Arg→Ser	1	<1	80	63
B-69	Exon 2	2.1	A→C	Asn→Asp	1	6	-	39
B-70	Exon 2	7.2	A→C	Glu→Ala	1	6	5	33
B-71	Exon 2	18.1	T-→C	Cys→Arg	1	<1	100	64
B-72	Exon 2	27.1	G→A	Glu→Lys	1	<1	30	65
B-73	Exon 2	27.2	A→T	Glu→Val	1	<1	3	66
B-74	Exon 2	29.2	*G→A	Arg→Gln	3	20-37	70	18,39,43
B-75	Exon 2	33.3	A→C	Glu→Asp	1	4	-	18
B-76	Exon 2	37.1-.3	△AGA	△Arg	1	<1	12	32
B-77	Exon 3	47.1	G→A	Asp→Gly	1	10	100	67
B-78	Exon 4	50.2	A→C	Gln→Pro	1	<1	114	68
B-79	Exon 4	55.1	C→G	Pro→Ala	3	10-12	30-58	32,43,69
B-80	Exon 4	56.2	G→A	Cys→Tyr	1	1	2	36
B-81	Exon 4	60.1	*G→A	Gly→Ser	11	10-17	11-34	18(3),34,39,43, 65(2),70(2),71
B-82	Exon 4	60.2	G→A	Gly→Asp	1	1	2	39
B-83	Exon 4	64.2	A→G	Asp→Gly	1	8	87	32
B-84	Exon 5	92.1	A→C	Asn→His	1	2	66	72
B-85	Exon 5	99.1	T→C	Cys→Arg	1	<1	-	39
B-86	Exon 5	107.2	T→C	Val→Ala	1	20	120	43
B-87	Exon 5	114.2	G→C	Gly→Ala	1	5	4	33
B-88	Exon 5	120.1	A→T	Asn→Tyr	1	<1	<1	32
B-89	Exon 6	132.1	T→C	Cys→Arg	1	<1	<1	73
B-90	Exon 6	145.1	*C→T	Arg→Cys	5	1-2(<1)	30-66,	34(2),74,75,83
B-91	Exon 6	145.2	*G→A	Arg→His	9	4-11	60-160	18,34(5),76-78

MISSENSE MUTATIONS AND OTHER AMINO ACID SEQUENCE CHANGES

The eight exons of factor IX largely code for distinct structural and functional domains. Changes in the coding by point mutations (or in one case, a 3 bp codon deletion), can lead to 1) a dysfunctional protein that circulates at near normal to normal levels, 2) a dysfunctional protein that circulates at reduced levels or 3) low circulating levels in which the clotting activities and antigen levels are equivalent. In patients with mild or moderate severity, the latter would appear as a normal specific activity. The normal specific activity associated with hemophilia B could arise from protein instability either after secretion or intracellularly. Alternatively, cDNA or mRNA may be unstable leading to decreased protein production. Missense mutations are presented by exons in the balance of Table 3.

Table 3. (Continued)

Defect	Location	Position or Codon	Substitution Nucleotide	Amino Acid	Number Reported	Severity IX:C	IX:Ag	Reference (# of Families)
B-92	Exon 6	180.1	*C→T	Arg→Trp	2	<1	100,130	79,80
B-93	Exon 6	180.2	*G→A	Arg→Gln	3	<1	110	73,79,81
B-94	Exon 6	181.1	G→T	Val→Phe	1	<1	130	79
B-95	Exon 6	182.1	G→T	Val→Phe	1	<1	120	82
B-96	Exon 6	182.2	G→C	Val→Leu	1	15	132	83
B-97	Exon 6	191.2	A→T	Gln→Leu	1	<1	<.1	43
B-98	Exon 7	211.1	G→T	Val→Phe	1	7	-	43
B-99	Exon 7	216.2	T→C	Ile→Thr	1	4	-	43
B-100	Exon 7	220.2	C→T	Ala→Val	2	<1,4	3,-	39(2)
B-101	Exon 7	222.3	T→G	Cys→Trp	1	1	-	18
B-102	Exon 7	233.1	*G→A	Ala→Thr	7	12-22	12-15	18,34(4),43(2)
B-103	Exon 8	248.2	*G→A	Arg→Gln	5	2-4	2-4	39,43(2),65(2)
B-104	Exon 8	260.2	A→G	Asn→Ser	1	24	-	18
B-105	Exon 8	271.2	C→T	Ala→Val	1	1	4	39
B-106	Exon 8	287.2	C→T	Pro→Leu	1	<1	<1	43
B-107	Exon 8	291.1	G→C	Ala→Pro	1	2	3	33
B-108	Exon 8	291.1	G→A	Ala→Thr	1	4-16	-	35
B-109	Exon 8	296.2	*C→T	Thr→Met	7	2-6	5-15	18,34(2),43(4)
B-110	Exon 8	307.2	T→C	Val→Ala	1	5	18	84
B-111	Exon 8	309.2	G→T	Gly→Val	1	<1	58	85
B-112	Exon 8	311.1	G→A	Gly→Arg	1	3	-	18
B-113	Exon 8	320.2	C→A	Ala→Asp	1	<1	90	39
B-114	Exon 8	328.1	G→T	Val→Phe	1	4	4	86
B-115	Exon 8	333.2	*G→A	Arg→Gln	7	1-3	32-135	32,34(2),43, 44,87,88
B-116	Exon 8	336.1	T→C	Cys→Arg	2	<1,2	2-5	32,43
B-117	Exon 8	348.1	A→G	Met→Val	1	3	103	43
B-118	Exon 8	360.2	C→T	Ser→Leu	1	2	130	43
B-119	Exon 8	363.2	G→T	Gly→Val	1	1-5	100	89
B-120	Exon 8	364.2	A→T	Asp→Val	1	<1	130	43
B-121	Exon 8	365.1	T→A	Ser→Arg	1	<1	90	39
B-122	Exon 8	367.1	G→A	Gly→Arg	1	<1	14	43
B-123	Exon 8	368.1	C→A	Pro→Thr	1	<1	156	79
B-124	Exon 8	378.1	T→C	Phe→Leu	1	<1	<1	43
B-125	Exon 8	381.2**	G→A	Gly→Glu	1	<1	<1	90
B-126	Exon 8	390.2	C→A	Ala→Glu	1	2	30	44
B-127	Exon 8	390.2	C→T	Ala→Val	2	<1,1-4	100,140	91,92
B-128	Exon 8	396.1	G→A	Gly→Arg	1	<1	90	53
B-129	Exon 8	397.2	T→C	Ile→Thr	21	(<1)1-5	45-96	18(5),43(2) 53(3),93-97(11)
B-130	Exon 8	402.2	C→T	Ser→Phe	1	2	-	39
B-131	Exon 8	407.1	T→C	Trp→Arg	1	1	-	18

Factor IX levels (ranges where >1) as in Table 2; where values were not reported in the description, they were either not available (-) or from a separately published data base.[98] B-125 (**) is the mutation in a colony with canine hemophilia B.

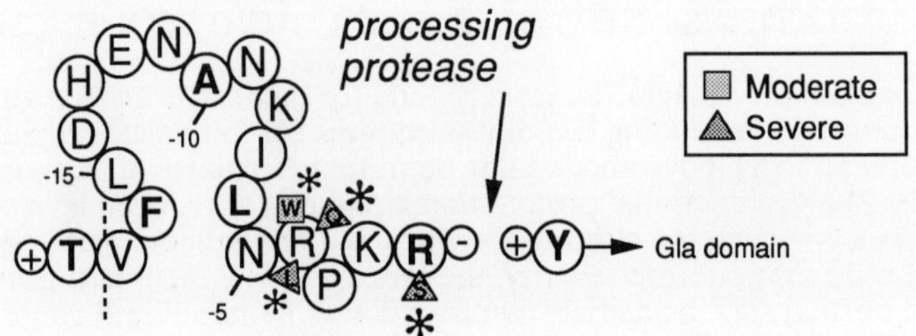

Figure 1. <u>Propeptide Sequence</u>. The propeptide provides gamma-carboxylase recognition. Basic side chains (-1 through -4) allow binding of a processing protease, which cleaves the Arg -1 to Tyr 1 bond prior to secretion. Missense mutations are indicated by severity (shaded symbols) with the substituted amino acid (single letter code, see Fig. 6 legend). Asterisks (*) are for dysfunctional factor IX with normal levels (largest asterisk), reduced levels (middle size asterisk) or low levels (smallest asterisk) in which the antigen level is at least twice the clotting activity.

<u>Propeptide Mutations</u>. As yet, there are no known mutations within exon 1 or the signal peptide leading to hemophilia B. In exon 2, three mutations at Arg -4 (B-65 through B-67)[32,39,59] and one at Arg -1 (B-68)[63] are associated with more severe hemophilia B with low normal to normal antigen levels (Fig. 1). Propeptide processing, however, is inhibited leaving an extended 17 amino acids attached to the normal amino terminus of the mature, circulating factor IX. This peptide interferes with the Gla domain's calcium-dependent conformation, accounting for its dysfunction.

<u>Gla Domain</u>. The balance of exon 2 and exon 3 code for the amino-terminus (Tyr 1) through the first base of the 47th codon (Asp). Defects of specific Gla residues (Fig. 2) include Glu 7 to Ala (B-70)[33] and Glu 33 to Asp (B-75)[18] associated with mild and moderately severe hemophilia B, respectively, and two severe defects in codon 27, B-72[65] and B-73.[66] Another dysfunctional defect is Cys 18 to Arg (B-71).[64] In this case, the hemophilic protein circulates with a peptide that is disulfide bound to Cys 23.[101] It is not known whether dysfunctional protein is present with defects in codons 2 or 33, but a patient with an in frame deletion of the codon for Arg 37 (B-76)[32] has severe hemophilia with detectable antigen (12 U/dl).[32]

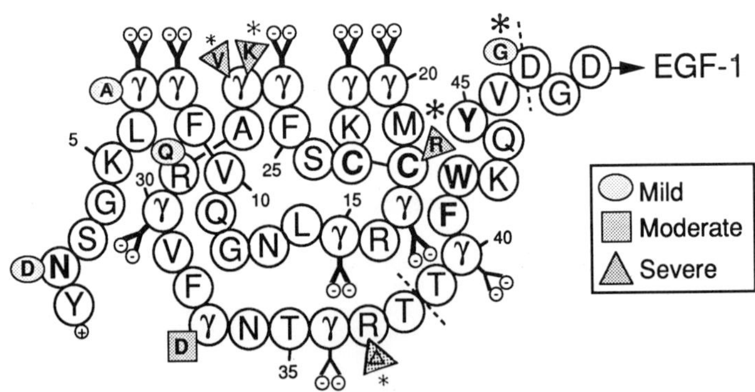

Figure 2. <u>Gla-Domain</u>. The amino acid residues are aligned to reflect the three-dimensional structure of the Gla-domain of prothrombin in which the three paired Gla residues are oriented on one surface.[30] Missense mutations and antigen levels are as in Fig. 1; △ is a 3 bp deletion.

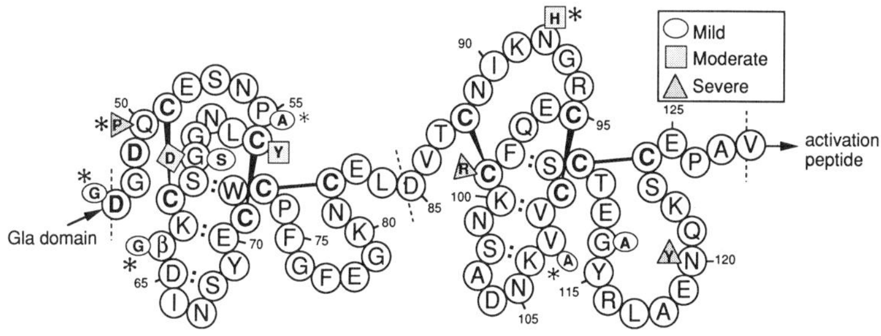

Figure 3. <u>Growth-Factor-Like Domains</u>. The amino acids of these regions are arranged according to structural data from epidermal growth factor in which the second disulfide loop is present as an anti-parallel, beta-pleated sheet.[30] Missense mutations and antigen levels are as in Fig. 1.

<u>Growth Factor-Like Regions</u>. Exons 4 and 5 code for two regions, each with 6 Cys residues, homologous with epidermal growth factor, and missense mutations are shown in Fig. 3. It appears that negatively charged side chains of Asp 47, 49 and 64 contribute to a high affinity calcium-binding site that is independent of the Gla domain. When the codons for the first and third of these are changed to Gly, the patients have mild hemophilia B with normal circulating antigen (B-77 and B-83).[67,32] At residue 50, introduction of a Pro for a Gln creates a severe defect (B-78) which has a normal level of circulating, dysfunctional antigen.[68] Some excess antigen is found in the mild defect Pro 55 to Ala (B-79)[32] and less with Gly 60 to Ser (B-81).[43,70] In a second growth factor-like region, the Asn 92 to His 72 and Val 107 to Ala[43] mutations have dysfunctional protein (B-84 and B-85).

<u>Activation Peptide</u>. Exon 6 defects are shown in Fig. 4. A Cys 132 to Arg mutation (B-89) involves the disulfide bond which normally binds the heavy and light chains of factor IXa together. The patient has severe hemophilia and no detectable antigen.[73] Mutations at the "alpha" cleavage site, Arg 145 (B-90 and B-91)[74,76] produce either moderately severe or mild hemophilia despite an inability of factor XIa to cleave the bond between the light chain and the activation peptide. Cleavage of the Arg 180-Val 181 bond, however, is essential for activity; defects of either residue are severe despite normal antigen levels (B-92 through B-94).[79-81] Additionally, less severe changes at residue 182 (B-95 and B-96) are known.[82,83]

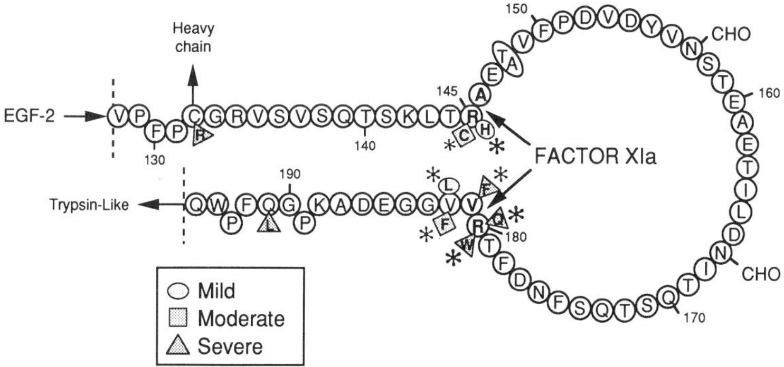

Figure 4. <u>Activation Peptide and Flanking Sequences</u>. There is no direct structural evidence for the conformation of this region and, indeed, the activation peptide itself varies considerably in both sequence and length among factor IXs from different species. Missense mutations and antigen levels are as in to Fig. 1.

<u>Tryptic Region</u>. Defects in factor IXa's heavy chain region (exons 6-8) are summarized in Fig. 5. In order for a trypsin-like serine protease like factor IXa to express proteolytic activity, it is essential for an *ion pair* to form from the amino terminus of the residue which is created by the beta-activation cleavage and the side chain of Asp 364, adjacent to the active center Ser 365.[30] Thus, when the Arg 180-Val 181 bond cannot be cleaved, ion pair formation cannot occur. Likewise, in a patient with severe hemophilia B and high normal antigen levels, an Asp 364 to Val mutation would be inactive and, indeed, is severe (B-120).[13] Another defect which appears to disrupt ion pair formation was found in a patient with severe hemophilia (B-111) with normal antigen levels due to Gly 309 to Val.[85] Although not close in the linear sequence, this residue is within a few Å of the ion pair and molecular models introducing any side chain for the H of Gly would prevent the ion pair interaction necessary for activation, and thus inhibit the hemophilic "factor IXa". Val 309 does not, however, interfere with factor XIa cleavages.[85]

Once the active center Ser-OH is labelized by the ion pair, it is able to form an unstable acyl intermediate with a specific substrate peptide bond. Catalysis is facilitated by two charged residues which, with Ser 365, make up the *catalytic triad*; the two are His 221 and Asp 269. As expected, mutations of Ser or adjacent, conserved residues produce severe hemophilia B with dysfunctional antigen (B-119 through B-123). Of interest, the Cys 222 to Trp mutation (adjacent to the His; B-91) is only moderately severe, but it is not known to what extent antigen circulates.[18] Less severe defects (B-98 through B-100) are within this disulfide loop (Cys 206-Cys-222).[13,39]

The third element of catalysis is *substrate binding*. The positively charged Arg side chain from the substrate's peptide bond to be cleaved binds ionically in a pocket in which Asp 359 presents its negative charge at the base. The Gly 396 to Arg mutation (B-128) is severe and this dysfunctional protein, by molecular models, binds its Arg 396 side chain in the pocket, inhibiting substrate binding.[53] In another defect, a model predicts a hydrogen bond from the new hydroxyl group in the Ile 397 to Thr mutation (B-129) could form with Trp 385 and thus disrupt secondary substrate binding.[93] In this defect, the dysfunctional factor IXa will cleave small molecular weight substrates,[95] but not factor X.[93] Two mutations of Ala 390 (B-126 and B-127)[44,91] may also interfere with substrate binding.

Defects at and around the disulfide loop from Cys 336 to Cys 350 (B-115 to B-118) are moderately severe and associated with detectable but low to normal dysfunctional antigen levels, but the role of this subregion in serine protease function is less well understood. On an opposite surface, the Ala 233 to Thr (B-102) and Arg 248 to Gln (B-103) missense mutations are of interest as each have essentially normal specific activities[43] suggesting instability of either their mRNAs or proteins. The protein present in patients with each of these defects fails to react to a specific monoclonal antibody[65] whose conformational epitope resides in what is probably a looped region on the surface by molecular models. Of several other defects within the heavy chain region, specific activities (where known) are near normal, suggesting that mRNA or protein stability is predominantly affected.

Patients who develop alloantibodies to factor IX have had either a deletion, frameshift or splicing, or nonsense mutation (Tables 1-2). The only exception is patient B-95[102] whose Gly 191 to Val substitution[43] was the only predicted change after sequencing his entire coding, splice junction and 5' regulator regions. It remains possible that an unknown, more distant 5' or 3' regulator, or even an intron mutation could account for his severe defect as a second mutation; alternatively, single base changes can occasionally make the mRNA severely unstable leading to undetectable circulating antigen. In this regard, other missense mutations at 287 (B-106), 378 (B-124) and 381 (B-125) are found in patients (human or canine) with undetectable to trace circulating antigen levels[43,90] but without inhibitors.

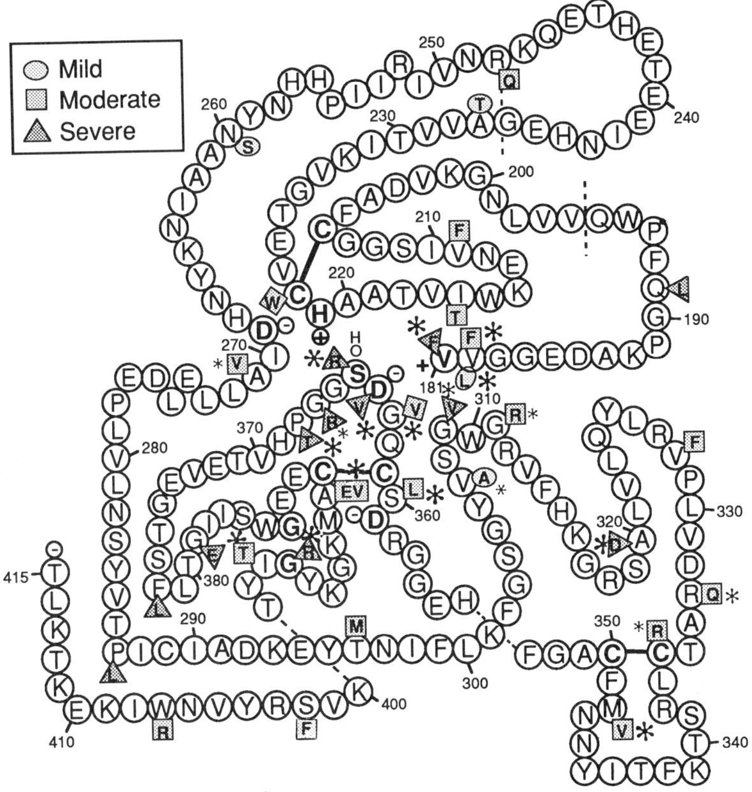

Figure 5. Heavy Chain of Factor IXa (Trypsin-Like Region). The sequence of the carboxy-terminal half of factor IXa is arranged to reflect the three-dimensional structure based on molecular models of trypsin-like enzymes.[30] The active center Ser 365 is shown with its side chain hydroxyl group (OH) as is the ion pair between the amino terminus of Val 181 (+) and the carboxyl side chain of Asp 364 (-). The two charged residues of the catalytic triad (with Ser 365 being the third) are His 221 (+) and Asp 269 (-). Asp 359 (-) is at the base of the substrate's P1' binding site. Missense mutations and antigen levels of dysfunctional protein are indicated as described in Fig. 1.

Hougie and Twomey[103] described a patient with hemophilia B associated with a prolonged prothrombin time (extrinsic system) but only when bovine (not human or rabbit) thromboplastin was used as a source of tissue factor. This B_m prolongation has varied in different families, even when the same molecular defect is present.[95] B_m patients have all been severe or moderately severe with normal levels of dysfunctional factor IX antigen and defects within the catalytic regions of factor IXa. The mutations are heterogeneous, however, and include missense changes near the beta activation cleavage site (B-92 through B-96),[13,79-81] the active center (B-123)[79] and residues involved in a secondary substrate binding site (B-127 through B-129).[53,90-95] In purified systems, high levels of the dysfunctional factor IX appear to competitively inhibit human factor VIIa, delaying direct (factor VIIa mediated) and/or indirect (through factor IXa) activation of factor X with bovine tissue factor.[95] This explanation predicts that human or rabbit tissue factor would bind factor VIIa more efficiently and allow normal extrinsic factor X activation *in vitro*, even in the absence factor IX.

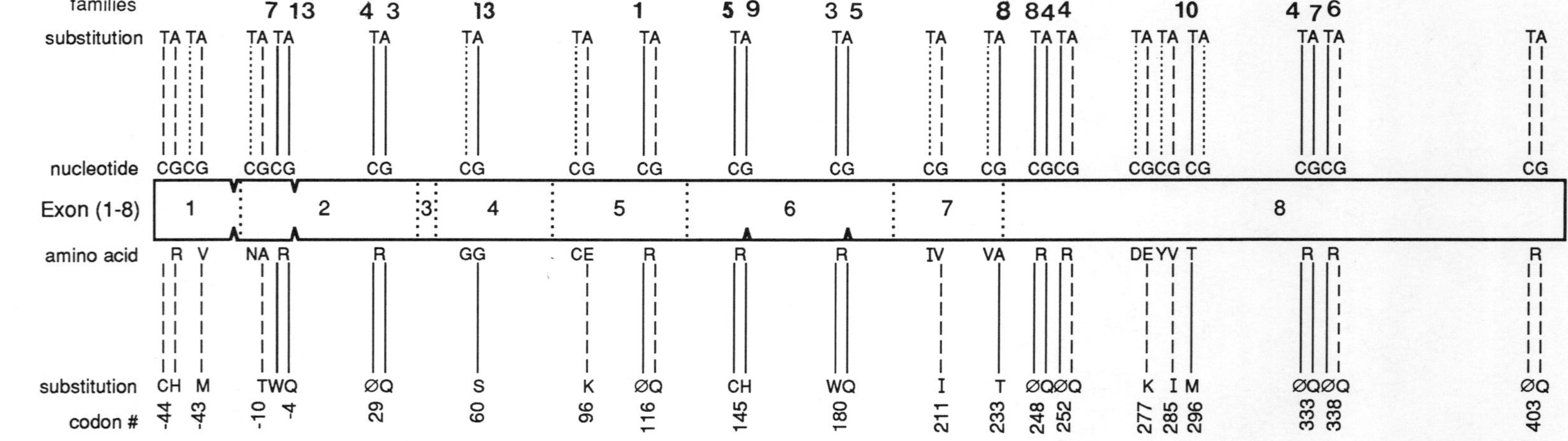

Figure 6. <u>Point Mutations Within CG-Dinucleotides in the Coding Regions of the Factor IX Gene</u>. The 20 CGs are displayed above the central bar which indicates the nucleotide sequence of the eight exons. The amino acids coded for are beneath the bar with the codon number provided. Solid lines indicate nucleotide and amino acid substitutions that have been found and the number of families described (Tables 2 and 3) and additional unpublished reports,[98] are provided along the top. Dotted lines are for transition mutations that would not change the amino acid code. Dashed lines are for mutations that have not been reported. The single letter amino acid code is indicated (Ø represents a new Stop codon): A=Ala, C=Cys, D=Asp, E=Glu, F=Phe, G=Gly, H=His, I=Ile, K=Lys, L=Leu, M=Met, N=Asn, P=Pro,Q=Gln, R=Arg, S=Ser, T=Thr, V=Val, W=Trp, Y=Tyr.

RECURRENT MUTATIONS

<u>CG Dinucleotides Defects</u>. Of 20 CG dinucleotides within the factor IX coding regions, 31 base substitutions would produce either missense or nonsense mutations.[18,43] These account for 91 defects in separate families at 18 of the 31 potential sites (Tables 2 and 3); 30 additional unpublished cases have been noted in a data base,[98] as shown in Fig. 6. The presence of a rare variant in two of the three families with the Arg 29 to Gln defect (B-64)[18,43] strongly argues for their being distant, yet unknown, relatives with the same mutation. This effect may be operative in some of the other, particularly milder, defects as well. In several defects it is possible to show independent, *de novo* origin in transition mutations within CG dinucleotides and the majority of severe defects are probably independent mutations.

<u>"Founder" Effect</u>. Although occasionally, a non-CG dinucleotide defect has been found in a second and possibly unrelated family, other sites of recurrent mutations have been transitions within CG dinucleotides (Fig. 6). A striking exception is 21 distinct families with the Ile 397 to Thr defect. On haplotype analysis, however, of the 18 tested to date,[13,96,97] all share a BamHI polymorphic site just 0.5 kb 5' to exon 1 which is present in only about 3% of Caucasian factor IX genes. The haplotype analyses represent strong indirect evidence for a common, distant (unknown) ancestor who proliferated the gene to a large number of families. Families with Thr 397 have either been sporadic or, where familial, have come from the Pacific Northwest, the mid-West, California, New England or Ontario. It will be of interest to see if the three patients described in different families in France[53] also share this rare haplotype.

SUMMARY

Of the known mutations in hemophilia B, the 29 gross gene alterations are probably overrepresented as they have been relatively easy to identify by Southern blotting. The increased incidence of inhibitor phenotype in patients with deletions appears real, but again represents those inhibitor phenotypes in whom defects have been found and therefore selection bias. Within a fairly representative series,[43] 2% of patients had gross alterations and 1%, microdeletions. Nonsense mutations are clearly more frequent than splicing or frameshift changes; these three types occur in about one-fourth of all defects described to date. The overall data are also biased to some degree by the fact that of those in whom defects have been found, about one-half have had severe hemophilia B; within a given population, however, only about one-third of the hemophilic patients are severe. "Regulatory" defects and missense mutations account for over half of the defects described and the majority of these are in the trypsin-like region.

Strategies can be offered in estimating where a given hemophilic defect might have occurred in the family of a patient whose familial or sporadic mutation has not been defined. For some patients, decreased reactivity to a monoclonal antibody known to react to a localized epitope will provide the necessary clue.[65] At other times, recurrent mutations within CG dinucleotides are so common that from the known examples of patients with different mutations, one can initiate a search. Potential sites are from known specific defects in which patients have similar absolute levels and specific activities. For severely affected patients with no detectable antigen, the majority have had either new nonsense codons (at five of the six potential C to T transitions within CG dinucleotide sites), or they have had gross gene alterations. In any event, identification of the specific defect can provide information about the relationship of structure to function, particularly in those missense defects in which dysfunctional antigen circulates. In addition, it can provide a direct means for carrier testing and prenatal diagnosis for that family.[84] This is especially important in patients in whom the defect is either sporadic or in whom key family members are not available to complete screening for inheritance of linked polymorphisms.

ACKNOWLEDGEMENTS

The collaborative contributions of Shi-Han Chen, Ph.D. (Pediatric Genetics) include the sequencing of hemophilia B defects in the Seattle series and are gratefully acknowledged. Professor George G. Brownlee, Oxford, UK, is thanked for sharing a preprint of the data base manuscript.[98] This work was supported in part by the National Institutes of Health (HL-31193) and American Heart Association (88-805).

REFERENCES

1. Pavlovsky A: Contribution to the pathogenesis of hemophilia. Blood 2:185-187, 1947
2. Aggeler PM, White SG, Glendening MB, Page EW, Leake TB, Bates G: Plasma thromboplastin component (PTC) deficiency: a new disease resembling hemophilia. Proc Soc Exper Biol Med 79:692-694, 1952
3. Schulman I, Smith CH: Hemorrhagic disease in an infant due to a deficiency of a previously undescribed clotting factor. Blood 7:794-807, 1952
4. Biggs R, Douglas AS, Macfarlane RG, Dacie JV, Pitney WR, Merskey C, O'Brien JR: Christmas disease, a condition previously mistaken for hemophilia. Br Med J ii:1378-1382, 1952
5. Thompson AR: Factor IX antigen by radioimmunoassay. Abnormal factor IX protein in patients on warfarin therapy and with hemophilia B. J Clin Invest 59:900-910, 1977
6. Yoshitake S, Schach BG, Foster DC, Davie EW, Kurachi K: Nucleotide sequence of the gene for human factor IX (antihemophilic factor B). Biochemistry 24:3736-3750, 1985
7. Giannelli F, Choo KH, Rees DJG, Boyd Y, Rizza CR, Brownlee GG: Gene deletions in patients with haemophilia B and anti-factor IX antibodies. Nature 303:181-182, 1983
8. Anson DS, Blake DJ, Winship PR, Birnbaum D, Brownlee GG: Nullisomic deletion of the *mcf.2* transforming gene in two haemophilia B patients. EMBO J 7:2795-2799, 1988
9. Taylor SAM, Lillicrap DP, Blanchette V, Giles AR, Holden JJA, White BN: A complete deletion of the factor IX gene and new TaqI variant in a hemophilia B kindred. Hum Genet 79:273-276, 1988
10. Matthews RJ, Anson DS, Peake IR, Bloom AL: Heterogeneity of the factor IX locus in nine hemophilia B inhibitor patients. J Clin Invest 79:746-753, 1987
11. Matthews RJ, Peake IR, Bloom AL, Anson DS: Carrier detection through the use of abnormal deletion junction fragments in a case of haemophilia B involving complete deletion of the factor IX gene. J Med Genet 25:779-780, 1988
12. Ludwig M, Schwaab R, Eigel A, Horst J, Egli H, Brackmann H-H, Olek K: Identification of a single nucleotide C-to-T transition and five different deletions in patients with severe hemophilia B. Am J Hum Genet 45:115-122, 1989
13. Chen S-H, Thompson AR: 1989 & 1990, unpublished results
14. Wadelius C, Blomback M, Pettersson U: Molecular studies of haemophilia B in Sweden. Hum Genet 81:13-17, 1988
15. Bernardi F, Del Senno L, Barbieri R, Buzzoni D, Gambari R, Marchetti G, Conconi F, Panicucci F, Positano M, Pitruzzello S: Gene deletion in an Italian haemophilia B subject. J Med Genet 22:305-307, 1985
16. Mikami S, Nishino M, Nishimura T, Fukui H: RFLPs of factor IX gene in Japanese haemophilia B families and gene deletion in two high-responder-inhibitor patients. Jpn J Human Genet 32:21-31, 1987
17. Tanimoto M, Kojima T, Kamiya T, Takamatsu J, Ogata K, Obata Y, Inagaki M, Iizuka A, Nagao T, Kurachi K, Saito H: DNA analysis of seven patients with hemophilia B who have anti-factor IX antibodies: Relationship to clinical manifestations and evidence that the abnormal gene was inherited. J Lab Clin Med 112:307-313, 1988

18. Koeberl DD, Bottema CDK, Buerstedde J-M, Sommer SS: Functionally important regions of the factor IX gene have a low rate of polymorphism and a high rate of mutations in the dinucleotide CpG. Am J Hum Genet 45:448-457, 1989

19. Hassan HJ, Leonardi A, Guerriero R, Chelucci C, Cianetti L, Ciavarella N, Ranieri P, Pilolli D, Peschle C: Hemophilia B with inhibitor: Molecular analysis of the subtotal deletion of the factor IX gene. Blood 66:627-630, 1985

20. Vidaud M, Chabret C, Gazengel C, Grunebaum L, Cazenave JP, Goossens M: A *de novo* intragenic deletion of the potential EGF domain of the factor IX gene in a family with severe hemophilia B. Blood 68:961-963, 1986

21. Chen S-H, Scott CR, Edson JR, Kurachi K: An insertion within the factor IX gene: Hemophilia B$_{El\ Salvador}$. Am J Hum Genet 42:581-584, 1988

22. Peake IR, Matthews RJ, Bloom AL: Haemophilia B Chicago: Severe haemophilia B caused by two deletions and an inversion within the factor IX gene. Br J Haematol 71(Supp I):1, 1989

23. Chen S-H, Yoshitake S, Chance PF, Bray GL, Thompson AR, Scott CR, Kurachi K: An intragenic deletion of the factor IX gene in a family with hemophilia B. J Clin Invest 76:2161-2164, 1985

24. Bray GL, Thompson AR: Partial factor IX protein in a pedigree with hemophilia B due to a partial gene deletion. J Clin Invest 77:1194-1200, 1986

25. Chen S-H, Scott CR: Uneven crossing-over between two homologous sequences (14bps) as the mechanism for the gene deletion in factor IX$_{Seattle\ 1}$. (Am J Hum Genet, in press)

26. Vidaud M, Vidaud D, Siguret V, Lavergne JM, Goossens M: Mutational insertion of an Alu sequence causes hemophilia B. Am J Hum Genet 45:A226, 1989

27. Trent RJ, Wallace RC, Rickard KA: Deletion/insertion of DNA in an intron of the factor IX gene produces severe hemophilia B-. Blood 72:312a, 1988

28. Green PM, Bentley DR, Mibashan RS, Giannelli F: Partial deletion by illegitimate recombination of the factor IX gene in a haemophilia B family with two inhibitor patients. Mol Biol Med 5:95-106, 1988

29. McGraw RA, Davis LM, Lundblad RL, Stafford DW, Roberts, HR: Structure and function of factor IX: Defects in haemophilia B. Clin Haematol 14:359-383, 1985

30. Thompson AR: Structure and biology of factor IX, in Benz EJ, Cohen JH, Furie B, Hoffman R, Shatil SJ (eds): Hematology - Basic Principles and Practice, chapter 102, New York, NY, Churchill Livingston, 1990, 1308-1316

31. Frazier D, Smith KJ, Ware J, Lin S-W, Thompson AR, Reisner H, Bajaj SP, Stafford DW: Mapping of monoclonal antibodies to factor IX. Blood 74:971-977, 1989

32. Green PM, Bentley DR, Mibashan RS, Nilsson IM, Giannelli F: Molecular pathology of haemophilia B. EMBO J 8:1067-1072, 1989

33. Winship PR: Characterisation of the molecular defect in haemophilia B patients using the polymerase chain reaction procedure. Thromb Haemostas 62:465, 1989

34. Green PM, Montandon AJ, Bentley DR, Ljung R, Nilsson IM, Giannelli F: The incidence and distribution of CpG->TpG transitions in the coagulation factor IX gene. A fresh look at CpG mutation hotspots. Nucl Acids Res 18:3227-3231, 1990

35. Montandon AJ, Green PM, Giannelli F, Bentley DR: Direct detection of point mutations by mismatch analysis: application to haemophilia B. Nucl Acids Res 17:3347-3358, 1989

36. Koeberl DD, Bottema CDK, Sarkar G, Ketterling RP, Chen S-H, Sommer SS: Recurrent nonsense mutations at arginine residues cause severe hemophilia B in unrelated hemophiliacs. Hum Genet 84:387-390, 1990

37. Matsushita T, Kamiya T, Tanimoto M, Yamamoto K, Sugiura I, Hamaguchi M, Takamatsu J, Saito H: DNA sequence analysis of three inhibitor positive hemophilia B patients (Jpn). Acta Haematol Jpn 53:452, 1990

38. Winship PR: Carrier detection and patient studies in haemophilia B. D.Phil-Thesis, Oxford University, 1986

39. Koeberl DD, Bottema CDK, Ketterling RP, Sarkar G, Sommer SS: Mutations causing hemophilia B: Direct estimates of the underlying of spontaneous germ line transitions, transversions and deletions in a human gene. Am J Hum Genet 47:202-217,1990

40. Schach BG, Yoshitake S, Davie EW: Hemophilia B (factor IX$_{Seattle\ 2}$) due to a single nucleotide deletion in the gene for factor IX. J Clin Invest 80:1023-1028, 1987

41. Montandon AJ, Green PM, Bentley DR, Ljung R, Nilsson IM, Giannelli F: Two factor IX mutations in the family of an isolated hemophilia B patient -- direct carrier detection by amplified mismatch detection (AMD). Hum Genet 85:200-204, 1990

42. Rees DJG, Rizza CR, Brownlee GG: Haemophilia B caused by a point mutation in a donor splice junction of the human factor IX gene. Nature 316:643-645, 1985

43. Chen S-H, Zhang M, Lovrien EW, Scott CR, Thompson AR: CG dinucleotide transitions in the factor IX gene account for about half of the point mutations in hemophilia B patients. (Hum Genet, submitted)

44. Wang NS, Thompson AR, Chen S-H: Point mutations in four hemophilia B patients from China. (Thrombos Haemostas, in press)

45. Bottema CDK, Ketterling RP, Koeberl DD, Taylor SA, Sommer SS: Mutations at arginine residues in two Asian hemophilia B patients. Nuc Acids Res 18:1924, 1990

46. Siguret V, Amselem S, Vidaud M, Assouline Z, Kerbiriou-Nabias D, Pietu G, Goossens M, Larrieu MJ, Bahnak B, Meyer D, Lavergne JM: Identification of a CpG mutation in the coagulation factor-IX gene by analysis of amplified DNA sequences. Br J Haematol 70:411-416, 1988

47. Chen S-H, Scott CR, Schoof J, Lovrien EW, Kurachi K: Factor IX-Portland: A nonsense mutation (CGA to TGA) resulting in hemophilia B. Am J Hum Genet 44:567-569, 1989

48. Taylor SAM: The molecular analysis of mutations of the human factor IX gene. PhD Thesis, Queen's University, Kingston, Ontario, 1990

49. Driscoll MC, Bouhassira E, Aledort LM: A codon 338 nonsense mutation in the factor IX gene in unrelated hemophilia B patients: Factor IX-338$_{New\ York}$. Blood 74:737-742, 1989

50. Freedenberg DL, Chen S-H, Scott R: A C to T mutation in the second TaqI site of exon VIII in the factor IX gene: detection by PCR amplification and direct sequence. Am J Hum Genet 45:A186, 1989

51. Rao KJ, Lyman G, Hamsabhushanam K, Scott JP, Jagadeeswaran P: Human factor IX$_{Lincoln\ Park}$: A molecular characterization. (Mol Cell Probes, in press)

52. Bottema CDK, Ketterling RP, Cho HI, Sommer SS: Hemophilia B in a male with a four-base insertion that arose in the germline of his mother. Nucl Acids Res 17:10139, 1989

53. Attree O, Vidaud D, Vidaud M, Amselem S, Lavergne J-M, Goossens M: Mutations in the catalytic domain of human coagulation factor IX: Rapid characterization by direct genomic sequencing of DNA fragments displaying an altered melting behavior. Genomics 4:266-272, 1989

54. Reitsma PH, Bertina RM, Ploos van Amstel JK, Riemens A, Briet E: The putative factor IX gene promoter in hemophilia B Leyden. Blood 72:1074-1076, 1988

55. Hirosawa S, Fahner JB, Salier JP, Wu CT, Lovrien EW, Kurachi K: Structural and functional basis of the developmental regulation of human coagulation factor IX gene: factor IX Leyden. Proc Natl Acad Sci (USA) 87:4421-4425, 1990

56. Gispert S, Vidaud M, Vidaud D, Gazengel C, Boneu B, Goossens M: A promoter defect correlates with an abnormal coagulation factor IX gene expression in a French family (hemophilia B Leyden). Am J Hum Genet 45:A189, 1989

57. Reitsma PH, Mandalaki T, Kasper CK, Bertina RM, Briet E: Two novel point mutations correlate with an altered developmental expression of blood coagulation factor IX (hemophilia B Leyden phenotype). Blood 73:743-746, 1989

58. Crossley PM, Winship PR, Black A, Rizza CR, Brownlee GG: Unusual case of haemophilia B. Lancet i:960, 1989

59. Bentley AK, Rees DJG, Rizza C, Brownlee GG: Defective propeptide processing of blood clotting factor IX caused by mutation of arginine to glutamine at position -4. Cell 45:343-348, 1986

60. Ware J, Diuguid DL, Liebman HA, Rabiet M-J, Kasper CK, Furie BC, Furie B, Stafford DW: Factor IX-San Dimas substitution of glutamine for Arg-4 in the propeptide leads to incomplete gamma-carboxylation and altered phospholipid binding properties. J Biol Chem 264:11401-11406, 1989

61. Liddell MB, Lillicrap DP, Peake IR, Bloom AL: Defective propeptide processing and abnormal activation underlie the molecular pathology of factor IX Troed-y-Rhiw. Br J Haematol 72:208-215, 1989

62. Sugimoto M, Miyata T, Kawabata S, Yoshioka A, Fukui H, & Iwanaga S: Factor IX Kawachinagano: Impaired function of the Gla-domain caused by attached propeptide region due to substitution of arginine by glutamine at position -4. Br J Haematol 72:216-221, 1989

63. Diuguid DL, Rabiet MJ, Furie BC, Liebman HA, Furie B: Molecular basis of hemophilia B: A defective enzyme due to an unprocessed propeptide is caused by a point mutation in the factor IX precursor. Proc Natl Acad Sci USA 83:5803-5807, 1986

64. Reitsma PH, Bertina RM: 1989, personal communication

65. Chen S-H, Thompson AR, Zhang M, Scott CR: Three point mutations in the factor IX genes of five hemophilia B patients: Identification strategy using localization by altered epitopes in their hemophilic proteins. J Clin Invest 84:113-118, 1989

66. Wang NS, Zhang M, Thompson AR, Chen S-H: Factor IX-Chongqing: A new mutation in the calcium-binding domain of factor IX resulting in severe hemophilia B. Thrombos Haemostas 63:24-26, 1990

67. Davis LM, McGraw RA, Ware JL, Roberts HR, Stafford DW: Factor IX$_{Alabama}$: A point mutation in a clotting protein results in hemophilia B. Blood 69:140-143, 1987

68. Lozier JN, Monroe DM, Stanfield-Oakley S, Lin S-W, Smith KJ, Roberts HR, High KA: Factor IX New London: substitution of proline for glutamine at position 50 causes severe hemophilia B. Blood 75:1097-1104, 1990

69. Spitzer S, Katzman D, Kasper C, Bajaj SP: Factor IX Hollywood: Substitution of 55 Pro to Ala in the first EGF domain. Thromb Haemostas 62:203, 1989

70. Denton PH, Fowlkes DM, Lord ST, Reisner HM: Hemophilia B-Durham: A mutation in the first EGF-like domain of factor IX that is characterized by polymerase chain reaction. Blood 72:1407-1411, 1988

71. Poort SR, Briet E, Bertina RM, Reitsma PH: A Dutch pedigree with mild hemophilia B with a missense mutation in the first EGF domain (factor IX$_{Oud en Nieuw Gastel}$). Nucl Acids Res 17:5869, 1989

72. Suehiro K, Okamura T, Murakawa M, Niho Y, Takeya H, Nishimura H, Iwanaga S: Factor IX-Fukuoka. Blood & Vessel 20:397, 1989

73. Vidaud M: Thesis, Paris VII University, 1990

74. Toomey J, Stafford D, Smith K: Factor IX Albuquerque (arginine 145 to cysteine) is cleaved slowly by factor XIa and has reduced coagulant activity. Blood 72:312a, 1988

75. Liddell MB, Peake IR, Taylor SAM, Lillicrap DP, Giddings JC, Bloom AL: Factor IX Cardiff: A variant factor IX protein that shows abnormal activation is caused by an arginine to cysteine substitution at position 145. Br J Haematol 72:556-560, 1989

76. Noyes CM, Griffith MJ, Roberts HR, Lundblad RL: Identification of the molecular defect in factor IX-Chapel Hill: Substitution of histidine for arginine at position 145. Proc Natl Acad Sci USA 80:4200-4202, 1983

77. Diuguid DL, Rabiet M-J, Furie BC, Furie B: Molecular defects of factor IX Chicago-2 (Arg 145—→ His) and prothrombin Madrid (Arg 271—→ Cys): arginine mutations that preclude zymogen activation. Blood 74:193-200, 1989

78. Suehiro K, Miyata T, Takeya H, Takamatsu J, Saito H, Niho Y, Iwanaga S: Blood clotting factor IX Nagoya 3: substitution of arginine-145 by histidine (Jpn). Acta Haematol Jpn 53:452, 1990

79. Bertina RM, van der Linden IK, Mannucci PM, Reinalda-Poot HH, Cupers R, Poort SR, Reitsma PH: Mutations in haemophilia B_m occur at the Arg^{180}-Val activation site or in the catalytic domain. J Biol Chem 265:876-883, 1990

80. Suehiro K, Kawabata S, Miyata T, Takeya H, Takamatsu J, Ogata K, Kamiya T, Saito H, Niho Y, Iwanaga S: Blood clotting factor IX B_m Nagoya: Substitution of arginine-180 by tryptophan and its activation by chymotrypsin. J Biol Chem 264:21257-21265, 1989

81. Huang M-N, Kasper CK, Roberts HR, Stafford DW, High KA: Molecular defect in factor IX_{Hilo}, a hemophilia B_m variant: Arg—→ Gln at the carboxyterminal cleavage site of the activation peptide. Blood 73:718-721, 1989

82. Sakai T, Yoshioka A, Yamamoto K, Niinomi K, Fujimura Y, Fukui H, Miyata T, Iwanaga S: Blood clotting factor IX Kashihara: Amino acid substitution of valine-182 by phenylalanine. J Biochem 105:756-759, 1989

83. Taylor SAM, Liddell MB, Peake IR, Bloom AL, Lillicrap DP: A mutation adjacent to the beta cleavage site of factor IX (valine 182 to leucine) results in mild hemophilia B. Br J Haematol 75:217-221, 1990

84. Bottema CDK, Koeberl DD, Sommer SS: Direct carrier testing in 14 families with haemophilia B. Lancet ii:526-529, 1989

85. Thompson AR, Chen S-H, Brayer GD: Severe hemophilia B due to a G to T transversion changing Gly 309 to Val and inhibiting active protease conformation by preventing ion pair formation. Blood 74:134a, 1989

86. Winship PR: Haemophilia B caused by mutation of a potential thrombin cleavage site in factor IX. Nucl Acids Res 18:1310, 1990

87. Tsang TC, Bentley DR, Mibashan RS, Giannelli F: A factor IX mutation, verified by direct genomic sequencing, causes haemophilia B by a novel mechanism. EMBO J 7:3009-3015, 1988

88. Poort SR, Briet E, Bertina RM, Reitsma PH: A Dutch family with moderately severe hemophilia B (Factor IX_{Heerde}) has a missense mutation identical to that of factor $IX_{London 2}$). Nucl Acids Res 17:3614, 1989

89. Bajaj SP, Spitzer SG, Welsh WJ, Warn-Cramer BJ, Kasper CK, Birktoft JJ: Experimental and theoretical evidence supporting the role of Gly^{363} in blood coagulation factor IXa (Gly^{193} in chymotrypsin) for proper activation of the proenzyme. J Biol Chem 265:2956-2961, 1990

90. Evans JP, Brinkhous KM, Brayer GD, Reisner HM, High KA: Canine hemophilia B resulting from a point mutation with unusual consequences. Proc Natl Acad Sci (USA) 86:10095-10099, 1989

91. Spitzer SG, Pendurthi UR, Kasper CK, Bajaj SP: Molecular defect in factor $IX_{Bm Lake Elsinore}$. Substitution of Ala^{390} by Val in the catalytic domain. J Biol Chem 263:10545-10548, 1988

92. Sugimoto M, Miyata T, Kawabata S, Yoshioka A, Fukui H, Takahashi H, Iwanaga S: Blood clotting factor IX Niigata: Substitution of alanine-390 by valine in the catalytic domain. J Biochem 104:878-880, 1988

93. Geddes VA, Le Bonniec BF, Louie GV, Brayer GD, Thompson AR, MacGillivray RTA: A moderate form of hemophilia B is caused by a novel mutation in the protease domain of factor IX-Vancouver. J Biol Chem 264:4689-4697, 1989

94. Ware J, Davis L, Frazier D, Bajaj SP, Stafford DW: Genetic defect responsible for the dysfunctional protein: Factor IX-Long Beach. Blood 72:820-822, 1988

95. Spitzer SG, Warn-Cramer BJ, Kasper CK, Bajaj SP: Replacement of 397 Ile by Thr in the clotting protease factor IXa (Los Angeles and Long Beach variants) affects macromolecular catalysis but non L-tosylarginine methyl ester hydrolysis. Lack of correlation between the ox-brain prothrombin time and the mutation site in the variant proteins. Biochem J 265:219-225, 1990

96. Thompson AR, Bajaj SP, Chen S-H, MacGillivray RTA: "Founder" effect in different families with hemophilia B mutation. Lancet i:418, 1990

97. Bottema CDK, Ketterling RP, Koeberl DD, Bowie EJW, Taylor SAM, Lillicrap D, Shapiro A, Gilchrist G, Sommer SS: A past mutation at isoleucine 397 now is a common cause of moderate-mild hemophilia B. Br J Haematol 75:212-216, 1990

98. Giannelli F, Green PM, High KA, Lozier JN, Lillicrap DP, Ludwig M, Olek R, Reitsma PH, Goosens M, Yoshioka A, Sommer S, Brownlee GG: Haemophilia B data base of point mutations and short additions and deletions. Nucl Acids Res 18:4053-4059, 1990

99. Briet E, Bertina RM, van Tilburg NH, Veltkamp JJ: A sex-linked hereditary disorder that improves after puberty. N Engl J Med 306:788-790, 1982

100. Salier J-P, Hirosawa S, Kurachi K: Functional characterization of the 5'-regulatory region of human factor IX gene. J Biol Chem 265:7062-7068, 1990

101. Bertina RM, van der Linden IK: Factor IX Zutphen. A genetic variant of blood coagulation factor IX with an abnormally high molecular weight. J Lab Clin Med 100:695-704, 1982

102. Thompson AR: Alloantibodies in hemophilia B binding to multiple factor IX epitopes. Thromb Res 46:169-174, 1987

103. Hougie C, Twomey JJ: Haemophilia Bm: A new type of factor IX deficiency. Lancet i:698-700, 1965

104. Crossley M, Winship PR, Austen DEG, Rizza CR, Brownlee GG: A less severe form of haemophilia B Leyden. Nucl Acids Res 18:4633, 1990

MOLECULAR DEFECTS IN HUMAN ANTITHROMBIN III DEFICIENCY

W.P. Sheffield, F. Fernandez-Rachubinski, R.C. Austin, and
M.A. Blajchman

The Canadian Red Cross Society Blood Transfusion Service
and Department of Pathology
McMaster University, Hamilton, Ontario, Canada

INTRODUCTION

The presence, and unopposed action, of activated coagulation factors
in the circulation would result in clot or thrombus formation within
intact blood vessels, with potentially life-threatening consequences.
Such events are prevented, in the non-pathological state, by the presence
in plasma of circulating inhibitors of coagulation. The most abundant of
these is a plasma glycoprotein known for historical reasons as antithrombin III (AT-III).[1,2] Genetic disorders that reduce functional
activity of AT-III leave individuals at risk for thromboembolism. The
majority of those affected suffer at least one such episode by their fifth
decade, with the proportion of AT-III-deficient patients clinically
affected rising by roughly 1% per year of life.[3-6] In this article we
summarize recent progress in defining the molecular defects involved in
hereditary antithrombin III deficiency, in the context of the normal
structure and function of this important serine protease inhibitor.

Antithrombin III: A heparin-activated inhibitor protein

Human AT-III is a single chain polypeptide of 432 amino acids[7]
containing four sites of N-linked glycosylation and three important
disulphide bonds linking cysteine residues 8 and 128, 21 and 95, and 247
and 430.[8] Reduction of the molecule, but not enzymatic deglycosylation,
impairs its ability to inhibit coagulation factors.[8,9] It is known to be
secreted from the liver,[10] and may also be synthesized in other
tissues.[11,12] A minor form, termed β-AT-III, representing approximately
10% of total plasma AT-III, is glycosylated at only three residues.[13]

Under different conditions, AT-III exhibits different activities.
Thrombin is a key coagulation enzyme responsible for the generation of the
fibrin clot from fibrinogen. AT-III inactivates thrombin when the latter
is added to plasma. This activity is known as progressive antithrombin
activity. Heparin greatly accelerates the inactivation of thrombin by
AT-III,[14] and this catalysis is termed AT-III heparin cofactor activity.
Since heparin _per se_ is present only in trace amounts in human blood, it
is thought that naturally occurring vessel wall glycosaminoglycans, such
as heparan sulfate, provide the _in vivo_ cofactor requirement.[15-17] Indeed,
AT-III has been shown to bind to the vessel wall both _in vitro_[15] and _in_
vivo.[17] Explants have been employed to demonstrate the specificity of the
binding, in that enzymatic degradation of heparin-like species of the

Recombinant Technology in Hemostasis and Thrombosis
Edited by L.W. Hoyer and W.N. Drohan, Plenum Press, New York, 1991

vessel wall abolishes binding,[16] while normal AT-III but not the Toyama mutant polypeptide binds to aortic endothelium in culture.[23]

The binding of heparin to AT-III most likely enhances its reactivity with thrombin by altering the inhibitor's conformation in a way that promotes attack by the protease. Heparin binding has been shown, for example, to activate Lys 236 in a way that enhances its ability to undergo chemical modification.[18] Heparin binding has also been shown to alter the overall fluorescence spectrum of the AT-III polypeptide.[19]. At least two sites on the molecule have been reported to constitute heparin/heparan binding sites. Amino acids 41-49 are implicated both by the mapping of altered residues in naturally occuring mutant AT-III molecules, and by the finding that chemical modification of Trp 49 blocks heparin binding.[20] Residues 107-156 are also thought to be relevant as chemical modification of lysine and arginine residues in this region have been shown to inhibit heparin binding,[21] and since a peptide derived from this region of AT-III by V8 protease treatment specifically binds heparin.[21] It should be noted that chemical modification at locations other than bona fide heparin binding sites may alter the structure of the molecule such that heparin binding is disrupted. Similarly, fragments of the AT-III molecule may behave differently alone rather than when incorporated into the full protein. Some uncertainty therefore remains concerning the precise location of the heparin binding domain(s). Two other considerations also further complicate the picture: firstly, the fact that thrombin also binds heparin;[22] and secondly, the demonstration that alterations to AT-III in the thrombin-binding region affect heparin binding (discussed below).

The reactive site of antithrombin III is easier to delineate. Attack by thrombin results in cleavage of the peptide bond between Arg 393 and Ser 394, and formation of a covalent ester linkage between the active serine residue of thrombin and Arg 393 of AT-III.[24-26] The carboxy-terminal peptide comprised of residues 394-432 is not released, but instead remains bound through a disulphide bridge between Cys 247 and 430. It has been suggested that this peptide may provide the signal that allows rapid uptake of the complex by a specific receptor in the liver.[27] Although serine proteases other than thrombin such as coagulation factors IXa,[28] Xa,[29] XIa,[30] XIIa,[31] kallikrein,[32] plasmin,[33] urokinase,[34] and trypsin[35] have all been shown to be inactivated by AT-III in the same manner as thrombin, kinetic considerations[36] suggest that thrombin is the primary physiological target of AT-III. Studies of mutant antithrombin III molecules from kindreds with familial venous thromboembolism have shown that alterations to specific residues between amino acids 382 and 407 abolish or reduce AT-III-thrombin complex formation (see below).

Austin et al. have developed a rapid cell-free expression system, in our laboratory, to aid in mapping the thrombin and heparin binding sites of antithrombin III.[37,38] cDNA sequences encoding the mature region of human AT-III (His1-Lys432) have been introduced into the pGEM transcrip-tion vector, and mRNA transcribed _in vitro_ employing a DNA-dependent RNA polymerase. Translation of the mRNA in a rabbit reticulocyte lysate in the presence of ^{35}S-methionine yields non-glycosylated AT-III that is capable of interacting with both heparin and thrombin. Cell-free derived AT-III variants mutated to any amino acid at positions 382, 393, and 430 are being investigated to assess the effect of changes at these positions to AT-III structure and function. In addition, truncated and deleted forms of the inhibitor have been synthesized. Preliminary results from one such engineered inhibitor lacking heparin binding region 1 (see Fig. 1), for instance, suggest that the absence of this region is associated with both impaired thrombin-AT-III complex formation and slightly reduced heparin affinity.

Computer searches of protein sequence data banks[40] have shown that AT-III is a member of a family of proteins dubbed the SERPINS, since the majority of its members are serine protease inhibitors. These include heparin cofactor II, α-1-antiplasmin, α-1-antitrypsin, and the inhibitors of protein C, C1, chymotrypsin, and plasminogen activator, respectively. Angiotensin, thyroxin-binding globulin, and ovalbumin are non-inhibitor members of the serpin family. Cleavage at the reactive centre of inhibitor serpins causes a dramatic conformational alteration of the protein, expressed as increased heat stability.[41] X-ray crystallography

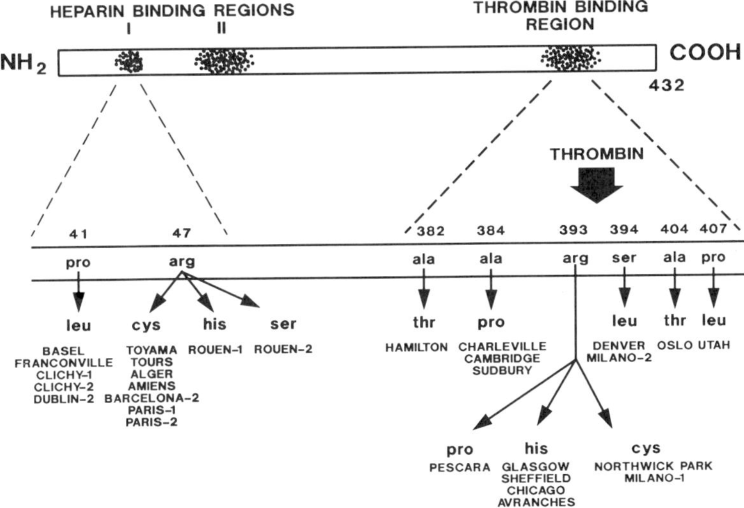

Fig. 1. A schematic diagram showing the two putative heparin binding regions and the thrombin-binding region of antithrombin III. The characterized antithrombin III mutants are shown, indicating the amino acid substitution in each case.

of α-1-antitrypsin cleaved at its reactive centre shows that the two amino acids at the reactive site, ordinarily contiguous, lie at opposite poles of the molecule.[42] These findings are summarised in the hypothesis that proposes that the inhibitor serpins circulate in a stressed conformation in which the reactive centre is exposed at the molecule's surface on a stressed peptide loop, with cleavage converting the inhibitor to a relaxed conformation.[40,41] The recent crystallization of a cleaved form of bovine AT-III[43] should demonstrate the degree to which AT-III fits this model. The apparent difficulty in crystallizing both intact α-1-antitrypsin and AT-III may also indicate that the stressed forms of the inhibitors are insufficiently stable to crystallize.

Although only 28% of the amino acid residues of AT-III and α-1-
antitrypsin are identical, the proteins share considerable structural
homology. This point is dramatically demonstrated by the description of
α-1-antitrypsin Pittsburgh.[44] Conversion of Met 358 of the antitrypsin's
reactive centre to Arg in this naturally occurring mutant produces a
potent antithrombin whose expression results in a lethal bleeding
disorder. Moreover, the high affinity of the mutant protein for thrombin
in the absence of heparin has general implications. It is possible that
protein engineering of serpins by interchanging features of one family
member with another may be one avenue towards producing novel proteins of
therapeutic value.

<u>Human Antithrombin III: mRNA and gene structure</u>

Molecular cloning has allowed the determination of the complete
nucleotide sequence of the mRNA and much of the gene encoding antithrombin
III. Analysis of overlapping cDNA clones from human liver[7] revealed that
the mRNA contains an open reading frame of 1392 nucleotides encoding a
polypeptide containing a 32 amino acid secretory signal peptide and a 432
amino acid mature form corresponding to plasma AT-III. Codon 432 and the
poly A tail are separated by 87 bases of 3' untranslated sequence containing a standard polyadenylation signal. Subsequent analysis of revealed
that the 5' untranslated region of the mRNA stretches from the transcriptional start site 70 bases upstream of the initiator methionine.[45]

The human AT-III gene extends over approximately 19 kb and is
interrupted by six intervening sequences (see Fig. 2).[46,48] Polymorphic
sites have been described, one resulting from sequence heterogeneity, a
translationally silent transition in codon 305,[48] and the other from a
length heterogeneity.[49] The latter polymorphism is characterized by the
presence of either a 32 or 108 base pair nonhomologous sequence at -275
relative to the start of transcription. The high frequency of these
polymorphisms (approaching 50% in some populations) has greatly aided
family studies of inherited AT-III deficiency.[50-52]

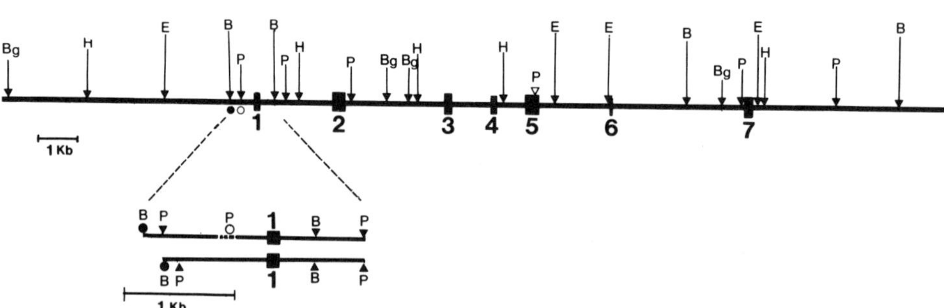

Fig. 2.   Structure of the human antithrombin III gene showing 7 exons
           and six introns, and selected restriction endonuclease
           cleavage sites. PstI polymorphic sites are designated by open
           symbols, an open triangle for the sequence polymorphism (48)
           and an open circle for the length polymorphism (49) (also
           shown in expanded area). B, BamHI; Bg, BglII; E, EcoRI; H,
           HindIII; P. PstI.

To date, only two studies have addressed the question of what
genetic regulatory regions control AT-III transcription. In the first,[53]
the presence of two sequence motifs, one homologous to the human Jk-Ck
intron enhancer, and the other to its murine counterpart, were noted
within the first 60 bases of the non-transcribed portion of the AT-III
gene, immediately upstream of the transcriptional start site. DNA
fragments encompassing this region, when introduced into enhancerless
reporter CAT constructs and transfected into cells, produced an increase
in CAT activity in liver and kidney cell lines, but not in lines derived
from lymphoid tissues. Unlike the AT-III gene, which does not contain a
recognizable TATA box, the test constructs provided a TATA box as well as
a truncated portion of the SV40 upstream region between the putative
enhancers and the cap site.[53]

The other investigation examined the binding of liver nuclear
proteins to upstream DNA sequences in a number of genes expressed in
liver.[54] This study provided evidence that the same factor binds both to
a region between bases -89 and -68 of the antithrombin III gene and to a
homologous site in the transferrin gene. If this region is critical to
the control of AT-III gene expression, then its deletion from reporter
constructs or in vitro transcription templates should decrease or abolish
promoter activity. However, such experiments have not yet been reported.

Linkage analysis has shown that the AT-III gene is loosely linked to
the Duffy gene, which encodes an erythrocyte membrane antigen, and is
found on chromosome 1.[55] In situ hybridization has further defined the
chromosomal location of the AT-III gene to the region 1q23-q25,[56] a
location consistent with analysis of a patient with 50% AT-III antigen and
function with a deletion of this region of chromosome 1 observed in
karyotype analysis.[57]

<u>Inherited antithrombin III deficiency</u>

Mutations and deletions of the antithrombin III gene result in
inherited deficiencies of the coagulation inhibitor that, in the majority
of cases, lead to clinically evident thromboembolic events. This heterogenous group of disorders may be placed into three classes.[58] Type I
AT-III deficiency is an autosomal dominant disorder in which the gene
product of one AT-III allele is absent from the patient's plasma (see
Table 1). This condition arises either through deletion of the gene or as
a consequence of mutations within the gene that prevent transcription,
translation, secretion, or circulation of the inhibitor. The other types
are mutant disorders in which an altered AT-III molecule is evident in
plasma; if the mutation primarily affects thrombin binding, it is
classified Type II, while if heparin binding is impaired, it is grouped
with Type III deficiencies. Known mutations are shown diagramatically, in
Figure 1, and listed in Tables II and III.

Table I. Type I Antithrombin III Deficiency

Number	Type of Gene Deletion	Reference
18	No reduction in size observed	59,60,61
1	Complete	59
1	Partial	62

Table II. Type II Antithrombin III Deficiency

Heparin Affinity of Mutant Polypeptides		
Normal	Increased	Reduced
Aalborg	Avranches (393)	Budapest
Cambridge (384)	Chicago (393)	Tokyo
Charleville (384)	Glasgow (393)	Malmo
Denver (394)	Sheffield (393)	Oslo (404)
Hamilton (382)	Milano-1 (393)	
Hvidorve	Northwick Park (393)	
Milano-2 (394)		
Pescara (393)		
Sudbury (384)		
Trento		
Utah		
Vincenza	[where known, the position of mutant residue is given in brackets]	

Table III.   Type III Antithrombin III Deficiency

Antigen Concentration of Mutant Polypeptides	
Normal	Reduced
Alger (47)	Barcelona
Amiens (47)	Johannesburg
Ann Arbor	Roma
Basel (41)	
Clichy-1 (41)	
Clichy-2 (41)	
Dublin-2 (41)	
Fontainebleau	
Franconville (41)	
Padua-1	
Padua-2	
Paris-1 (47)	
Rouen-1 (47)	
Rouen-2 (47)	
Tours (47)	
Toyama (47)	[where known, the position of the mutant residue is given in brackets]

As is implied by the long list of potential steps resulting in Type I deficiency cited above, its molecular pathology is somewhat unclear. Of 20 kindreds examined[59-62] (see Table I), only one is completely lacking in one AT-III allele; all others, save one, yield Southern blot patterns indistinguishable from normal individuals. Either the deletions involved are too small to be seen by this technique, or the deficiency stems from point mutations at key positions within the gene. These changes could lie, for instance, in the promoter, at intron/exon boundaries, or in the secretory signal sequence. In our laboratory, Fernandez-Rachubinski et al. have recently characterized a family with a partial, but extensive, gene deletion of one AT-III allele.[62] All AT-III genetic information upstream of a point between exons 2 and 3 has been deleted, as is illustrated by the restriction map shown in Figure 3. Work in progress is aimed at characterizing the nucleotide sequence of the abnormal allele at and around the break point.

The molecular defects involved have been easier to define in mutant
AT-III disorders (Types II and III), particularly since the methodologies
involved have undergone rapid evolution. For example, two years ago, we
reported the mutation responsible for the AT-III Hamilton phenotype, after
the relatively laborious process of cloning and sequencing the mutant gene
of the propositus.[63] In the interim, we have determined the mutations
involved in the AT-III Amiens[64] and Sudbury[65] kindreds, by sequencing DNA
obtained after polymerase chain reaction (PCR)[66] amplification of portions
of patients' genomic DNA. The latter technique has greatly reduced the
amount of time required to identify a given mutation, and may soon make
necessary a change in the naming of mutations. Typically, kindreds are
given a topynym, usually the place of residence of the propositus, since
in the past kindreds were reported before the definition of their
molecular defect. We now have instances of the same mutation being known
by up to seven topynyms (see Fig. 1). If kindreds continue to be
characterized at the current rate, a new nomenclature placing more
emphasis on the mutation involved may be required.

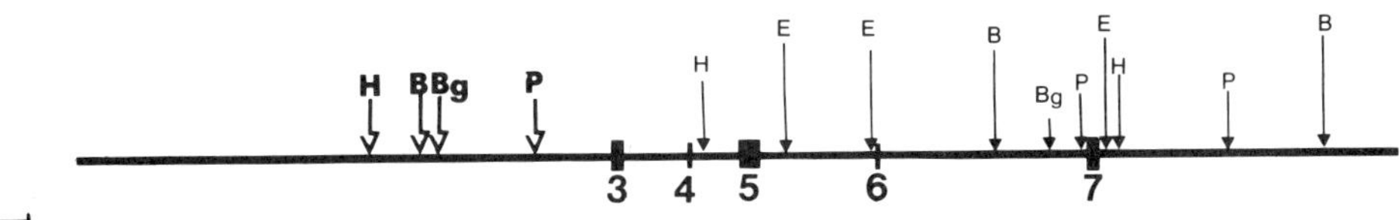

Fig. 3.   Structure of the partially deleted antithrombin III gene of the
           kindred described by Fernandez-Rachubinski et al (62).
           Restriction endonuclease cleavage sites are shown as in Fig. 2.
           Open arrowheads show cleavages in areas unrelated to the
           antithrombin III gene, upstream of the break point; closed
           arrowheads show cleavages within the antithrombin III gene
           common to the mutant and normal alleles.

     All Type II mutations characterized to date are single nucleotide
changes that alter codons between amino acids 382 and 407, a region of the
polypeptide bracketing the thrombin-reactive site (see Fig. 1 and Table
II). The primary laboratory abnormality in these cases is impaired
progressive antithrombin activity. In the simplest cases, heparin
affinity is unaltered. Substitutions of His or Cys for Arg at position
394 (termed the Pl' site of bond scission), however, result in an
antithrombin with greater than normal affinity for heparin;[67-70] the latter
Milano-1 and Northwick Park mutants circulate disulphide-bonded to albumin
and other plasma proteins through the novel cysteine residue. On the
other hand, three kindreds have been described in which heparin binding,
in addition to thrombin reactivity, is reduced;[71-75] in two, the Budapest
and Tokyo mutations, the propositii are severely affected, with little or

no antithrombin or heparin cofactor activities, and both arterial and venous thrombosis. The site of the mutation in these cases is as yet unknown, although the detection of high molecular weight forms of the Budapest protein suggest an additional cysteine in the molecule.

Alterations to the polypeptide chain of AT-III may both impair function and change the stability of the inhibitor in the circulation. The first kindred described with AT-III deficiency, a Norwegian family, was originally classified as suffering from a quantitative, Type I deficiency;[76] subsequent analysis detected a trace amount of an altered AT-III protein in the plasma, and the deficiency was described as AT-III Oslo (404 Ala to Thr).[39,77] Similarly, AT-III Utah[50] is characterized by the presence in the circulation of small quantities of a mutant antithrombin III.

Type III mutations involve single nucleotide changes that alter codons between residues 41 and 47 and impair heparin binding (see Fig. 1 and Table III). Patients with such disorders have normal progressive antithrombin activity and their mutant antithrombin III exhibits normal complex formation with thrombin in the absence of heparin; however, heparin cofactor activity is substantially impaired. Nevertheless, it seems that only homozygous individuals (e.g. Alger)[78] are severely affected, with a clear cut history of thrombosis. According to one study, only 6% of heterozygous Type III mutants suffer thromboemboli.[80] One AT-III variant with decreased heparin binding due to an alteration outside of the amino acid 41-47 region has recently been reported. In this kindred, Ile 7 has been converted to Asn, and the new residue is glycosylated.[81] The reduction in heparin affinity by glycosylation at this site contrasts with the naturally occurring minor β form of AT-III, whose lack of oligosaccharide at Asn 135 correlates with increased heparin affinity.[13]

It has been widely assumed that homozygosity of either Type I or Type II deficiency was incompatible with life, as no patient of this kind had been characterized. Recently, however, a Turkish family was described in which marriage between first cousins, both deficient in both AT-III antigen and function, resulted in the birth of two affected infants, neither of whom survived their first month of life.[82] Both AT-III antigen and functional levels, assayed with a peptide substrate, were roughly 10% normal. Autopsy of the second infant revealed massive infarcts of the lung and brain, disseminated thrombosis of the small vessels with attendant necrosis and gangrene, and recanalysed cerebral thrombi; interpreted by the authors as evidence of intrauterine thrombosis. Normal neonatal levels of AT-III are roughly 50% of adult levels, with no apparent deleterious consequences.[83] This case report implies that further reductions in AT-III levels can be disastrous in the early development of the infant.

<u>Therapeutic considerations and production of AT-III</u>

The reduction in functional AT-III levels seen in individuals with a hereditary deficiency of the inhibitor must cause relatively subtle changes in the hemostatic equilibrium, since most patients remain clinically unaffected for one or two decades. Their hypercoagulable state often goes undetected until a trauma such as injury, surgery, pregnancy, or parturition occurs. Patients with known AT-III deficiency, if asymptomatic, are typically placed on a temporary prophylactic regimen of anticoagulants, either heparin or coumadin, prior to surgery or delivery. In the latter instance, careful monitoring of the patient is essential, since coumadin in the first trimester can be teratogenic,[84] while prolonged administration of heparin has been suggested to predispose the patient to osteoporosis.[85] Some patients may also develop heparin-induced thrombocytopenia.[86] Where pregnancy is not a concern, and following a thrombotic episode, life-long coumadin treatment is indicated. Patients with a genetic deficiency of AT-III have been estimated to represent 2-4% of all patients presenting with venous thromboembolism (VTE), with the frequency of inherited AT-III deficiency in North American populations estimated to lie between 1 in 2,000 and 1 in 5,000.[87,88]

Replacement therapy with AT-III concentrates has become possible in the last few years. These preparations are obtained from plasma by chromatography on immobilized heparin, followed by pasteurization to eliminate blood-borne viruses. The American Red Cross has recently conducted a two-phase study[89] on the efficacy and safety of AT-III concentrates, and found the preparation efficacious as judged both by the absence of thrombotic complications after surgery or parturition in patients genetically deficient in AT-III, and by the non-extension or non-recurrence of thrombosis in patients experiencing VTE. No tranmission of either hepatitis B or HIV-1 was attributed to the administration of the concentrates. Others have reported on the usefulness of AT-III concentrates for the prevention of VTE in the clinical management of patients undergoing hip replacement,[90] or hemodialysis.[91]

Another potentially important means of producing antithrombin III is by expression of its cloned gene in eukaryotic cells maintained in culture. Three groups[92-94] have reported production of AT-III by tissue culture cells transformed with appropriate expression plasmids. Expression of AT-III in yeast lead to the secretion of small quantities of a product with near normal progressive antithrombin activity but with only 10% heparin cofactor activity.[95] This shortcoming may be due to the fact that yeast cells add only high mannose sugar groups when glycosylating proteins, instead of the more complex sugars found in plasma AT-III. Two of three groups employing higher eukaryotic cells in culture reported obtaining preparations of recombinant AT-III that were very similar in structure and function to the natural serpin. Stephens et al. (93) used COS (African green monkey kidney) cells to produce normal AT-III as well as eight variants mutated at the P1' residue, and obtained 50 ng/10^6 cells/24 hrs. Zettlemeissl et al.[94] were able to obtain yields of 10 μg/10^6 cells/24 hrs by using CHO (Chinese hamster ovary) cells in which AT-III expression plasmids were integrated into the genome and amplified. The product was indistinguishable from plasma AT-III in terms of activity, circular dichroism, and ultraviolet and fluoresence spectroscopy, but differed somewhat in its pattern of complex sugars. Nevertheless, no differences in clearance time between recombinant and natural AT-III were seen when the recombinant product was injected into rabbits. Wasley et al.[92] also expressed AT-III in CHO cells, but found that only 5% of the secreted product was functional. This discrepancy can most likely be accounted for by the fact that the AT-III cDNA employed differed in three codons compared to that reported by Bock et al.,[7] and employed by Zettlemeissl et al.[94]

The ability to produce recombinant AT-III without resorting to concentrates made from large pools of plasma has obvious advantages in terms of safety. Additionally, production of recombinant molecules in culture offers the possibility of engineering AT-III molecules with extended half-lives, increased activity, or reduced susceptibility to inactivation in vivo by agents other than coagulation factors. At the same time, the ability to produce pure mutant antithrombins will allow for continued insights into the molecular mechanisms of inherited antithrombin deficiency.

ACKNOWLEDGEMENTS

This work was supported, in part, by a Research and Development Grant (HA-04-89) from the Canadian Red Cross Society (CRCS). WPS holds a Career Fellowship award from the CRCS; FFR holds a Canadian Medical Research Council Research Fellowship; and RCA holds a Scholarship Award from the Heart and Stroke Foundation of Canada.

REFERENCES

1. Abilgaard U: Purification of two progressive antithrombins of human plasma. Scand J Clin Lab Invest 19:190-195, 1967
2. Rosenberg RD, Damus PS: The purification and mechanism of action of human antithrombin-heparin cofactor. J Biol Chem 248:6490-6505, 1973
3. Thaler E, Lechner K: Antithrombin III deficiency and thromboembolism. Clin Haematol 10:369-390, 1981
4. Bick RL: Clinical relevance of antithrombin III. Sem Thromb Hemost 8:276-287, 1982
5. Roka L, Eckhardt T: Antithrombin III: Properties and function, in Seegers WH, Walz DA (eds): Prothrombin and Other Vitamin K Proteins. Vol. 2. Boca Raton, FL CRC, 1986 pp 1-15
6. Beresford CH: Antithrombin III deficiency. Blood Rev 2:239-250, 1988
7. Bock SC, Wion KL, Vehar GA, et al: Cloning and expression of the cDNA for human antithrombin III. Nucleic Acids Res 10:8113-8125, 1982
8. Sun X-J, Chang J-Y: Heparin binding domain of human antithrombin III inferred from the sequential reduction of its three disulfide linkages. J Biol Chem 264:11288-11293, 1989
9. Rosenfeld L, Danishefsky I: Effects of enzymatic doglycosylation on the biological activities of human thrombin and antithrombin. Arch Biochem Biophys 229:359-367, 1984
10. Hedner U, Nilsson IM: Antithrombin III in a clinical material. Thromb Res 3:631-641, 1973
11. Lee AKY, Chan V, Chan TK: The identification in and localization of antithrombin III in human tissues. Thromb Res 14:209-217, 1979
12. D'Souza SE, Mercer JFB: Antithrombin III mRNA in adult liver and kidney and in rat liver during development. Biochem Biophys Res Comm 142:417-421, 1987
13. Brennan SO, George PM, Jordan RE: Physiological variant of antithrombin III lacks carbohydrate side chain at Asn 135. FEBS Lett 219:431-436, 1987
14. Rosenberg RD: Role of heparin and heparin-like molecules in thrombosis and atherosclerosis. Fed Proc 44:404-409, 1985
15. Hatton MW, Moar SL, Richardson M: Evidence that rabbit ^{125}I-antithrombin binds to proteoheparin sulphate at the subendothelium of the rabbit aorta in vitro. Blood Vessels 25:12-27, 1988
16. Stern D, Nawroth P, Marcum J, et al: Interaction of antithrombin III with bovine aortic segments. Role of heparin in binding and enhanced anticoagulant activity. J Clin Invest 75:272-279, 1985
17. Marcum JA, McKenney JB, Rosenberg RD: Acceleration of thrombin-antithrombin complex formation in rat hindquarters via heparin-like molecules bound to the endothelium. J Clin Invest 74:341-350, 1984.
18. Chang J-Y: Binding of heparin to human antithrombin III activities selective chemical modification at Lys 236. J Biol Chem 264:3111-3115, 1989
19. Olson ST, Shore JD: Binding of high affinity heparin to antithrombin III. J Biol Chem 256:11065-11072, 1982

20. Blackburn MN, Smith RL, Carson J, et al: The heparin-binding site of antithrombin III. Identification of a critical tryptophan in the amino acid sequence. J Biol Chem 259:939-941, 1984

21. Smith JW, Knauer DJ: A heparin binding site in antithrombin III. J Biol Chem 262:11964-11972, 1987

22. Fenton JW II, Witting JI, Pouliott C, et al: Thrombin anion-binding exosite interactions with heparin and various polyanions. Ann NY Acad Sci 556:158-165, 1989

23. Saito S, Takahashi K, Sakuragawa N: Interaction of abnormal antithrombin-III Toyama with cultured porcine aortic endothelial cells. Thromb Res 50:19-25, 1988

24. Fish WW, Bjork I: Release of a two-chain form of antithrombin from the antithrombin-thrombin complex. Eur J Biochem 101:31-38, 1979

25. Bjork I, Jackson CM, Jornvall H, et al: The active site of antithrombin. J Biol Chem 257:2406-2411, 1982

26. Owen WG: Evidence for the formation of an ester between thrombin and heparin cofactor. Biophys Acta 405:380-387, 1975

27. Pizzo SV: Serpin Receptor 1: A hepatic receptor that mediates the clearance of antithrombin III protease complexes. Am J Med 87 (Suppl. 3B):10S-14S, 1989

28. Kurachi K, Fujikawa K, Schmer G, et al: Inhibition of bovine factor IXa and factor Xaβ by antithrombin III. Biochemistry 15:373-377, 1976

29. Biggs R, Denson KWE, Akman N, et al: Antithrombin III, antifactor Xa and heparin. Br J Haematol 19:283-305, 1970

30. Scott CF, Colman RW: Factors influencing the acceleration of human factor Xa inactivation by antithrombin III. Blood 73:1873-1879, 1989

31. Stead N, Kaplan AP, Rosenberg RD: Inhibition of activated Factor XII by antithrombin-heparin cofactor. J Biol Chem 251:6481-6488, 1976

32. Lahiri B, Rosenberg R, Talamo RC, et al: Antithrombin III: An inhibitor of human plasma kallikrein. Fed Proc 33:642, 1974 (Abstr)

33. Highsmith RF, Rosenberg RD: The inhibition of human plasmin by human antithrombin-heparin cofactor. J Biol Chem 249:4335-4338, 1974

34. Clemmensen I: Inhibition or urokinase by complex formation with human antithrombin III in the absence and presence of heparin. Thromb Haemost 39:616-623, 1978

35. Abildgaard U, Egeberg O: Thrombin inhibitory activity of fractions obtained by gel filtration of antithrombin III deficient plasma. Scand J Haemat 5:155-157, 1968

36. Buchanan MR, Boneu B, Ofosu F, et al: The relative importance of thrombin inhibition and factor Xa inhibition to the antithrombotic effects of heparin. Blood 65:198-201, 1985

37. Austin R, Rachubinski R, Fernandez-Rachubinski F, et al: Expression of biologically active antithrombin III in a cell-free system to define its thrombin-binding domain. Blood 72(Suppl 1):362a, 1988 (abstr)

38. Austin RC, Rachubinski R, Fernandez-Rachubinski F, et al: Expression in a cell-free system of normal and variant forms of human antithrombin III: Ability to bind heparin and react with α-thrombin. (Blood in press)

39. Hultin MB, McKay J, Abildgaard U: Antithrombin Oslo: Type 1b classification of the first report antithrombin deficient family, with a review of hereditary antithrombin variants. Thrombos Haemost 59:468-473, 1988

40. Hunt LT, Dayhoff MO: A surprising new protein superfamily containing ovalbumin, antithrombin-III, and alpha 1-protease inhibitor. Biochem Biophys Res Commun 95:864-871, 1980

41. Carrell RW, Owen MC: Plakalbumin, α-antitrypsin, antithrombin and the mechanism of inflammatory thrombosis. Nature 317:730-732, 1985

42. Loeberman H, Tokuoka R, Deisenhofer J, et al: Human α-1-proteinase inhibitor crystal structure analysis of two crystal modifications,

molecular model and preliminary analysis of the implications for function. J Mol Biol 177:531-556, 1984

43.   Samama J, Delarue M, Muvrey L, et al: Crystallization and preliminary crystallographic data for bovine antithrombin-III. J Mol Biol 210:877-879, 1989

44.   Owen MC, Brennan SO, Lewis JH, et al: Mutation of antitrypsin to antithrombin. Alpha 1-antitrypsin Pittsburgh (358 Met leads to Arg) a fatal bleeding disorder. N Engl J Med 390:694-698, 1983

45.   Prochownik EV, Orkin SH: In vivo transcription of a human antithrombin III "minigene". J Biol Chem 259:15386-15392, 1984

46.   Bock SC, Marrinan JA, Radziejewska E: Antithrombin III Utah: Proline-407 to leucine mutation in a high conserved region near the inhibitor reactive site. Biochem 27:6171-6178, 1988

47.   Prochownik EV, Bock SC, Orkin SH: Intron structure of the human antithrombin III gene differs from that of other members of the serine protease inhibitor superfamily. J Biol Chem 260:9608-9612, 1985

48.   Prochownik EV, Markham AF, Orkin SH: Isolation of a cDNA clone for human antithrombin III. J Biol Chem 258:8389-8394, 1983

49.   Bock SC, Levitan DJ: Characterization of an unusual DNA length polymorphism 5' to the human antithrombin III gene. Nucleic Acids Res 11:8569-8582, 1983

50.   Bock SC, Harris JF, Schwartz CE, et al: Hereditary thrombosis in a Utah kindred is caused by a dysfunctional antithrombin III gene. Am J Hum Genet 37:32-41, 1985

51.   Le Paslier D, Rochu D, Lucotte G: Pst polymorphism of the antithrombin III gene in a French population. Vox Sang 49:168-170, 1985

52.   Prochownik EV, Antonarakis S, Bauer KA, et al: Molecular heterogeneity of inherited antithrombn III deficiency. N Engl J Med 308:1549-1552, 1983

53.   Prochownik EV: Relationship between an enhancer element in the human antithrombin III gene and an immunoglobulin light-chain gene enhancer. Nature 316:845-848, 1985

54.   Ochoa A, Brunel F, Mendelzon D, et al: Different liver nuclear proteins bind to similar DNA sequences in the 5'-flanking regions of three hepatic genes. Nucleic Acids Res 17:116-133, 1989

55.   Lovrien EW, Magenis Re, Rivas ML, et al: Linkage study of antithrombin III. Cytogenet Cell Genet 22:319-323, 1978

56.   Bock SC, Harris FJ, Balazx I, et al: Assignment of human antithrombin III structural gene to chromosome 1q23-25. Cytogenet Cell Genet 39:67-69, 1985

57.   Winter JH, Bennett B, Watt JL, et al: Confirmation of linkage between antithrombin III and Duffy blood group and assignment of AT3 to 1q22 → q25. Ann Hum Genet 46:29-34, 1982

58.   Sas G, Peto I, Banhegyi D, et al: Heterogeneity of the 'classical' antithrombin III deficiency. Thromb Haemost 43:133-136, 1980

59.   Bock SC, Prochownik EV: Molecular genetic survey of sixteen kindreds with hereditary antithrombin III deficiency. Blood 70:1273-1278, 1987

60.   Fernandez-Rachubinski F, Blajchman MA: Unpublished observations.

61.   Sacks SH, Mold J, Reader ST, et al: Evidence linking familial thrombosis with a defective antithrombin III gene in two British kindreds. J Med Genet 25:20-24, 1988

62.   Fernandez F, Rachubinski R, Brill-Edwards P, et al: Evidence for a partial deletion of the antithrombin III gene in a family with recurrent thrombosis and type 1 antithrombin III deficiency. Proc 4th Int Cong of Cell Biol, Montreal, Quebec, 1988, p 424 (abstr)

63. Devraj-Kizuk R, Chui DHK, Prochownik EV, et al: Antithrombon-III-Hamilton: A gene with a point mutation (guanine to adenine) in codon 382 causing impaired serine protease reactivity. Blood 72:1518-1523, 1988

64. Fernandez-Rachubinski F, Eng B, Murray WW, et al: Incorporation of 7-deaza-2' dGTP during amplification by the polymerase chain reaction improves subsequent direct sequencing. Sequence - J DNA Mapping and Sequencing (in press)

65. Pewarchuk W, Fernandez-Rachubinski F, Blajchman MA: Manuscript in preparation.

66. Saiki RJ, Scarf S, Faloma F, et al: Enzymatic amplification of β-globin genomic sequences and restriction site analysis for detection of sickle cell anemia. Science 230:1350-1354, 1985

67. Thein SL, Lane DA: Use of synthetic oligonucleotides in the characterization of antithrombin III Northwick Park (393 CGT→TGT) and antithrombin III Glasgow (393CGT→CAT). Blood 72:1817-1821, 1988

68. Lane DA, Erdjument A, Glynn V, et al: Antithrombin Sheffield: Amino acid substitution at the reactive site (Arg393 to His) causing thrombosis. Br J Haematol 71:91-96, 1989

69. Erdjument H, Lane DA, Ireland H, et al: Antithrombin III Milano. Single amino acid substitution at the reactive site, Arg393 to Cys. Thromb Haemost 60:471-475, 1988

70. Erdjument H, Lane DA, Ireland H, et al: Formation of a covalent disulfide-linked antithrombin-albumin complex by an antithrombin variant, antithrombin "Northwick Park". J Biol Chem 262:13381-13384, 1987

71. Sas G, Pepper DS, Cash JD: Further investigations on antithrombin III in the plasmas of patients with the abnormality of antithrombin III Budapest. Thromb Diath Haemorrh 33:564-572, 1975

72. Murayama H, Matsuda M: Abnormal antithrombin III with defective biological and immunological functions found in a thrombophilic patient "Antithrombin III Tokyo". Thromb Haemost 50:358, 1983 (abstr)

73. Murayama H, Matsuda M: Abnormal properties and behaviors of antithrombin III found in a thrombophilic patient: Defective biological functions and dissimilar antigenic determinants. Thromb Haemost 56:165-171, 1986

74. Uratani Y, Murayama H, Matsuda M, et al: Conformation of antithrombin III with defective biological functions derived from a thrombophilic patient. Thromb Res 49:591-600, 1988

75. Tengborn L, Frohm B, Nilsson LE, et al: A Swedish family with abnormal antithrombin III. Scand J Haematol 34:412-416, 1985

76. Egeberg O: Inherited antithrombin deficiency causing thrombophilia. Thromb Diath Haemorrh 13:516-530, 1965

77. Bock SC, Silbermann JA, Wikoff W, et al: Identification of a threonine for alanine substitution at residue 404 of antithrombin III Oslo suggests integrity of the 404-407 region is important for maintaining normal plasma inhibitor levels. Thromb Haemostas 62:494 (abstr)

78. Brunel F, Buchange N, Fischer AM, et al: Antithrombin III Alger: A new case of Arg 47→Cys mutation. Am J Hematol 25:223-224, 1987

80. Finazzi G, Caccia R, Barbui T: Different prevalence of thromboembolism in the subtypes of congenital antithrombin III deficiency: Review of 404 cases. Thromb Haemost 58:1094, 1987 (letter)

81. Brennan SO, Borg J-Y, George PM, et al: New carbohydrate site in mutant antithrombin (7 Ile→Asn) with decreased heparin affinity. FEBS Lett 237:118-122, 1988

82. Hakten M, Deniz U, Ozbay G, et al: Two cases of homozygous antithrombin III deficiency in a family with congenital deficiency of AT-III. In Senzinger H and Vinazzer H (eds) Thrombosis and Haemorrhagic Disorders. Proc of the 6th International Meeting of the Danubian League Against Thrombosis and Haemorrhagic Disorders. Schmitt and Meyer GmbH, Wurzburg, Germany, 1989, pp 177-181

83. Peters M, Jansen E, Ten Cate JW, et al: Neonatal antithrombin III. Brit J Haemat 58:579-587, 1984

84. Ruthnum P, et al: Atypical malformations in an infant exposed to warfarin during first trimester of pregnancy. Teratology 36:299-301, 1987

85. Howell R, Fidler J, Letsky E, et al: The risks of antenatal subcutaneous heparin prophylaxis: A controlled trial. Br J Obstet Gynaecol 90:1124-1128, 1983

86. Warkentin TE, Kelton JG: Heparin-induced thrombocytopenia. Ann Rev Med 40:31-44, 1989

87. Rosenberg RD: Actions and interactions of antithrombin and heparin: N Engl J Med 292:146-152, 1975

88. Abildgaard V: Antithrombins and related inhibitors of coagulation, in Poller L (ed): Recent Advances in Blood Coagulation. Edinburgh, Scotland, Churchill Livingstone, 1981 pp 151-173

89. Menacho D, O'Malley JP, Schorr JB, et al: Evaluation of the safety, recovery, half-life and clinical efficacy of antithrombin III (human) in patients with hereditary antithrombin III deficiency. Blood 75:33-39, 1990

90. Francis CW, Pellegrini VD, Marder VJ, et al: Antithrombin III/low-dose heparin in prevention of venous thrombosis following total hip arthroplasty. Circulation 78 (Suppl II):312, 1988

91. Schrader J, Kostering H, Kramer P, et al: Antithrombin-III-Substitution bei dialysepflichtiger Niereninsuffizienz. Dtsch med Wschr 107:1847-1850, 1982

92. Wasley LC, Attra DH, Bauer KA, et al: Expression and characterization of human antithrombin III synthesized in mammalian cells. J Biol Chem 262:14766-14772, 1987

93. Stephens AW, Siddiqui A, Hirs CHW: Expression of functionally active human antithrombin III. Proc Natl Acad Sci USA 84:3886-3890, 1987

94. Zettlemeissl G, Cenrodt HS, Nimtz M, et al: Characterization of recombinant human antithrombin III synthesized in Chinese Hamster Ovary Cells. J Biol Chem 264:21153-21159, 1989

95. Broker M, Ragg H, Kargis HE: Expression of human antithrombin III in Saccharomyces cerevisiae and Schizosaccharomyces pombe. Biochim Biophys Acta 908:203-213, 1987

THE BIOLOGIC IMPACT OF HEREDITARY DEFECTS THAT CAUSE THROMBOSIS

Kenneth A. Bauer

Harvard Medical School
Brockton-West Roxbury Veterans Administration Medical
Center,1400 VFW Parkway, West Roxbury, MA 02132 and
Beth Israel Hospital,330 Brookline Avenue, Boston, MA
02215

INTRODUCTION

Considerable effort has been devoted to investigating the natural anticoagulant mechanisms that regulate the hemostatic mechanism. Two of these are the heparan sulfate-antithrombin III and protein C-thrombomodulin-protein S mechanisms that inactivate the serine proteases and activated cofactors of the coagulation cascade, respectively.[1,2] Antithrombin III neutralizes the activity of thrombin as well as factors Xa, IXa, XIa, and XIIa. Heparan sulfate associated with endothelial cells of the vessel wall accelerates the inactivation of these enzymes. To perform its anticoagulant function, protein C, a vitamin-K dependent glycoprotein, is converted to activated protein C when thrombin binds to thrombomodulin on vascular endothelial cell membranes. Activated protein C inhibits the platelet-dependent conversion of prothrombin to thrombin via factor Xa by destroying factor VIIIa and platelet-bound factor Va. Protein S enhances the binding of activated protein C to phospholipid-containing membranes and accelerates the proteolytic inactivation of factor VIIIa and factor Va. The association of severe thrombotic disease in persons with hereditary deficiencies of antithrombin III, protein C, or protein S demonstrates the physiologic importance of these natural anticoagulants in vivo. This chapter describes the clinical features, laboratory evaluation, and management of patients with these familial disorders.

INHERITED THROMBOTIC DISORDERS

The prevalence of an inherited basis for a thrombotic disorder has been evaluated in several cross-sectional investigations of thrombotic patients with a positive family history or a personal history of spontaneous venous thrombosis.[3-5] In a cohort of 141 consecutive unrelated German patients under age 45, deficiencies of protein C and protein S were present in 4

and 5 percent, respectively, of the individuals.[5] Deficiencies
of antithrombin III as well as identifiable plasminogen and
fibrinogen abnormalities were identified in 3, 2 and 1 percent,
respectively, of subjects. Similar results have been obtained in
such populations by investigators in the Netherlands[3] and
Austria.[4] The prevalence of these genetic abnormalities has not
yet been evaluated in large groups of unselected patients, but is
likely to be considerably less if one includes a large number of
older subjects.

With currently available tests, one can diagnose a
hereditary disorder in only 10 to 20 percent of young patients
presenting with a documented episode of venous thromboembolism.
The following clinical features however should suggest the
presence of one of these diagnoses and prompt a screening
laboratory evaluation: thrombosis occurring at an early age, a
family history of thrombotic disease, thrombosis occurring at
unusual sites (e.g., mesenteric venous thrombosis, cerebral
venous thrombosis), or recurrent thrombosis without apparent
precipitating factors.

Most individuals with thrombophilia in association with
congenital deficiencies of the proteins of the natural
anticoagulant mechanisms (i.e., antithrombin III, protein C,
protein S) are primarily afflicted with venous thromboembolic
disease. The reasons for the absence of arterial thromboses in
these disorders are unknown though the traditional view is that
these events occur in the setting of gross defects in platelet
and/or endothelial cell function.

<u>Antithrombin III Deficiency</u>

In 1965, Egeberg[6] reported a Norwegian family with a history
of repeated thrombotic events whose members exhibited plasma
antithrombin III concentrations that were 40 to 50% of normal.
Subsequently, numerous additional families with a similar
constellation of clinical and laboratory abnormalities have been
desscribed.[6-25] Antithrombin III deficiency is inherited in an
autosomal dominant fashion and affects both sexes equally. The
true prevalence of antithrombin III deficiency in the general
population is unknown, but estimates reported in the literature
range from about one in 2000 to 5000.[26,27] Thaler and Lechner[28]
compiled the published cases of familial antithrombin III
deficiency and observed that approximately 55 percent of
biochemically affected patients had experienced at least one
thrombotic event. However, patients had rarely manifested
thrombotic episodes prior to puberty and increasing numbers of
patients reported such events as they reached the age of 50. The
initial clinical manifestations of this disorder occurred
spontaneously in about 42 percent of subjects, but appeared to be
related to pregnancy, delivery, contraceptive pill ingestion,
surgery, or trauma in the remaining 58 percent of patients. The
most common sites of disease were the deep veins of the leg, the
iliofemoral veins, and the mesenteric veins. Approximately 60
percent of individuals developed recurrent thrombotic episodes,
and clinical signs of pulmonary embolism are evident in 40
percent.

Two major types of inherited antithrombin III deficiency
have been delineated. The classic deficiency state (type I)
results from the reduced synthesis of biologically normal
protease inhibitor molecules.[24,29] In these cases, the
immunological and biological activities of antithrombin III
within the blood are reduced to the same extent. The second type
of antithrombin III deficiency is produced by a discrete
molecular defect within the protease inhibitor (type II). Under
these circumstances, the plasma levels of antithrombin III are
reduced as judged by biologic activity measurements, whereas the
immunologic determinations of this inhibitor are normal.

The first family with a functional deficiency of
antithrombin III was reported by Sas in 1974.[30] Many families
with this type of deficiency state have now been reported, and
they have been further subcategorized on the basis of two
different functional assays of antithrombin III activity.[31] The
first is the antithrombin III-heparin cofactor assay which
measures the ability of heparin to bind to lysyl residues on the
inhibitor and catalyze the neutralization of coagulation enzymes
such as thrombin and factor Xa. Another protein in human plasma
exhibits heparin cofactor activity.[32-34] This protein has been
termed heparin cofactor II; however in contrast to antithrombin
III, this inhibitor requires concentrations of heparin of at
least 1 unit/ml in the reaction mixture in order to function as
an effective inhibitor of thrombin and therefore probably plays a
minimal role when heparin is utilized clinically as an
anticoagulant.[34] Heparin cofactor II does not interact with other
serine proteases generated during blood clotting, and another
mucopolysaccharide, dermatan sulfate, dramatically accelerates
the neutralization of thrombin by this inhibitor. Several
patients have been described with inherited deficiencies of
heparin cofactor II and thrombotic phenomena,[35,36] but the causal
relationship is uncertain.[37] The second test is the progressive
antithrombin III activity assay, which quantifies the capacity of
this inhibitor to neutralize the enzymatic activity of thrombin
in the absence of heparin. These two functional assays have
identified antithrombin III-deficient patients with reductions in
heparin cofactor activity with or without concordant decrements
in progressive antithrombin III activity. The application of the
techniques of classical protein chemistry or molecular biology
has led to the identification of the mutation in many of these
abnormal antithrombin III molecules.

The literature suggests that the prevalence of thrombosis is
different in patients with the two types of functional
antithrombin III deficiency. Unlike the type I subjects who
comprised all but a few of the cases summarized by Thaler and
Lechner,[28] heterozygous persons with reductions in plasma
antithrombin III-heparin cofactor activity to approximately 50
percent of normal with normal progressive antithrombin III
activity appear infrequently to experience thrombotic episodes.[38-42]
Several of these cases were brought to clinical attention when
children of these heterozygous subjects presented with severe
venous or arterial thromboses at a young age,[38,40,43] and were
determined to have plasma antithrombin III-heparin cofactor
levels of less than 10 percent. This latter group of patients
were determined to be homozygous for an antithrombin III

molecular defect. In each instance, there was a history of
consanguinity in the family. The clinical histories of the
heterozygotes with this type of antithrombin III abnormality are
also seemingly different from type II patients with both
diminished progressive antithrombin III activity and antithrombin
III-heparin cofactor activity. This latter group of subjects
sustain venous thromboembolism as often as type I patients.[44]

The mean concentration of antithrombin III in normal pooled
plasma is 140 µg/ml. In the plasmas from normal persons, the
range of antithrombin III concentrations as determined by
immunologic or functional tests is quite narrow, 80 to 120
percent for antithrombin III-heparin cofactor determinations[45] and
a somewhat wider range for immunoassay results.[46] Given the
various subtypes of the familial deficiency state that might be
encountered, the best single screening test for the disorder is
the antithrombin III-heparin cofactor assay. Healthy newborns
have about half the normal adult concentration[47,48] and gradually
reach this level by 6 months of age.[48] The levels may be
considerably lower in infants born after 30 to 36 weeks of
gestation.[49]

A variety of pathophysiologic conditions can reduce the
concentration of antithrombin III within the blood. While acute
thrombosis will infrequently lower antithrombin III levels
substantially,[50] disseminated intravascular coagulation may result
in reductions in the level of this inhibitor.[51] Lowered
antithrombin III concentrations occur in patients with liver
disease (mainly cirrhosis) due to decreased protein synthesis.[52]
Decreased antithrombin III levels are also observed in
individuals with the nephrotic syndrome as a consequence of
urinary excretion.[53] Furthermore, the use of estrogens is
associated with modest reductions in plasma antithrombin III
concentrations.[28] This reduction can be seen in women on oral
contraceptives, as well as in patients receiving estrogens for
other purposes. Infusions of L-asparaginase, a chemotherapeutic
agent employed in the treatment of acute lymphocytic leukemia,
can substantially lower the plasma concentration of this
inhibitor.[54] In addition, the administration of heparin can
result in significant decrements in plasma antithrombin III
levels,[55] presumably on the basis of accelerated in vivo clearance
of the inhibitor. Evaluation of plasma samples from patients
suspected of having congenital antithrombin III deficiency during
a period of heparinization can therefore potentially lead to an
erroneous diagnosis of the disorder.

Because of the number of clinical conditions associated with
reductions in the plasma concentration of antithrombin III,
definitive diagnosis of the hereditary deficiency state is
oftentimes difficult in the setting of an acute thrombotic event.
While an antithrombin III level in the normal range drawn upon
clinical presentation is usually sufficient to exclude the
presence of the disorder, low levels should be confirmed by a
second determination. This is ideally performed when the patient
is no longer receiving oral anticoagulants as these medications
have occasionally been reported to raise plasma antithrombin III
concentrations into the normal range in patients with the
hereditary deficiency state;[9] clinical assessment as to the

individual's risk of recurrent thrombosis will determine whether
this approach is feasible. In most antithrombin III-deficient
subjects, however, oral anticoagulants do not obscure the
diagnosis.[56] Confirmation of the hereditary nature of the
disorder requires investigation of other family members.
Diagnosis of other biochemically affected family members also
allows for appropriate counseling regarding the need for
prophylaxis against venous thrombosis.

Patients with antithrombin III deficiency can usually be
treated successfully with intravenous heparin. In some
situations, unusually high doses of the drug are required to
achieve adequate anticoagulation. Indeed, the diagnosis of
antithrombin deficiency is usually considered in the differential
diagnosis of heparin resistance and an occasional patient with
this disorder is diagnosed in this manner. In antithrombin III-
deficient patients receiving heparin for the treatment of acute
thrombosis, the adjunctive role of antithrombin III concentrate
purified from human plasma is not clearly defined as controlled
trials have not been performed. However, on the basis of
available data, administration of this product should be
considered when difficulty is encountered in achieving adequate
heparinization or recurrent thrombosis is observed despite
adequate anticoagulation. Antithrombin III-deficient patients may
be treated with concentrate before major surgery or in obstetric
cases in which the risk of bleeding in the presence of full
anticoagulation is deemed unacceptable. Antithrombin III
concentrate is more than 95 percent pure and the viruses
responsible for hepatitis B and the acquired immune deficiency
syndrome are inactivated by commercial manufacturing
processes;[57,58] it is therefore preferable to administer
antithrombin III concentrate rather than fresh frozen plasma in
these clinical settings.

Protein C Deficiency

In 1981, Griffin et al.[59] described the first kindred in
which several patients had plasma levels of protein C antigen of
approximately 50 percent of normal in association with a history
of recurrent thrombotic events. Subsequently, other
investigators[60-65] have reported numerous other families with this
disorder. This biochemical deficiency state is inherited as an
autosomal dominant gene and has clinical features quite similar
to those of hereditary antithrombin III deficiency. About 75
percent of these persons have experienced one or more thrombotic
events. The initial episode occurred spontaneously in
approximately 70 percent of patients, with the remaining 30
percent of these persons having the usual associated risk factors
at the time they developed acute thrombotic events (e.g.
pregnancy, delivery, oral contraceptive ingestion, surgery, or
trauma). However, these patients are not usually symptomatic
until the age of 20, with increasing numbers of persons
experiencing thrombotic events as they reach the age of 50. The
most common sites of disease are the deep veins of the legs, the
iliofemoral veins, and the mesenteric veins. Approximately 63
percent of affected patients developed recurrent venous
thrombosis, with about 40 percent exhibiting signs of pulmonary
embolism.[62] Investigators from the Netherlands have noted a high

frequency of superficial thrombophlebitis of the leg veins[61] as well as several cases of cerebral venous thrombosis in their protein C deficient patients.[66]

Other kindreds have been reported in which patients heterozygous for protein C deficiency have minimal symptoms. First, a number of case reports have described a homozygous or doubly heterozygous state in which newborns develop purpura fulminans and laboratory evidence of disseminated intravascular coagulation in association with protein C antigen levels less than 1 percent of normal.[67-76] However, the heterozygous parents of these infants have only infrequently exhibited thrombotic manifestations, in contrast to the the patients from thrombophilic families with a partial deficiency of protein C. Second, Miletich et al.[77] observed that the frequency of heterozygous protein C deficiency is as high as one per 200 in a healthy adult population and that biochemically affected persons had not yet exhibited thrombotic manifestations. These data suggest that other, as yet undefined, factors modulate the phenotypic expression of heterozygous protein C deficiency.

The occurrence of coumarin-induced skin necrosis has been associated with the presence of heterozygous protein C deficiency.[78-81] This syndrome typically occurs during the first several days of warfarin therapy, often in association with the administration of large loading doses of the medication. The skin lesions occur on the extremities, breasts, and trunk as well as the penis and marginate over a period of hours from an initial central erythematous macule. If vitamin K or or a product containing protein C is not rapidly administered, the affected cutaneous areas become edematous, develop central purpuric zones, and ultimately become necrotic. Biopsies demonstrate fibrin thrombi within cutaneous vessels with interstitial hemorrhage. The dermal manifestations of coumarin-induced skin necrosis are clinically and pathologically similar to those seen in infants with purpura fulminans due to severe protein C deficiency.

The pathogenesis of warfarin-induced skin necrosis is attributable to the emergence of a transient hypercoagulable state. The initiation of the drug at standard doses leads to a decrease in protein C anticoagulant activity levels to approximately 50 percent of normal within one day.[82] While factor VII activity measurements follow a pattern similar to that of protein C, the levels of the other vitamin K-dependent factors decline at slower rates consistent with their longer half-lives. Increased thrombin generation has been documented in patients during this early phase of warfarin therapy using a sensitive immunochemical assay for fragment F_{1+2}, an index of the in vivo activation of prothrombin mediated by factor Xa.[83] During this period, it therefore appears that the drug's suppressive effect on protein C has a greater influence on the hemostatic mechanism as contrasted to its action on factor VII. These effects are likely to be augmented when loading dose schedules are used, or the patient has an underlying hereditary deficiency of protein C. Only approximately one-third of patients with warfarin-induced skin necrosis have an inherited deficiency of protein C;[84] this complication is a rather infrequently reported event among persons with the heterozygous deficiency state.

Two major subtypes of heterozygous protein C deficiency have been delineated utilizing immunologic and functional assays. The classic, or type I, deficiency state appears to be the most common form, characterized by a parallel reduction in the immunologic and biologic activity of the zymogen in the blood.[62] The majority of the families with this disorder have nonsense or missense point mutations in their protein C genes.[85,86] In families with a type II deficiency state, affected individuals exhibit normal protein C levels on immunologic examination, yet possess lowered functional levels of the zymogen.[87-92]

A clinical disorder has been reported in which newborns develop purpura fulminans in association with protein C antigen levels that are less than 1 percent of normal.[67-76] In some instances, there was a history of consanguinity in the family, making it highly likely that the affected infants were homozygous for the deficiency state.[68,71,74] However, a number of case reports have documented the existence of a form of severe protein C deficiency in which neonatal purpura fulminans is not present. These individuals generally have protein C levels of under 20 percent of normal in the absence of oral anticoagulant therapy, and their clinical presentation is generally similar to that of severely affected subjects from thrombophilic kindreds with the heterozygous deficiency state.[93-98] The parents of these subjects have a type I deficiency state, as do the parents of babies with purpura fulminans. In addition, patients who are doubly heterozygous for both type I and type II deficiency states that were inherited separately from each of the parents have been described.[99,100] In one of these cases, the type II defect resulted from the replacement of arginine by tryptophan at position 12 of the heavy chain; this residue is the site at which protein C is activated by the thrombin-thrombomodulin complex.[100]

A variety of immunologic and functional techniques have been developed to measure protein C levels in plasma samples. The most commonly used procedures for antigen determinations are electroimmunoassay,[59,60] enzyme-linked immunosorbent assay,[93,101] or radioimmunoassay.[102,103] Functional assays have utilized either thrombin[87,104] or the thrombin-thrombomodulin complex[88,105] to activate protein C after adsorbing the zymogen with aluminum hydroxide[85] or barium citrate.[104,105] With the exception of the method of Francis and Patch,[104] which measures the ability of activated protein C to prolong the activated partial thromboplastin time, the endpoint of these assays is the cleavage of a synthetic substrate by the enzyme. Other methods have utilized the thrombin-thrombomodulin complex to first activate protein C in plasma followed by immunoadsorption of the enzyme with goat antihuman protein C IgG-agarose.[88] Alternatively, the protein C is first adsorbed with a calcium-dependent monoclonal antibody and subsequently activated by the thrombin-thrombomodulin complex.[82] The activity of the enzyme is then assessed using a chromogenic substrate[87,88,105] or by measuring its anticoagulant activity in a factor Xa one-stage clotting assay.[82]

The development of simpler functional assays has been facilitated by the observation that the venom from the Southern copperhead snake (*Agkistrodon Contortrix*) is able to activate

protein C in plasma.[106-108] This is followed by measuring the
amidolytic activity of the enzyme toward a suitable chromogenic
substrate or its anticoagulant activity in a clotting assay.
Functional assays using amidolytic and clotting endpoints may
give useful information regarding the nature of the molecular
defect in patients with type II protein C deficiency.[82,90,92]

Protein C circulates in human plasma at a concentration of 4
µg/ml. The levels of protein C antigen in healthy adults are
log-normally distributed; 95 percent of the values range from 70
to 140 percent.[77] There is no significant gender dependence, but
mean protein C antigen concentrations increase by approximately 4
percent per decade. This is largely a reflection of the tendency
for some older subjects to have higher values rather than
individual change at the lower end of the normal range. Similar
results are seen with functional assays. The relatively wide
normal range of protein C measurements in the general population
will frequently make it more difficult to identify a given
patient definitively as having heterozygous protein C deficiency.
If medical and pharmacologic causes of low levels are excluded
(see below), protein C values of less than 55 percent lead to a
high likelihood that the person has the genetic abnormality.
Levels from 55 to 65 percent are consistent with either a
deficiency state or the lower end of the normal distribution.[77]
To document this disorder, it is useful to perform repeat
determinations on the patient suspected of being protein C
deficient as well as performing family studies to document the
presence of an autosomal dominant inheritance pattern for the
abnormality.

Protein C levels in newborns are 20 to 40 percent of normal
adult levels[109,110]; preterm infants have even lower levels.[111,112]
Acquired protein C deficiency occurs in a number of clinical
disorders, including liver disease,[82,109,113,114] disseminated
intravascular coagulation,[82,109,113,115,116] adult respiratory
distress syndrome,[109] the post-operative state,[109] and in
association with L-asparaginase therapy.[117] In contradistinction
to antithrombin III, the antigenic concentrations of vitamin K-
dependent plasma proteins including protein C are oftentimes
elevated in patients with the nephrotic syndrome.[118,119]

Warfarin therapy reduces functional[82,87,105] and to a lesser
extent immunologic measurements of protein C.[59,60] This makes it
quite difficult to diagnose patients with heterozygous protein C
deficiency who are being treated with oral anticoagulants.
Several research laboratories have used a reduced ratio of
protein C antigen to prothrombin antigen to identify patients
with a type I deficiency state.[59,60] This approach however can
only be used in subjects in a stable phase of oral
anticoagulation; also, the diagnostic criteria for the disorder
vary with the intensity of warfarin therapy.[60] In practice, it is
preferable to investigate patients suspected of having the
deficiency state after oral anticoagulation has been discontinued
for at least a 1-week period and to perform family studies. If
it is not possible to discontinue warfarin due to the severity of
the thrombotic tendency, such patients can be studied while
receiving heparin therapy. Heparin has not been shown to alter
plasma protein C levels.

Management of thromboembolic events in heterozygous protein C-deficient patients is similar to that of subjects without this disorder. It is advisable to keep the subject fully anticoagulated with heparin during the initiation of oral anticoagulation; large loading doses of warfarin must be avoided. Oral anticoagulants are effective in managing individuals with protein C deficiency; the recommendations for its use in such patients who have either sustained recurrent venous thrombosis or are asymptomatic are similar to those in patients with congenital antithrombin III deficiency.

The infrequent occurrence of warfarin-induced skin necrosis and the diagnostic difficulty in making a rapid definitive laboratory diagnosis of the deficiency state are arguments against the routine measurement of plasma protein C levels for all patients with venous thrombosis before the initiation of oral anticoagulants. If, however, one is starting oral anticoagulants in a patient who is already known or likely to be protein C deficient, it would seem prudent from the biochemical standpoint to start the drug under the cover of full heparinization and also to increase the dose of warfarin gradually starting from a relatively low level (e.g., 2 mg for the first three days, and then increasing in increments of 2-3 mg until therapeutic anticoagulation is achieved). The successful oral anticoagulation of a subject with heterozygous protein C deficiency and a prior history of warfarin-induced skin necrosis has been reported.[81] Therapeutic doses of heparin as well as protein C replacement in the form of fresh frozen plasma or purified concentrate can be used to prevent the development of this complication.

The management of neonatal purpura fulminans in association with severe protein C deficiency is more complicated. Neither heparin therapy or antiplatelet agents have been shown to be effective.[67,69-71] The administration of a source of protein C appears to be critical in the initial treatment of these patients. Fresh frozen plasma has therefore been used successfully in these infants. However, the half-life of protein C in the circulation is short, at about 6 to 16 hours,[82,120] and the administration of plasma on a frequent basis is limited by the development of hyperproteinemia, hypertension, loss of venous access, and the potential for exposure to infectious viral agents. Protein C concentrate has recently been used successfully in this disorder,[121] and highly purified concentrates of this protein are currently undergoing clinical trials for this indication. Warfarin has been administered to these infants without the redevelopment of skin necrosis during the phased withdrawal of fresh frozen plasma infusions[67,72,74,122,123]; this medication has been used chronically to control the thrombotic disorder. A 20-month-old child with liver failure and homozygous protein C deficiency has been successfully treated by liver transplantation.[124] This approach resulted in a normalization of protein C levels as well as resolution of the thrombotic tendency.

<u>Protein S Deficiency</u>

In 1984, Comp et al.[125] and Schwarz et al.[126] described members from several kindreds who exhibited reduced levels of protein S in association with a striking history of recurrent venous thrombotic disease. Subsequently, many additional families with this disorder have been reported.[127-139] The clinical presentation of patients with heterozygous protein S deficiency is similar to that of antithrombin III deficiency or protein C deficiency. Among 71 protein S-deficient members from 12 Dutch pedigrees,[135] 74 percent, 72 percent, and 38 percent of patients sustained deep venous thrombosis, superficial thrombophlebitis, or pulmonary emboli, respectively. The mean age of the first thrombotic event was 28 years, with a range of 15 to 68 years; 56 percent of the episodes were spontaneous and the remainder were precipitated by an identifiable factor. Thromboses have also been reported in the axillary, mesenteric, and cerebral veins. A case report has appeared describing warfarin-induced skin necrosis in a patient with heterozygous protein S deficiency.[140]

Under normal conditions, approximately 60 percent of the total protein S antigen in plasma is complexed to a complement component, C4b-binding protein. Only the free 40 percent is functionally active as a cofactor in mediating the anticoagulant effects of activated protein C.[127] This observation has led to the development of methods for measuring total[126,141,142] and free protein S antigen.[132] The most reliable measurements of total protein S antigen are by radioimmunoassay[141-143] or enzyme-linked immunosorbent assay techniques, which involve dilution of plasma samples and thereby favor dissociation of the protein S-C4b-binding protein complexes. After removal of protein S-C4b-binding protein complexes from plasma by polyethylene glycol precipitation,[132,144] free protein S may be quantified by immunoassay of the supernatant fractions. Functional assay methods usually require immunodepleted protein S-deficient plasma or immobilized antibodies to protein S.[125,145-149] They are based on the ability of protein S to serve as a cofactor for the anticoagulant effect of activated protein C.

The classic deficiency state is associated with approximately 50 percent of the normal total S antigen level[126,129] and with decrements in free protein S antigen and protein S functional activity to about 40 percent or less of normal values.[132] Another type of hereditary deficiency state has been described in which total protein S antigen measurements are within the normal range but in which the levels of free protein S and protein S functional activity are disproportionately reduced. A thrombotic patient with normal concentrations of total and free protein S, but diminished functional activity, on a hereditary basis has recently been described.[150]

A family has been reported in which two individuals had doubly heterozygous or homozygous protein S deficiency with recurrent venous thromboembolic disease.[127] The parents of these children were asymptomatic and had laboratory studies consistent with the classic protein S-deficient state. In addition, neonatal purpura fulminans in association with homozygous protein S deficiency has been described.[151]

The concentration of total protein S antigen in normal adults is approximately 23 μg/ml.[141] The levels have been observed to increase significantly with advancing age and are significantly lower and more variable in females than males.[152,153] These factors have confounded the reliable estimation of the prevalence of heterozygous protein S deficiency in the normal population. These factors complicate the diagnosis of heterozygous protein S deficiency by performing assays on a given patient at a single point in time. The resampling of patients as well as family studies are usually required to establish a firm diagnosis.

Acquired protein S deficiency occurs during pregnancy[143,144] and in association with the use of oral contraceptives.[153,154] Reduced protein S levels have been noted in association with L-asparaginase therapy,[117] and in patients with disseminated intravascular coagulation[146,155] and acute thromboembolic disease.[146] C4b-binding protein is an acute phase protein, and the decline in protein S activity in the latter two conditions as well as in other inflammatory disorders is attributable to a shift of the protein to the complexed, inactive form.[146] Total protein S antigen measurements are generally increased in patients with the nephrotic syndrome[118,119,156] although functional assays give reduced values, partly explained by the loss of free protein S in the urine and elevations in C4b-binding protein levels. Total and free protein S antigen concentrations are moderately decreased in liver disease.[142,146] Total protein S antigen values in healthy newborns at term are 15 to 30 percent of normal, while C4b-binding protein is markedly reduced to less than 20 percent. Thus, the free form of the protein predominates in this setting, and functional levels are only slightly reduced as compared to those in normal adults.[149,157,158] Interpretation of protein S measurements in individuals on oral anticoagulants is complicated inasmuch as the antigenic and functional levels of the protein drop substantially in this setting. Reductions in the ratio of total protein S antigen to prothrombin antigen may be used to infer a diagnosis of the classic type of protein S-deficient state, as for protein C-deficient subjects.[126,129]

CONCLUSION

Several inherited deficiencies or abnormalities of coagulation proteins have been described which predispose patients to venous thromboembolism. Laboratory tests have been developed to screen for these defects, although these assays are currently able to provide diagnoses in fewer than 20 percent of young patients with venous thromboembolism. This diagnostic armamentarium provides relatively little assistance in the evaluation of individuals with acquired risk factors for venous thrombosis or those with arterial vascular disease.

Advances in our understanding of the biochemistry of the hemostatic mechanism has led to the development of sensitive immunochemical methods for measuring peptides or enzyme-inhibitor complexes that are liberated with the activation of the coagulation and fibrinolytic systems in vivo.[31,103,159-165] Two of these techniques are specific assays for fragment F_{1+2}[159,160] and fibrinopeptide A[161-163] that measure the cleavage of the prothrombin molecule by factor Xa and the proteolysis of fibrinogen by thrombin, respectively. All persons exhibit measurable amounts of these markers under normal conditions, and elevated levels have been reported in patients with acute thromboembolism,[162,163] and disseminated intravascular coagulation.[160,162,163,166] Clinical investigations have also demonstrated that significant increments in factor Xa activity (as measured by the F_{1+2} assay), but not thrombin activity (as measured by the fibrinopeptide A assay), regularly occur in the blood of asymptomatic patients with several of the inherited thrombotic disorders.[97,167] The continuing application of biochemical, molecular biologic, and clinical approaches to investigations of the hemostatic system should lead to more precise identification of patients who are entering a hypercoagulable state and intervene with appropriate therapy before the onset of overt thrombotic disease.

ACKNOWLEDGMENTS

This work was supported in part by grant PO1 HL33014 from the National Institutes of Health. Dr. Bauer is a recipient of an Established Investigatorship from the American Heart Association.

REFERENCES

1. Rosenberg RD: Regulation of the hemostatic mechanism, in Stomatoyannopoulos G, Nienhuis AW, Leder P, Majerus PW (eds): The Molecular Basis of Blood Diseases. Philadelphia, PA, Saunders, 1987, p 534

2. Esmon CT: Protein C: the regulation of natural anticoagulant pathways. Science 235:1348, 1987

3. Briet E, Engesser L, Brommer EJP, Broekmans AW, Bertina RM: Thrombophilia: its causes and a rough estimate of its prevalence. Thromb Haemost 58:39, 1987 (abstr)

4. Vikydal R, Korninger C, Kyrle PA, Niessner H, Pabinger I, Thaler E, Lechner K: The prevalence of hereditary antithrombin-III deficiency in patients with a history of venous thromboembolism. Thromb Haemost 54:744, 1985

5. Gladson CL, Scharrer I, Hach V, Beck KH, Griffin JH: The frequency of type I heterozygous protein S and protein C deficiency in 141 unrelated young patients with venous thrombosis. Thromb Haemost 59:18, 1988

6. Egeberg O: Inherited antithrombin deficiency causing thrombophilia. Thromb Diath Haemorrh 13:516, 1965

7. Van der Meer J, Stoepman-van Dalen EA, Jansen JMS: Anti-thrombin-III deficiency in a Dutch family. J Clin Path 26:532, 1973

8. Shapiro SS, Prager D, Martinez J: Inherited antithrombin III deficiency associated with multiple thromboembolic phenomena. Blood 42:1001, 1973 (abstr)

9.   Marciniak E, Farley CH, DeSimone PA: Familial thrombosis due
     to antithrombin III deficiency. Blood 43:219, 1974
10. Gruenberg JC, Smallridge RC, Rosenberg RD: Inherited anti-
     thrombin-III deficiency causing mesenteric venous
     infarction: a new clinical entity. Ann Surg 181:791, 1975
11. Zucker ML, Metz J, Gomperts ED: Inherited antithrombin III
     deficiency as a cause of multiple venous thromboses. S Afr
     Med J 49:1425, 1975
12. Filip DJ, Eckstein JD, Veltkamp JJ: Hereditary antithrombin
     III deficiency and thromboembolic disease. Am J Hematol
     2:343, 1976
13. Carvalho A, Ellman L: Hereditary antithrombin III
     deficiency: effect of antithrombin III deficiency on
     platelet function. Am J Med 61:179, 1976
14. Odegard OR, Abildgaard U: Antifactor Xa activity in
     thrombophilia. Studies in a family with AT-III deficiency.
     Scand J Haematol 18:86, 1977
15. Stathakis NE, Papayannis AG, Antonopoulos M, Gardikas C:
     Familial thrombosis due to antithrombin III deficiency in a
     Greek family. Acta Haematol (Basel) 57:47, 1977
16. Mackie M, Bennett B, Ogston D, Douglas AS: Familial
     thrombosis: inherited deficiency of antithrombin III. Br Med
     J 1:136, 1978
17. Johansson L, Hedner U, Nilsson IM: Familial antithrombin III
     deficiency as pathogenesis of deep venous thrombosis. Acta
     Med Scand 204:491, 1978
18. Gyde OHB, Middleton MD, Vaughan GR, Fletcher DJ:
     Antithrombin III deficiency, hypertriglyceridaemia and
     venous thrombosis. Br Med J 1:621, 1978
19. Matsuo T, Ohki Y, Kondo S, Matsuo O: Familial antithrombin
     III deficiency in a Japanese family. Thromb Res 16:815, 1979
20. Pitney WR, Manoharan A, Dean S: Antithrombin III deficiency
     in an Australian family. Br J Haematol 46:1479, 1980
21. Boyer C, Wolf M, Lavergne JM, Larrieu MH: Thrombin
     generation and formation of thrombin-antithrombin III
     complexes in congenital antithrombin deficiency. Thromb Res
     20:207, 1980
22. Beukes CA, Heyns ADuP: A South African family with
     antithrombin III deficiency. S Afr Med J 58:528, 1980
23. Ambruso DR, Jacobson LJ, Hathaway WE: Inherited antithrombin
     III deficiency and cerebral thrombosis in a child.
     Pediatrics 65:125, 1980
24. Scully MF, De Haas H, Chan P, Kakkar VV: Hereditary
     antithrombin III deficiency in an English family. Br J
     Haematol 47:235, 1981
25. Winter JH, Fenech A, Ridley W, Bennett B, Cumming AM, Mackie
     M, Douglas AS: Familial antithrombin III deficiency. Quart J
     Med 204:373, 1982
26. Rosenberg RD: Actions and interactions of antithrombin and
     heparin. N Engl J Med 292:146, 1975
27. Odegard OR, Abildgaard U: Antithrombin III: critical review
     of assay methods. Significance of variations in health and
     disease. Haemostasis 7:127, 1978
28. Thaler E, Lechner K: Antithrombin III deficiency and
     thromboembolism, in Prentice CRM (ed): Clinics in
     Haematology, vol 10. London, Saunders, 1981, p 369

29. Ambruso DR, Leonard BD, Bies RD, Jacobson L, Hathaway WE, Reeve EB: Antithrombin III deficiency: decreased synthesis of a biochemically normal molecule. Blood 60:78, 1982

30. Sas G, Blasko G, Banhegyi D, Jako J, Palos LA: Abnormal antithrombin III (antithrombin III "Budapest") as a cause of familial thrombophilia. Thromb Diath Haemorrh 32:105, 1974

31. Bauer KA, Teitel JM, Rosenberg RD: Assays for the quantitation of antithrombin III thrombin-antithrombin complex and prothrombin activation fragments, in Colman RW (ed): Disorders of thrombin formation. New York, Churchill Livingstone, New York, 1983, p 142

32. Tollefsen DM, Blank MK: Detection of a new heparin-dependent inhibitor of thrombin in human plasma. J Clin Invest 68:589, 1981

33. Tollefsen DM, Majerus DW, Blank MK: Heparin cofactor II. Purification and properties of a heparin-dependent inhibitor of thrombin in human plasma. J Biol Chem 257:2162, 1982

34. Wunderwald P, Schrenk WJ, Port H: Antithrombin BM from human plasma: an antithrombin binding moderately to heparin. Thromb Res 25:177, 1982

35. Tran TH, Marbet GA, Duckert F: Association of hereditary heparin cofactor II deficiency with thrombosis. Lancet II:413, 1985

36. Sie P, Dupouy D, Pichon J, Boneu B: Constitutional heparin cofactor II deficiency associated with recurrent thrombosis. Lancet II:414, 1985

37. Bertina RM, van der Linden IK, Engesser L, Muller HP, Brommer EJP: Hereditary heparin cofactor II deficiency and the risk of development of thrombosis. Thromb Haemost 57:196, 1987

38. Sakuragawa N, Takahashi K, Kondo S, Koide T: Antithrombin III Toyama: a hereditary abnormal antithrombin III of a patient with recurrent thrombophlebitis. Thromb Res 31:305, 1983

39. Chasse JF, Esnard F, Guitton JD, Mouray H, Perigois F, Fauconneau G, Gauthier F: An abnormal plasma antithrombin with no apparent affinity for heparin. Thromb Res 34:297, 1984

40. Fischer AM, Cornu P, Sternberg C, Meriane F, Dautzenberg MD, Chafa O, Beguin S, Desnos M: Antithrombin III Alger: a new homozygous AT III variant. Thromb Haemost 55:218, 1986

41. Owen MC, Borg JY, Soria C, Soria J, Caen J, Carrell RW: Heparin binding defect in a new antithrombin III variant: Rouen, 47 Arg to His. Blood 69:1275, 1987

42. Borg JY, Owen MC, Soria C, Soria J, Caen J, Carrell RW: Proposed heparin binding site in antithrombin based on arginine 47. A new variant Rouen-II, 47 arg to ser. J Clin Invest 81:1292, 1988

43. Boyer C, Wolf M, Vedrenne J, Meyer D, Larrieu MJ: Homozygous variant of antithrombin III: AT III Fountainebleau. Thromb Haemost 56:18, 1986

44. Finazzi G, Caccia R, Barbui T: Different prevalence of thromboembolism in the subtypes of congenital antithrombin deficiency: review of 404 cases. Thromb Haemost 58:1094, 1987 (letter)

45. Odegard OR, Lie M, Abildgaard U: Heparin cofactor activity measured with an amidolytic method. Thromb Res 6:287, 1975

46. Fagerhol MK, Abilgaard U: Immunological studies on human antithrombin III. Influence of age, sex and use of oral contraceptives on serum concentration. Scand J Haematol 1:10, 1970

47. McDonald MM, Hathaway WE, Reeve EB, Leonard BD: Biochemical and functional study of antithrombin III in newborn infants. Thromb Haemost 47:56, 1982

48. Andrew M, Paes B, Milner R, Johnston M, Mitchell L, Tollefsen DM, Powers P: Development of the human coagulation system in the full-term infant. Blood 70:165, 1987

49. Andrew M, Paes B, Milner R, Johnston M, Mitchell L, Tollefsen DM, Castle V, Powers P: Development of the human coagulation system in the healthy premature infant. Blood 72:1651, 1988

50. de Boer AC, van Riel LAM, den Ottolander GJH: Measurement of antithrombin III, α_2-macroglobulin and α_1-antitrypsin in patients with deep venous thrombosis and pulmonary embolism. Thromb Res 15:17, 1979

51. Damus PS, Wallace GA: Immunologic measurement of antithrombin III-heparin cofactor and α_2-macroglobulin in disseminated intravascular coagulation and hepatic failure coagulopathy. Thromb Res 6:27, 1989

52. von Kaulla E, von Kaulla KN: Antithrombin III and diseases. Am J Clin Path 48:69, 1967

53. Kauffman RH, Veltkamp JJ, Van Tilburg NH, Van Es LA: Acquired antithrombin III deficiency and thrombosis in the nephrotic syndrome. Am J Med 65:607, 1978

54. Buchanan GR, Holtkamp CA: Reduced antithrombin III levels during L-asparaginase therapy. Med Pediatr Oncol 8:7, 1980

55. Marciniak E, Gockemen JP: Heparin-induced decrease in circulating antithrombin III. Lancet II:581, 1978

56. Kitchens CS: Amelioration of antithrombin III deficiency by coumarin administration. Am J Med Sci 293:403, 1987

57. Hoffman DL: Purification and large-scale preparation of antithrombin III. Am J Med 87 (suppl 3B):23S, 1989

58. Menache D, O'Malley JP, Schorr JB, Wagner B, Williams C, Alving BM, Ballard JO, Goodnight SH, Hathaway WE, Hultin MB, Kitchens CS, Lessner HE, Makary AZ, Manco-Johnson M, McGehee WG, Penner JA, Sanders JE: Evaluation of the safety, recovery, half-life, and clinical efficacy of antithrombin III (human) in patients with hereditary antithrombin III deficiency. Blood 75:33, 1990

59. Griffin JH, Evatt B, Zimmerman TS, Kleiss AJ, Wideman C: Deficiency of protein C in congenital thrombotic disease. J Clin Invest 68:1370, 1981

60. Bertina RM, Broekmans AW, van der Linden IK, Mertens K: Protein C deficiency in a Dutch family with thrombotic disease. Thromb Haemost 45:237, 1982

61. Broekmans AW, Veltkamp JJ, Bertina RM: Congenital protein C deficiency and venous thromboembolism: a study of three Dutch families. N Engl J Med 309:340, 1983

62. Broekmans AW, Bertina RM: Protein C, in Poller L (ed): Recent Advances in Blood Coagulation, no 4. New York, Churchill Livingstone, 1985, p 117

63. Horellou MH, Conard J, Bertina RM, Samama M: Congenital protein C deficiency and thrombotic disease in nine French families. Br Med J 289:1285, 1984

64. Pabinger-Fasching I, Bertina RM, Lechner K, Niessner H, Korninger C: Protein C deficiency in two Austrian families. Thromb Haemost 50:810, 1983

65. Bovill EG, Bauer KA, Dickerman JD, Callas P, West B: The clinical spectrum of heterozygous protein C deficiency in a large New England kindred. Blood 73:712, 1989

66. Wintzen AR, Broekmans AW, Bertina RM, Briet E, Briet PE, Zecha A, Vielvoye GJ, Bots GThAM: Cerebral hemorrhagic infarction in young patients with hereditary protein C deficiency: evidence for "spontaneous" cerebral venous thrombosis. Br Med J 290:350, 1985

67. Branson HE, Katz J, Marble R, Griffin JH: Inherited protein C deficiency and coumarin-responsive chronic relapsing purpura fulminans in a newborn infant. Lancet II:1165, 1983

68. Seligsohn U, Berger A, Abend M, Rubin L, Attias D, Zivelin A, Rapaport SI: Homozygous protein C deficiency manifested by massive venous thrombosis in the newborn. N Eng J Med 310:559, 1984

69. Sills RH, Marlar RA, Montgomery RR, Desphande GN, Humbert JR: Severe homozygous protein C deficiency. J Pediatr 105:409, 1984

70. Estelles A, Garcia-Plaza I, Dasi A, Aznar J, Duart M, Sanz G, Perez-Requejo JL, Espana F, Jimenez C, Abeledo G: Severe inherited 'homozygous' protein C deficiency in a newborn infant. Thromb Haemost 52:53, 1984

71. Marciniak E, Wilson HD, Marlar RA: Neonatal purpura fulminans: a genetic disorder related to the absence of protein C in blood. Blood 65:15, 1985

72. Yuen P, Cheung A, Lin HJ, Ho F, Mimuro J, Yoshida N, Aoki N: Purpura fulminans in a Chinese boy with congenital protein C deficiency. Pediatrics 77:670, 1986

73. Rappaport ES, Speights VO, Helbert B, Trowbridge A, Koops B, Montgomery RR, Marlar RA: Protein C deficiency. South Med J 80:240, 1987

74. Peters C, Casella JF, Marlar RA, Montgomery RR, Zinkham WH: Homozygous protein C deficiency: observations on the nature of the molecular abnormality and the effectiveness of warfarin therapy. Pediatrics 81:272, 1988

75. Tarras S, Gadia C, Meister L, Roldan E, Gregorios JB: Homozygous protein C deficiency in a newborn. Clinicopathologic correlation. Arch Neurol 45:214, 1988

76. Auletta MJ, Headington JT. Purpura fulminans. A cutaneous manifestation of severe protein C deficiency. Arch Dermatol 124:1387, 1988

77. Miletich J, Sherman L, Broze G Jr: Absence of thrombosis in subjects with heterozygous protein C deficiency. N Engl J Med 317:991, 1987

78. Broekmans AW, Bertina RM, Loeliger EA, Hofmann V, Klingemann HG: Protein C and the development of skin necrosis during anticoagulant therapy. Thromb Haemostas 49:244, 1983 (letter)

79. McGehee WG, Klotz TA, Epstein DJ, Rapaport SI: Coumarin necrosis associated with hereditary protein C deficiency. Ann Intern Med 100:59, 1984

80. Samama M, Horellou MH, Soria J, Conard J, Nicolas G: Successful progressive anticoagulation in a severe protein C deficiency and previous skin necrosis at initiation of oral

anticoagulant treatment. Thromb Haemost 51:132, 1984
(letter)

81. Zauber NP, Stark MW: Successful warfarin anticoagulation
despite protein C deficiency and a history of warfarin
necrosis. Ann Intern Med 104:659, 1986

82. D'Angelo SV, Comp PC, Esmon CT, D'Angelo A: Relationship
between protein C antigen and anticoagulant activity during
oral anticoagulation and in selected disease states. J Clin
Invest 77:416, 1986

83. Conway EM, Bauer KA, Barzegar S, Rosenberg RD: Suppression
of hemostatic system activation by oral anticoagulants in
the blood of patients with thrombotic diatheses. J Clin
Invest 80:1535, 1987

84. Broekmans AW, Teepe RGC, v d Meer FJM, Briet E, Bertina RM:
Protein C (PC) and coumarin-induced skin necrosis. Thromb
Res 6:137, 1986 (abstr)

85. Crabtree GR, Plutzky J, Marlar R, Bertina RM, Broekmans AW,
Griffin J, Zacharski L, Gruppo R, Sala N, Long G. The range
of genotypes underlying human protein C deficiency. Thromb
Haemost 54:56, 1985 (abstr)

86. Romeo G, Hassan HJ, Staempfli S, Roncuzzi L, Cianetti L,
Leonardi A, Vicente V, Mannucci PM, Bertina R, Peschle C,
Cortese R: Hereditary thrombophilia: identification of
nonsense and missense mutations in the protein C gene. Proc
Natl Acad Sci USA 84:2829, 1987

87. Bertina RM, Broekmans AW, Krommenhoek-van Es C, van
Wijngaarden A: The use of a functional and immunologic assay
for plasma protein C in the study of the heterogeneity of
congenital protein C deficiency. Thromb Haemost 51:1, 1984

88. Comp PC, Nixon RR, Esmon CT: Determination of functional
levels of protein C, an antithrombotic protein, using
thrombin-thrombomodulin complex. Blood 63:15, 1984

89. Barbui T, Finazzi G, Mussoni L, Riganti M, Donati MB,
Colucci M, Collen D: Hereditary dysfunctional protein C
(protein C Bergamo) and thrombosis. Lancet II:819, 1984
(letter)

90. Sala N, Borrell M, Bauer KA, D'Angelo SV, Fontcuberta J,
Felez J, Rutllant ML. Dysfunctional activated protein C (PC
Cadiz) in a patient with thrombotic disease. Thromb Haemost
57:183, 1987

91. Tirindelli MC, Franchi F, Tripodi A, Mariani G, Mannucci PM:
Familial dysfunctional protein C. Thromb Res 44:893, 1986

92. Faioni EM, Esmon CT, Esmon NL, Mannucci PM: Isolation of an
abnormal protein C molecule from the plasma of a patient
with thrombotic diathesis. Blood 71:940, 1988

93. Soria J, Soria C, Samama M, Nicolas G, Kisiel W: Severe
protein C deficiency in congenital thrombotic disease-
description of an immunoenzymological assay for protein C
determination. Thromb Haemost 53:293, 1985

94. Manabe S, Matsuda M: Homozygous protein C deficiency
combined with heterozygous dysplasminogenemia found in a 21-
year-old thrombophilic man. Thromb Res 39:333, 1985

95. Sharon C, Tirindelli MC, Mannucci PM, Tripodi A, Mariani G:
Homozygous protein C deficiency with moderately severe
clinical symptoms. Thromb Res 41:483, 1986

96. Kakkar S, Melissari E, Kakkar VV: Congenital severe protein
C deficiency in adults. Thromb Haemost 58:410, 1987 (abstr)

97. Bauer KA, Broekmans AW, Bertina RM, Conard J, Horellou MH, Samama MM, Rosenberg RD: Hemostatic enzyme generation in the blood of patients with hereditary protein C deficiency. Blood 71:1418, 1988

98. Triplett DA, Sandquist DS, Musgrave KA: Clinical application of a functional assay for protein C. Hematol Pathol 1:239, 1988

99. Gruppo RA, Leimer P, Francis RB, Marlar RA, Silberstein E: Protein C deficiency resulting from possible double heterozygosity and its response to danazol. Blood 71:370, 1988

100. Matsuda M, Sugo T, Sakata Y, Murayama H, Mimuro J, Tanabe S, Yoshitake S: A thrombotic state due to an abnormal protein C. N Eng J Med 319:1265, 1988

101. Boyer C, Rothschild C, Wolf M, Amiral J, Meyer D, Larrieu MJ: A new method for the estimation of protein C by ELISA. Thromb Res 36:579, 1984

102. Epstein DJ, Begum PW, Bajaj SP, Rapaport SI: Radioimmunoassays for protein C and factor X. Plasma antigen levels in abnormal hemostatic states. Am J Clin Pathol 82:573, 1984

103. Bauer KA, Kass BL, Beeler DL, Rosenberg RD: Detection of protein C activation in humans. J Clin Invest 74:2033, 1984

104. Francis RB Jr, Patch MJ: A functional assay for protein C in human plasma. Thromb Res 32:605, 1983

105. Sala N, Owen WG, Collen D: Functional assay of protein C in human plasma. Blood 63:671, 1984

106. Martinoli JL, Stocker K: Fast functional protein C assay using Protac C, a novel protein C activator. Thromb Res 43:253, 1986

107. Francis RB Jr, Seyfert U: Rapid amidolytic assay of protein C in whole plasma using an activator from the venom of agkistrodon contortrix. Am J Clin Pathol 87:619, 1987

108. Walker PA, Bauer KA, McDonagh J. A simple, automated functional assay for protein C. Am J Clin Path 92:210, 1989

109. Mannucci PM, Vigano S: Deficiencies of protein C, an inhibitor of blood coagulation. Lancet II:463, 1982

110. Polack B, Pouzol P, Amiral J, Kolodie L: Protein C level at birth. Thromb Haemost 52:188, 1984

111. Karpatkin M, Mannucci P, Bhogal M, Vigano S, Nardi M: Low protein C in the neonatal period. Br J Haematol 62:137, 1986

112. Manco-Johnson MJ, Marlar RA, Jacobson LJ, Hays T, Warady BA: Severe protein C deficiency in newborn infants. J Pediatr 113:359, 1988

113. Griffin JH, Mosher DF, Zimmerman TS, Kleiss AJ: Protein C, an antithrombotic protein, is reduced in hospitalized patients with intravascular coagulation. Blood 60:261, 1982

114. Rodeghiero F, Mannucci PM, Vigano S, Barbui T, Gugliotta L, Cortellaro M, Dini E: Liver dysfunction rather than intravascular coagulation as the main cause of low protein C and antithrombin III in acute leukemia. Blood 63:965, 1984

115. Marlar RA, Endres-Brooks J, Miller C: Serial studies of protein C and its plasma inhibitor in patients with disseminated intravascular coagulation. Blood 66:59, 1985

116. Mimuro J, Sakata Y, Wakabayashi K, Matsuda M: Level of protein C determined by combined assays during disseminated intravascular coagulation and oral anticoagulation. Blood 69:1704, 1987

117. Pui CH, Chesney CM, Bergum PW, Jackson CW, Rapaport SI: Lack of pathogenetic role of proteins C and S in thrombosis associated with asparaginase-prednisone-vincristine therapy for leukaemia. Br J Haematol 64:283, 1986
118. Cosio FG, Harker C, Batard MA, Brandt JT, Griffin JH: Plasma concentrations of the natural anticoagulants protein C and protein S in patients with proteinuria. J Lab Clin Med 106:218, 1985
119. Vigano-D'Angelo S, D'Angelo A, Kaufman CE Jr, Sholer C, Esmon CT, Comp PC: Protein S deficiency occurs in the nephrotic syndrome. Ann Intern Med 107:42, 1987
120. Vigano S, Mannucci PM, Solinas S, Bottasso B, Mariani G: Decrease in protein C antigen and formation of an abnormal protein soon after starting oral anticoagulant therapy. Br J Haematol 57:213, 1984
121. Vukovich T, Auberger K, Weil J, Engelmann H, Knobl P, Hadorn HB: Replacement therapy for a homozygous protein C deficiency-state using a concentrate of human protein C and S. Br J Haematol 70:435, 1988
122. Garcia-Plaza I, Jimenez-Astorga C, Borrego D, Marty ML: Coumarin prophylaxis for fulminant purpura syndrome due to homozygous protein C deficiency. Lancet I:634, 1985 (letter)
123. Hartman KR, Manco-Johnson M, Rawlings JS, Bower DR, Marlar RA: Homozygous protein C deficiency: early treatment with warfarin. Am J Pediatr Hematol Oncol 11:395, 1989
124. Casella JF, Lewis JH, Bontempo FA, Zitelli BJ, Markel H, Starzl TE: Successful treatment of homozygous protein C deficiency by hepatic transplantation. Lancet I:435, 1988
125. Comp PC, Esmon CT: Recurrent venous thromboembolism in patients with a partial deficiency of protein S. N Engl J Med 311:1525, 1984
126. Schwarz HP, Fischer M, Hopmeier P, Batard MA, Griffin JH: Plasma protein S deficiency in familial thrombotic disease. Blood 64:1297, 1984
127. Comp PC, Nixon RR, Cooper MR, Esmon CT: Familial protein S deficiency is associated with recurrent thrombosis. J Clin Invest 74:2082, 1984
128. Szatkowski NS, Miller CM, Endres-Brooks JL, Madden RM, Marlar RA: Clinical studies of human protein S and identification of a patient with protein S deficiency and thrombotic complications. Circulation 70:II-204,1984 (abstr)
129. Broekmans AW, Bertina RM, Reinalda-Poot J, Engesser L, Muller HP, Leeuw JA, Michiels JJ, Brommer EJP, Briet E: Hereditary protein S deficiency and venous thrombo-embolism. A study in three Dutch families. Thromb Haemost 53:273, 1985
130. Sas G, Blasko G, Petro I, Griffin J: A protein S deficient family with portal vein thrombosis. Thromb Haemost 54:724, 1985 (letter)
131. Kamiya T, Sugihara T, Ogata K, Saito H, Suzuki K, Nishioka J, Hashimoto S, Yamagata K: Inherited deficiency of protein S in a Japanese family with recurrent venous thrombosis: a study of three generations. Blood 67:406, 1986
132. Comp PC, Doray D, Patton D, Esmon CT: An abnormal plasma distribution of protein S occurs in functional protein S deficiency. Blood 67:504, 1986
133. Mannucci PM, Tripodi A, Bertina RM: Protein S deficiency associated with "juvenile" arterial and venous thromboses. Thromb Haemaost 55:440, 1986 (letter)

134. Pabinger I, Bertina RM, Lechner K, Niessner H, Korninger C, Deutsch E: Protein S deficiency in 7 Austrian families. Thromb Res 42:VI-136, 1986 (abstr)

135. Engesser L, Broekmans AW, Briet E, Brommer EJP, Bertina RM: Hereditary protein S deficiency: clinical manifestations. Ann Intern Med 106:677, 1987

136. Boyer-Neumann C, Wolf M, Amiral J, Guyader AM, Meyer D, Larrieu MJ: Familial type I protein S deficiency associated with severe venous thrombosis-a study of five cases. Thromb Haemost 60:128, 1988 (letter)

137. Ploos van Amstel HK, Huisman MV, Reitsma PH, ten Cate JW, Bertina RM: Partial protein S gene deletion in a family with hereditary thrombophilia. Blood 73:479, 1989

138. Schwarz HP, Heeb MJ, Lottenberg R, Roberts H, Griffin JH: Familial protein S deficiency with a variant protein S molecule in plasma and platelets. Blood 74:213, 1989

139. Whitlock JA, Janco RL, Phillips JA III: Inherited hypercoagulable states in children. Am J Pediatr Hematol Oncol 11:170, 1989

140. Friedman KD, Marlar RA, Houston JG, Montgomery RR: Warfarin-induced skin necrosis in a patient with protein S deficiency. Blood 68:333a, 1986 (abstr)

141. Fair DS, Revak DJ: Quantitation of human protein S in the plasma of normal and warfarin-treated individuals by radioimmunoassay. Thromb Res 36:527, 1984

142. Bertina RM, van Wijngaarden A, Reinalda-Poot J, Roort SR, Bom VJJ: Determination of plasma protein S-the protein cofactor of activated protein C. Thromb Haemost 53:268, 1985

143. Comp PC, Thurnau GR, Welsh J, Esmon CT: Functional and immunologic protein S levels are decreased during pregnancy. Blood 68:881, 1986

144. Malm J, Laurell M, Dahlback B: Changes in the plasma levels of vitamin K-dependent proteins C and S and of C4b-binding protein during pregnancy and oral contraception. Br J Haematol 68:437, 1988

145. Van de Waart P, Preissner KT, Bechtold JR, Muller-Berghaus G: A functional test for protein S activity in plasma. Thromb Res 48:427, 1987

146. D'Angelo A, Vigano-D'Angelo S, Esmon CT, Comp PC: Acquired deficiencies of protein S. Protein S activity during oral anticoagulation, in liver disease, and in disseminated intravascular coagulation. J Clin Invest 81:1445, 1988

147. Suzuki K, Nishioka J: Plasma protein S activity measured using Protac, a snake venom derived activator of protein C. Thromb Res 49:241, 1988

148. Kobayashi I, Amemiya N, Endo T, Okuyama K, Tamura K, Kume S: Functional activity of protein S determined with use of protein C activated by venom activator. Clin Chem 35:1644, 1989

149. Schwarz HP, Muntean W, Watzke H, Richter B, Griffin JH: Low total protein antigen but high protein S activity due to decreased C4b-binding protein in neonates. Blood 71:562, 1988

150. Mannucci PM, Valsecchi C, Krachmalnicoff A, Faioni EM, Tripodi A: Thromb Haemost 62:763, 1989

151. Mahasandana C, Suvatte V, Marlar R, Manco-Johnson M, Jacobson L, Hathaway WE: Neonatal purpura fulminans associated with homozygous proteins S deficiency. Thromb Haemost 62:301, 1989 (abstr)

152. Miletich JP, Broze GJ Jr: Age and gender dependence of total protein S antigen in the normal adult population. Blood 72:371a, 1988 (abstr)

153. Boerger LM, Morris PC, Thurnau GR, Esmon CT, Comp PC: Oral contraceptives and gender affect protein S status. Blood 69:692, 1987

154. Gilabert J, Fernandez JA, Espana F, Aznar J, Estelles A: Physiological coagulation inhibitors (protein S, protein C and antithrombin III) in severe preeclamptic states and in users of oral contraceptives. Thromb Res 49:319, 1988

155. Heeb MJ, Mosher D, Griffin JH: Activation and complexation of protein C and cleavage and decrease of protein S in plasma of patients with disseminated intravascular coagulation. Blood 73:455, 1989

156. Gouault-Heilmann M, Gadelha-Parente T, Levent M, Intrator L, Rostokor G, Lagrue G: Total and free protein S in nephrotic syndrome. Thromb Res 49:37, 1988

157. Malm J, Bennhagen R, Holmberg L, Dahlback B: Plasma concentrations of C4b-binding protein and vitamin K-dependent protein S in term and preterm infants: low levels of protein S-C4b-binding protein complexes. Br J Haematol 68:445, 1988

158. Melissari E, Nicolaides KH, Scully MF, Kakkar VV: Protein S and C4b-binding protein in fetal and neonatal blood. Br J Haematol 70:199, 1988

159. Lau HK, Rosenberg JS, Beeler DL, Rosenberg RD: The isolation and characterization of a specific antibody population directed against the prothrombin activation fragments F2 and F_{1+2}. J Biol Chem 254:8751, 1979

160. Teitel JM, Bauer KA, Lau HK, Rosenberg RD: Studies of the prothrombin activation pathway utilizing radioimmunoassays for the $F2/F_{1+2}$ fragment and thrombin-antithrombin complex. Blood 59:1086, 1982

161. Nossel HL, Younger LR, Wilner GD, Procupez T, Canfield RE, Butler VP Jr: Radioimmunoassay of human fibrinopeptide A. Proc Natl Acad Sci USA 68:2350, 1971

162. Nossel HL, Yudelman I, Canfield RE, Butler VP Jr, Spanondis K, Wilner GD, Qureshi GD: Measurement of fibrinopeptide A in human blood. J Clin Invest 54:43, 1974

163. Nossel HL, Ti M, Kaplan KL, Spanondis K, Soland T, Butler VP Jr: The generation of fibrinopeptide A in clinical blood samples: Evidence for thrombin activity. J Clin Invest 58:1136, 1976

164. Nossel HL, Wasser J, Kaplan KL, LaGamma KS, Yudelman I, Canfield RE: Sequence of fibrinogen protolysis and platelet release after intrauterine infusion of hypertonic saline. J Clin Invest 64:1371, 1979

165. Harpel PC: α_2-Plasmin inhibitor and α_2-macroglobulin-plasmin complexes in plasma: quantification by an enzyme-linked differential antibody immunosorbent assay. J Clin Invest 68:46, 1981

166. Bauer KA, Rosenberg RD: Thrombin generation in acute promyelocytic leukemia. Blood 64:791, 1984

167. Bauer KA, Goodman TL, Kass BL, Rosenberg RD: Elevated factor Xa activity in the blood of asymptomatic patients with congenital antithrombin deficiency. J Clin Invest 76:826, 1985

PROTEIN PRODUCTION BY RECOMBINANT TECHNOLOGY

FACTORS LIMITING EXPRESSION OF SECRETED PROTEINS IN MAMMALIAN CELLS

Randal J. Kaufman, Robert J. Wise,
Louise C. Wasley, and Andrew J. Dorner

Genetics Institute
87 Cambridge Pk Dr.
Cambridge, MA 02140

INTRODUCTION

Gene expression may be controlled at the level of transcription, processing of precursor mRNAs, mRNA transport to the cytoplasm, mRNA stability and translational efficiency, and protein stability. Proteins which transit the secretory apparatus are subject to a variety of additional steps which may regulate the appropriate secretion of the mature polypeptide. Experience from studying the expression of a wide variety of proteins which transit the secretory apparatus demonstrates that the rate limiting step in secretion is transport from the endoplasmic reticulum (ER) to the Golgi apparatus[1]. It is now known that expression of heterologous secretory proteins from transfected genes introduced into heterologous cells may also be limited due to inefficient transport from the ER to Golgi. In addition, a wide variety of post-translational steps which are required for appropriate maturation and biological activity of a protein may be saturated as the expression of the specific protein is increased. Our work has focused on the limiting steps for the secretion of these heterologous proteins expressed at high level in Chinese hamster ovary (CHO) cells. In general, the factors limiting expression of these proteins are at the post-translational events which occur within the secretory pathway.

The highly compartmentalized structure of eukaryotic cells requires a mechanism for protein localization to ensure that proteins are directed to the appropriate organelle such as the ER, mitochondria, peroxisomes, lysosomes, etc. In many cases this mechanism may be saturated at high expression levels. The precursor forms of proteins transported through the ER generally contain a hydrophobic signal sequence at or near the amino-terminus. Translation of the mRNAs of secretory proteins is arrested at the early stage when the signal recognition particle (SRP) binds to this signal sequence on the nascent polypeptide chain. This arrested complex interacts with an SRP receptor (docking protein) on the rough ER and protein translation resumes[2]. The protein is cotranslationally translocated into the lumen of the ER via the docking protein interaction. The signal sequence is usually cleaved by the signal peptidase during translocation[3]. An important unanswered question is whether the translocation apparatus becomes saturated as the expression level of a secretory protein is increased. As the protein enters the ER, asparagine (N)-linked glycosylation occurs at appropriate recognition sites (Asn-X-Ser/Thr). The utilization of a particular N- linked site is determined by the structure of the growing polypeptide backbone and thus proteins expressed in heterologous cells most frequently exhibit occupancy of N-linked sites very similar to that of the native polypeptide [4]. As the protein is translocated into the lumen of the ER, it undergoes a complex series of reactions that result in attaining its final appropriate conformation (Figure 1). These steps include: protein folding[5]; disulphide bond formation[6]; oligomerization of

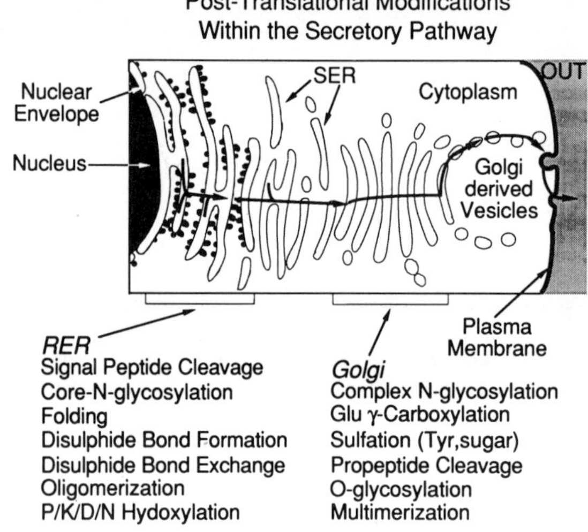

Fig. 1. Post-translational Modifications Within the
Secretory Pathway

multisubunit proteins[5]; hydroxylation of specific amino acids such as proline, aspartic acid, asparagine, and lysine[7]; trimming of glucose and mannose residues from the core N-linked oligosccharides[4]; and fatty acid acylation[8]. The culmination of these steps results in the vesicle mediated transport of the polypeptide from the ER to the cis-Golgi compartment[9,10]. As the protein transits through the Golgi complex, further modifications occur which include addition of N-acetylglucosamine, galactose, and sialic acid to N-linked carbohydrate[4], serine and threonine O-linked glycosylation, sulfation of tyrosine[11] and carbohydrate residues, phosphorylation, cleavage of propeptides[12], and perhaps gamma-carboxylation of vitamin K dependent proteins[13]. When the protein reaches the trans-Golgi network there exist three different pathways which can be taken[14]: 1) membrane-bound and secreted proteins are constitutively transported to the cell surface, 2) proteins may enter the regulated secretory pathway and be stored in vesicles for release upon stimulation; or 3) proteins may be targeted to lysosomal structures. All cells contain a constitutive secretory pathway. Only some cells have the capacity to store proteins in specialized granules and release them upon stimulation. This regulated pathway is required to mediate correct processing of proinsulin to insulin and proopiomelanocortin to ACTH. Thus, certain proteins require this specialized secretory pathway to mediate appropriate maturation of the precursor polypeptides.

FACTORS LIMITING TRANSPORT THROUGH THE ER

Different proteins exhibit dramatically different rates of transit through the ER. It is not known what structural features within a protein are responsible for the individual rates of transit. However, most results suggest that the rate by which a protein attains its appropriate conformation is a crucial limiting factor in the rate of exit from the ER [15,16]. Experiments with mutant proteins demonstrate that altered protein structures may result in improper folding. The improperly folded protein may exhibit altered solubility properties resulting in its aggregation and retention within the ER. Recently there has been much attention on the role that proteins within the ER may play to facilitate protein folding. The most extensively studied protein in this class is the glucose regulated protein of 78 kDa (GRP78), or also known as immunoglobulin binding protein (BiP). BiP is a lumenal protein of the ER homologous to hsp70[17]. BiP is a member of the glucose regulated protein family whose expression is induced by glucose starvation, treatment with inhibitors of N-linked glycosylation, or the presence of misfolded proteins within the ER[18-20]. BiP has been shown to bind ATP and has an ATPase activity[21].

The Role of BiP in ER to Golgi Transport

The potential importance of BiP in secretion is assumed since improperly glycosylated, misfolded, or incompletely folded or assembled proteins are detected in association with BiP[22-25]. Stable association correlates with retention in the ER. This stable association may serve to target secretion incompetent proteins for degradation. Other proteins may be detected in a transient association with BiP[24,26-28]. In this case, interaction with BiP may be part of the normal pathway leading to secretion. For example, vesicular stomatitis virus glycoprotein (VSV G) exhibits a transient association with BiP[26]. However, it is only the forms of VSV G which are intermediates in the disulphide bonding process that are detected in association with BiP. After appropriate disulphide bonds form, the BiP associated VSV G is released and then transits to the Golgi compartment.

We have studied the role of BiP association in secretion in a model system utilizing CHO cells expressing high levels of proteins which display different degrees of association with BiP[24]. Wildtype tissue plasminogen activator (tPA) expressed at high levels in CHO cells is only detected in a slight transient association with BiP. In contrast, expression of unglycosylated tPA, either by mutagenesis of asparagines at the 3 consensus N-linked sites or by treatment of the cells with tunicamycin, results in a greater proportion of unglycosylated tPA in both stable and transient association with BiP[24,27].

One approach to study the role of BiP in the secretion of these proteins was to ask what affect an increase or a decrease on the cellular level of BiP would have on secretion. To increase the level of BiP, cells were treated overnight with tunicamycin to elicit a 10-fold increase in the steady state levels of BiP[29]. In these cells the synthesis and secretion of unglycosylated tPA was studied. The results demonstrated that an increase in the steady state level of BiP extended the association of the unglycosylated tPA with BiP and severely inhibited secretion. To elicit a reduction of BiP within the cell, an antisense RNA strategy was employed. This approach yielded a 3-4-fold reduction in the steady state levels of BiP mRNA in an unglycosylated tPA expressing cell line. In these cells secretion of unglycosylated tPA was improved several fold. The secreted tPA exhibited a specific acitivity similar to the protein secreted in the absence of antisense BiP mRNA. Thus, it did not appear that unfolded tPA was being secreted by reducing BiP level. Most dramatically, in the cells with reduced BiP, the newly synthesized tPA exhibited a lesser and transient association with BiP. These results demonstrate that reduction of BiP level can improve secretion of an associated protein and implicate BiP as playing a role to prevent secretion of associated proteins.

The effect of expression level of the BiP associated protein was studied by evaluating secretion and BiP association as the expression level of wildtype and mutant tPA was increased[24]. Cells were obtained that expressed increasing amounts of the wildtype or mutant tPA through coamplification with a dihydrofolate reductase gene by selection for sequential increases in resistance to methotrexate. These results demonstrated that as expression of unglycosylated tPA increased a greater percentage of the newly synthesized tPA was retained in a complex with BiP within the cell. Thus, increased expression of a secretory protein resulted in a reduced efficiency of secretion and increased association with BiP.

To address whether it is possible to increase expression of a BiP associated protein to a point that newly synthesized protein will bypass association with BiP, we used an inducible vector system[30]. Factor VIII and vWF were co-expressed in the same cell using a vector which contains a promoter responsive to transcriptional activation by addition of sodium butyrate to the conditioned medium. In these cells, factor VIII exhibits both transient and stable association with BiP, whereas vWF is only detected in a slight transient association with BiP[24]. To further increase the factor VIII and vWF mRNA levels, sodium butyrate was added to the conditioned medium for 24 hr. At this time the factor VIII mRNA was induced 10-fold and the vWF mRNA was induced 4-fold. Analysis of the primary translation products demonstrated that factor VIII and vWF protein synthesis were elevated proportionally to the increase in mRNA. Although increased levels of vWF synthesis correlated with increased levels of vWF secreted into the conditoned medium, the majority of newly synthesized factor VIII was retained within the cell and was not secreted. Most dramatically, sodium butyrate induction of transcription from the factor VIII gene in CHO cells elicited an 80-fold induction of BiP mRNA. In contrast, the induction of vWF alone did not dramatically induce BiP mRNA. These results suggest that an increase in the expression of a BiP associated protein can not circumvent BiP association. In addition, within the same cell, significant association of factor VIII with BiP does not significantly affect association of vWF with BiP. This suggests that factor VIII and vWF transit independently through the secretory pathway and that a block to one protein does not interfere with others.

Since induction of factor VIII synthesis resulted in a dramatic BiP induction it appeared that factor VIII synthesis may result in a stress response within the ER. Therefore, we examined whether any obvious structural alterations within the cell occurred by transmission electron microscopy. Dramatically, CHO cells induced to synthesize factor VIII exhibited a dilated ER juxtaposed to many mitochondria (Fig. 2B) compared to control cells not treated with sodium butyrate (Fig. 2A). The presence of mitochondria in close proximity to the ER suggested that Factor VIII synthesis and secretion may require high levels of ATP.

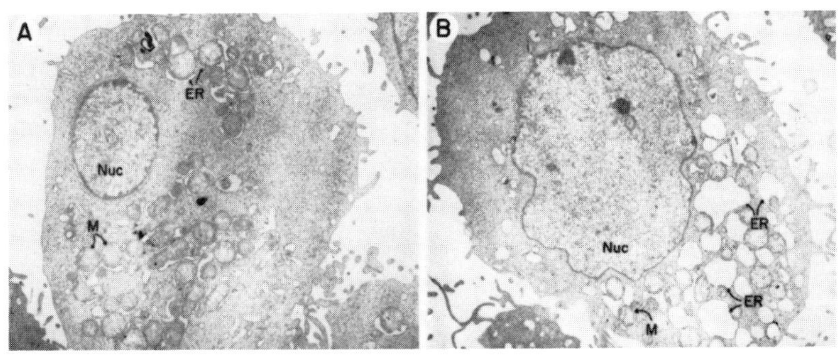

Fig. 2. Electron Microscopy of Chinese Hamster Ovary Cells
Induced to Synthesize Factor VIII.
    CHO cells in the absence (A) and presence (B) of induced
factor VIII secretion and were prepared for electron
microscopy.[30].

Is ATP Limiting for ER to Golgi Transport of Some Proteins?

The importance for ATP in dissociation of protein-BiP complexes has been shown _in vitro_[31]. Complexes of BiP with heavy chain immunoglobulin[17] or factor VIII (our unpub. obs.) exhibit an ATP dependent release from BiP upon incubation _in vitro_. The importance of ATP availability for the secretion of GRP78 associated proteins _in vivo_ was studied by examining the effect of depleting intracellular ATP levels through treatment of CHO cells with carbonylcyanide m-chlorophenylhydrazone (CCCP), an uncoupler of oxidative phosphorylation which reversibly inhibits transport from the ER[32-34]. Factor VIII and vWF coexpressing cells were pulse labeled with ^{35}S-methionine and then chased in the presence of various concentrations of CCCP[35]. Factor VIII and vWF secretion was monitored through analysis of cell extracts and conditioned medium samples by immunoprecipitation with anti-factor VIII and anti-vWF

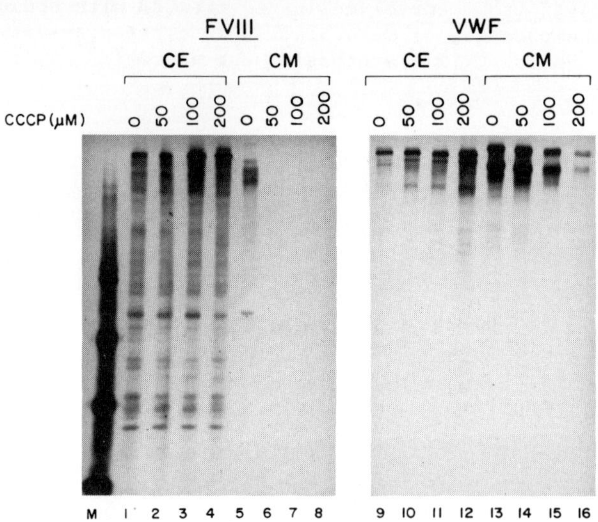

Fig. 3. Effect of Carbonylcyanide m-chlorophenylhydrazone (CCCP)
Concentration on Secretion of Factor VIII and vWF.
Factor VIII and vWF coexpressing cells were labeled with a 15
min pulse of ^{35}S-methionine followed by a 9 hr chase with
unlabeled complete medium containing the indicated
concentrations of CCCP. Autoradiogram of a 6% SDS-
polyacrylamide gel is shown. Lanes 1-4 represent cell
extracts (CE) and lanes 5-8 represent conditioned medium (CM)
of samples immunoprecipitated with anti-factor VIII
monoclonal antibody. Lanes 9-12 represent cell extracts
(CE) and lanes 13-16 represent conditioned medium (CM) of
samples immunoprecipitated with anti-vWF antiserum. M,
molecular weight markers, in kilodaltons, 200, 100, 92, 69,
46, and 30. Position of the 69 kDa marker is indicated by an
arrow. For details see [35].

specific antibodies and SDS-polyacrylamide gel electrophoresis. Factor VIII was detected in the cell extracts as a species migrating at 300 kDa. In addition, BiP was coprecipitated with factor VIII and detected as a species migrating at 78 kDa (Fig. 3, lane 1-4). Immunoprecipitation of the conditioned medium in the absence of CCCP revealed a 200 kDa heterogeneous smear corresponding to the heavy chain of factor VIII and an 80 kDa doublet corresponding to the light chain

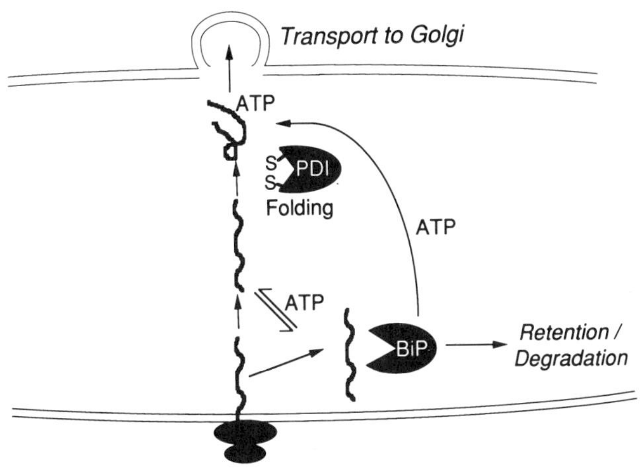

Fig. 4. ATP Requiring Steps for Transit of Secretory Proteins from
the Endoplasmic Reticulum.
Depicted are the processes of protein translocation into the
lumen of the ER, protein folding, and exit from the ER. ATP
dependent release from BiP is required for transport of
proteins which exhibit significant association with BiP. It
is not known if protein folding also requires ATP or is
facilitated by BiP. Vessicle mediated transport from the ER
is a second ATP requiring step. Proteins found in
significant association with BiP may be destined for
degradation.

(lane 5). Factor VIII was not detected in the conditioned medium in the presence of CCCP (10 uM and greater). Analysis of vWF in the same cell extracts detected the 330 kDa precursor polypeptide (Lanes 9-12). Coprecipitation of BiP was not detected. Analysis of vWF secreted into the conditioned medium showed that low concentrations (100 uM or less) of CCCP did not inhibit vWF secretion (Lanes 13-15). In the presence of 200 uM CCCP secretion of vWF was markedly reduced (lane 16). Thus, secretion of factor VIII and vWF synthesized in the same cells display different sensitivities to inhibition by CCCP. Low concentrations of CCCP inhibited secretion of factor VIII but not vWF coexpressed in the same cells. The finding that secretion of factor VIII but not vWF was inhibited at low CCCP concentration precludes a general disruption of secretion. Further studies demonstrated that this inhibition correlated with a block to dissociation of the BiP/factor VIII complex. Much higher concentrations of CCCP were required to block secretion of vWF in these cells. The finding that secretion of factor VIII but not vWF was inhibited at low CCCP concentration precludes a general disruption of secretion. In a similar manner, secretion of unglycosylated tPA was also inhibited by low levels of CCCP compared to wildtype tPA[35]. These results indicated that secretion of BiP associated proteins may be more sensitive to perturbation of intracellular ATP levels by CCCP treatment than unassociated proteins. In vivo dissociation from BiP may be a primary ATP dependent step for transport from the ER. These findings suggest that increased intracellular ATP levels may facilitate secretion of GRP78 associated proteins.

Two ATP requiring steps have been identified for transport through and out of the ER[36,37]: One step requires high levels of ATP for release from BiP and the other is likely at the stage of transitional vesicle formation upon exit from the ER (Fig. 4). Upon translocation into the lumen of the ER, proteins may interact with BiP. As the level of BiP increases or as the lumenal concentration of a secretory protein increases, more protein is observed in association with BiP. The influence of the intralumenal concentration of a secretory protein on promoting BiP association may result from the aggregation state of the protein transiting the ER. Proteins may aggregate dependent on their solubility properties and intralumen concentration. These aggregates may more avidly bind to BiP. Conversely, increased levels of BiP may promote association and aggregation of proteins transiting the ER which are incompletely folded or assembled. BiP associated protein requires ATP hydrolysis for dissociation. If ATP is limiting, proteins associated with BiP are trapped and blocked in their transit through the ER. Proteins associated stably with BiP are likely destined for degradation. Analysis of the kinetics of appropriate disulphide bond formation and BiP release of VSV G protein suggest that folding and disulphide bond formation must occur prior to release from BiP[23]. However, it is not known if ATP hydrolysis mediated by BiP actually assists in protein folding. Another lumenal protein, protein disulphide isomerase (PDI) has been shown to assist the formation of appropriate disulphide bond[38]. It is not known if all proteins exhibit BiP association since it is not possible to detect association of some proteins under the conditions used. Whether some proteins can transit the ER independent of BiP association is still an unanswered question.

LIMITING POST-TRANSLATIONAL MODIFICATIONS

Most proteins synthesized in heterologous cells frequently exhibit the same modifications that occur on the native protein. However, in a number of situations the post-translational modification may be inefficient. For example, the coagulation factor IX is a serine protease that exhibits gamma-carboxylation of 12 glutamic acid residues within the amino terminal region of the protein. These carboxyl groups are transferred by a vitamin K -dependent reaction and their presence is required for factor IX activity[13]. Expression of factor IX in CHO cells has resulted in high levels of factor IX antigen, but negligible factor IX activity due to saturation of the carboxylase activity[39]. Another post-translational modification that may be limiting for some proteins is tyrosine sulfation. Factor VIII contains at least 4 tyrosine residues which undergo post-translational addition of sulfate[40]. Factor VIII expressed in CHO cells does contain tyrosine sulfate, but evidence suggests that inefficient tyrosine sulfate addition may occur[41].

More significant is the observation that as expression of propolypeptides increases, saturation of the propeptide cleaving enzyme may occur[42-46]. Proteolytic processing of proproteins is an event that occurs late in both the regulated and constitutive secretory pathways of many different cell types. Expression of heterologous proteins in foreign cells has shown that the propeptide is usually cleaved correctly. However, cleavage is often incomplete. The inefficiency of cleavage may represent saturation of the endoproteolytic cleaving activity of the cell due to the large amounts of protein expressed. Alternativley, a portion of the expressed proprotein may bypass the compartment containing the endogenous protease, or its rate of transit through that compartment may be too rapid for complete cleavage to occur. As one approach to circumvent this limitation to high level expression, we have overexpressed a propeptide processing enzyme in order to improve processing of proproteins expressed in heterologous cells.

Very little is known about the enzyme(s) responsible for propeptide processing in mammalian cells. The best characterized proprotein processing enzyme is the Kex2 endoprotease which is responsible for appropriate processing of prepro-killer toxin and prepro-α-factor in Saccharomyces cerevisiae[47,48]. Kex2 cleaves after paired basic amino acid residues lys-arg and arg-arg[49,50]. It can accurately process proinsulin and proalbumin expressed in yeast[51,52] and can process proopiomelanocortin (POMC) upon transfection into mammalian cells[53]. Based on sequence homology to Kex2, a cDNA encoding a human homologue of Kex2 was isolated from a cDNA library prepared from the Human HepG2 cell line[54]. The deduced amino acid sequence demonstrated the protein to contain the conserved residues for the subtilysin type of serine protease. In addition, the protein contained a transmembrane domain, likely responsible for its presumed localization in the trans-Golgi. This protease has been given the acrynym PACE (Paired basic Amino acid Cleaving Enzyme).

To test the ability for PACE to improve processing of a heterolgous protein expressed in mammalian cells, the effect of PACE expression on processing of von Willebrand factor (vWF) was monitored. vWF is a multimeric plasma protein which is synthesized as a large precursor polypeptide that forms carboxy-terminal-linked disulphide bonded dimers after translocation into the lumen of the ER[55]. Upon transport to the Golgi and post-Golgi compartments, the protein is processed by formation of complex N-linked oligosaccharides, addition of O-linked oligosaccharides, sulfation, assembly of disulfide bonded miltimers, and propeptide cleavage. Transfection of a vWF cDNA expression vector into COS-1 monkey kidney cells directs the synthesis of prepro-vWF[56]. Although COS-1 cells can accurately cleave the vWF propeptide, the process is inefficient so that approximately 50% of the secreted protein is uncleaved pro-vWF. Thus, expression of PACE may result in more efficient processing of pro-vWF.

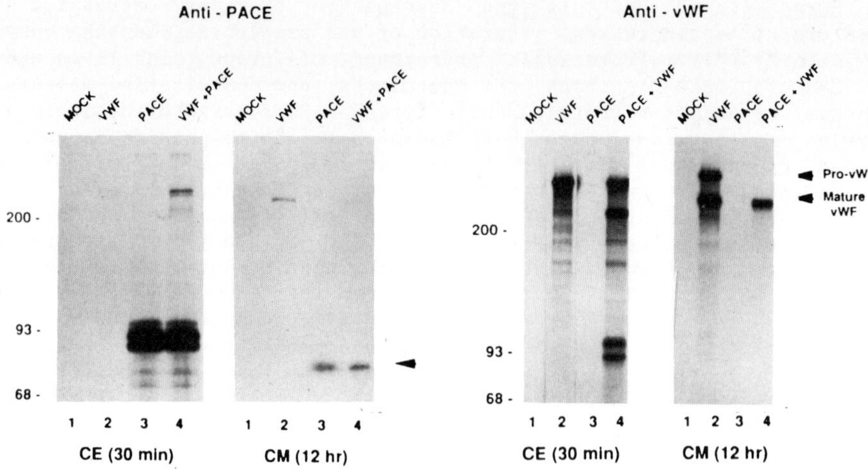

Fig. 5. Co-expression of PACE and pro-vWF in COS-1 Monkey Cells. SDS-polyacrylamide gel of radiolabeled proteins products immunoprecipitated with excess anti-PACE (left) or anti-vWF (right) antibody. Each lane represents an equivalent portion of total labeled cell extract (CE) or conditioned medium (CM). COS-1 cells were transfected with no DNA (lane 1), the vWF expression plasmid alone (lane 2), the PACE expression plasmid alone (lane 3), or the vWF and PACE expression plasmids together (lane 4). Cell extracts were prepared after a 30 min pulse with ^{35}S-methionine and conditioned medium samples were prepared after a 12 hr chase in the presence of cold methionine. Positions of the molecular weight markers are indicated on the left. The arrow on the right shows tha band of immunoreactive PACE protein in the conditioned medium. The migration of the pro-vWF and the mature vWF subunit from the conditoned medium is indicated on the right. The arrow in the center indicates the intracellular cleaved propeptide that appears as a doublet due to glycosylation. For details see [54].

A vWF expression plasmid was co-transfected in the presence and absence of a PACE expression vector into COS-1 monkey cells. Expression of PACE and of vWF was monitored by pulse labelling cells with ^{35}S-methionine and analysis of the cell extracts and conditioned medium by immunoprecipitation with anti-vWF and anti-PACE antibodies and SDS-polyacrylamide gel electrophoresis. In cells transfected with PACE, a complex doublet is observed migrating at 90 kDa within the cell extracts labeled for 30 min (Fig. 5A, CE: lanes 3,4). In the conditioned medium after a 12 hr chase, immunoprecipitable protein is detected migrating at 75 kDa (Fig. 5A, CM: lanes 3,4). This species may represent a truncated form of PACE generated by autoproteolysis at a lys-arg dibasic residue at amino acid 497-498. In the absence of PACE, no cleavage of pro-vWF is detected within the cell (Fig. 5B, CE: lane 2). Since this processing event occurs very late in the secretory pathway, very little processed vWF accumulates within the cell. However, in the presence of PACE a significant amount of propeptide cleavage is detected intracellularly as monitored by the appearance of the mature vWF as well as the presence of the propeptide migrating at 90 kDa (Fig. 5B, CE: lane 4). Although the anti-vWF antibody recognizes the mature vWF but not the propeptide, the propeptide is coimmunoprecipitated since it is disulphide bonded to the mature subunit within the cell. This disulphide bonded form of the propeptide and the mature vWF is an intermediate in the disulphide bond multimerizatrion process and is not present in the secreted vWF[56]. The identity of the vWF propeptide was confirmed by immunoprecipitation using antibody directed against the propeptide of vWF[54].

In the conditioned medium of cells transfected with vWF alone, both pro-vWF and mature vWF are detected (Fig. 5B, CM: lane 2). However, in the presence of PACE co-transfection, all the vWF is processed to the mature form (Fig. 5B, CM: lane 4). Further analysis by amino-terminal sequence analysis was consistent with cleavage occurring at the appropriate residue within the propeptide. In addition, a vWF mutant harboring a mutation at the cleavage site of the dibasic pair lys-arg to asp-glu was not cleaved either in the presence of absence of PACE expression[54]. A vWF mutant harboring a lys-lys pair in place of lys-arg was not cleaved in the absence of PACE, but was cleaved, albeit less efficiently than lys-arg, in the presence of PACE co-transfection. Thus, PACE cleavage of vWF appears to require a paired basic amino acid residue. Although further studies are required to define the substrate specificity for PACE, there is a recognition preference for lys-arg and/or arg-arg, compared to lys-lys, for cleavage mediated by Kex2[47,48].

These experiments demonstrate that expression of PACE in mammalian cells can enhance the processing of the propeptide of human von Willebrand factor. Thus, the expression of PACE in mammalian cells should allow the production of more completely processed protein products. In a similar manner, the identification and overexpression of other potentially rate limiting enzymes involved in post-translational modifications offers a solution to the inefficiency observed for these processes.

REFERENCES

1. Lodish, H.F., Kong, N., Snider, M., Strous, G.J.A.M. (1983) Hepatoma secretory proteins migrate from rough endoplasmic reticulum to Golgi at characteristic rates. Nature <u>304</u>: 80-83.

2. Bernstein, H.D., Poritz, M.A., Strub, K., Hoben, P.J., Brenner, S., and Walter, P. (1989) Model for signal sequence recognition from amino-acid sequence of 54 K subunit of signal recognition particle. Nature 340: 482-486.

3. Blobel, G. and Dobberstein, B (1975) Transfer of protein across membranes. I. Presence of proteolytically processed and unprocessed hascent immunoglobulin light chains on mambrane-bound ribosomes of murine myeloma. J. Cell Biol. 67: 835-851.

4. Kornfeld, R., and Kornfeld, S. (1985) Assembly of asparagine-linked oligosaccharides. Annu. Rev. Biochem. 54: 631-669

5. Hurtley, S.M., and Helenius, A. (1989) Protein oligomerization in the endoplasmic reticulum. Ann. Rev. Biochem. 57: 277-307.

6. Freedman, R.B., Bulleid, N.J., Hawkins, H.C., and Paver, J.L. (1989) Role of protein disulphide-isomerase in the expression of native proteins. Biochem. Soc. Symp. 55: 167-192.

7. Friedman, P.A., and Przysiecki, C.T. (1987) Vitamin K dependent carboxylation. Int. J. Biochem. 19: 1-7.

8. Towler, D.A., Gordon, J.I., Adams, S.P., Glaser, L. (1988) The biology and enzymology of eukaryotic protein acylation. Ann. Rev. Biochem. 57: 69-99.

9. Clary, D.O., Griff, I.C., and Rothman, J.E. (1990) SNAPs, a family of NSF attachment proteins involved in intracellular membrane fusion in animals and yeast. Cell 61: 709-721.

10. Kaiser, C.A., and Schekman, R. (1990) Distinct sets of SEC genes govern transport vesicle formation and fusion early in the secretory pathway. Cell 61: 723-733.

11. Huttner, W.B., Baeuerle, P.A. (1988) Protein sulfation on tyrosine. In: Modern Cell Biology. Birgit H Satiri (ed). Alan R. Liss, Inc N.Y., N.Y. 6: 97-140.

12. Thomas, G., Thorne, B.A., Hruby, D.E. (1988) Gene transfer techniques to study neuropeptide processing. Annu. Rev. Physiol. 50: 323-332.

13. Furie, B., and Furie, B.C. (1990) Molecular basis of vitamin K-dependent g-carboxylation. Blood 75: 1753-1762.

14. Burgess, T.L., and Kelley, R.B. (1987) Constitutive and regulated secretion of proteins. Ann. Rev. Cell Biol. 106: 629-639.

15. Rothman, J.E. (1987) Protein sorting by selective retention in the endoplasmic retriculum and Golgi stack. Cell 50: 1-23.

16. Pelham, H.R.B. (1989) Control of protein exit from the endoplasmic reticulum. Ann. Rev. Cell. Biol. 5: 1-23.

17. Munro, S., and Pelham, H.R.B. (1986) An hsp70-like protein in the ER: identity with the 78 kd glucose-regulated protein and immunoglobulin heavy chain binding protein. Cell 46: 291-300.

18. Chang, S.C., Wooden, S.K., Nakaki, T., Kim, T.K., Lin, A.Y., Kung, L., Attenello, J.W., and Lee, A.S. (1987) Rat gene encoding the 78-kDa glucose-regulated protein GRP78: Its regulatory sequences and the effect of protein glycosylation on its expression. Proc. Natl. Acad. Sci. 84: 680-684.

19. Kozutsumi, Y., Segal, M., Normington, K., Gething, M-J., and Sambrook, J. (1988) The presence of malfolded proteins in the endoplasmic reticulum signals the induction of glucose-regulated proteins. Nature 332: 462-464.

20. Watowich, S.S., and Morimoto, R.I. (1988) Complex regulation of heat shock- and glucose-responsive genes in human cells. Mol. Cell. Biol. 8: 393-405.

21. Kassenbrock, C.K., and Kelly, R.B. (1989) Interaction of heavy chain binding protein (BiP/GRP78) with adenine nucleotides. EMBO J. 8: 1461-1467.

22. Bole, D.G., Hendershot, L.M., and Kearney, J.F. (1986) Posttranslational association of immunoglobulin heavy chain binding protein with nascent heavy chains in nonsecreting and secreting hybridomas. J. Cell Biol. 102: 1558-1566.

23. Hurtley, S.M., Bole, D.G., Hoover-Litty, H., Helenius, A., and Copeland, C.S. (1989) Interactions of misfolded influenza virus hemagglutinin with binding protein (BiP). J. Cell. Biol. 108: 2117-2126.

24. Dorner, A.J., Bole. D.G., and Kaufman, R.J. (1987) The relationship of N-linked glycosylation and heavy chain binding protein association with the secretion of glycoproteins. J. Cell Biol. 105: 2665-2674.

25. Singh, I., Doms, R.W., Wagner, K.R., and Helenius, A. (1990) Intracellular transport of soluble and membrane-bound glycoproteins: Folding, assembly, and secretion of anchor-free influenza hemagglutinin. EMBO J. 9: 631-639.

26. Machamer, C.E., Doms, R.W., Bole, D.G., Helenius, A., and Rose, J.K. (1990) Heavy-chain binding-protein recognizes incompletely disulphide-bonded forms of vesicular stomatitis virus G protein. J. Biol. Chem. 265: 6879-6883.

27. Kaufman, R.J., Wasley, L.C., Davies, M.V., Wise, R.J., Israel, D.I., and Dorner, A.J. (1989) Effect of von Willebrand factor coexpression on the synthesis and secretion of factor VIII in Chinese hamster ovary cells. Mol. Cell. Biol. 9: 1233-1242.

28. Ng, D.T.W., Randall, R.E., and Lamb, R.A. (1989) Intracellular maturation and transport of the SV5 type II glycoprotein hemagglutinin-neuraminidase: Specific and transient association with GRP78-BiP in the endoplasmic reticulum and extensive internalization from the cell surface. J. Cell Biol. 109: 3273-3289.

29. Dorner, A.J., Krane, M.A.G., and Kaufman, R.J. (1988) Reduction of endogenous GRP78 levels improves secretion of a heterologous protein in CHO cells. Mol. Cell. Biol. 8: 4063-4070.

30. Dorner, A.J., Wasley, L.C., and Kaufman, R.J. (1989) Increased synthesis of secreted proteins induces expression of glucose-regulated proteins in butyrate-treated Chinese hamster ovary cells. J. Biol. Chem. 264: 20602-20607.

31. Flynn, G.C., Chappell, T.G., and Rothman, J.E. (1989) Peptide binding and release by proteins implicated as catalysts of protein assembly. Science 245: 385-390.

32. Jamieson, J.D., and Palade, G.E. (1968) Intracellular transport of secretory proteins in the pancreatic exocrine cell. IV. Metabolic requirements. J. Cell Biol. 39: 589-603.

33. Heytler, P.G. (1963) Uncoupling of oxidative phosphorylation by carbonylcyanamide phenyhydrazones. I. Some characteristics of m-Cl-CCP action on mitochondria and chloroplasts. Biochemistry 2: 357-361.

34. Argon, Y., Burkhardt, J.K., Leeds, J.M., and Milstein, C. (1989) Two steps in the intracellular transport of IgD are sensitive to energy depletion. J. Immunol. 142: 554-561.

35. Dorner, A.J., Wasley, L.C., and Kaufman, R.J. (1990) Protein dissociation from GRP78 and secretion is blocked by depletion of cellular ATP levels. Proc. Natl. Acad. Sci. (in Press).

36. Balch, W.E., Elliott, M.M., and Keller, D.S. (1986) ATP-coupled transport of vesicular stomatitis virus G protein between the endoplasmic reticulum and the Golgi. J. Biol. Chem. 261: 14681-14689.

37. Balch, W.E., and Keller, D.S.   (1986) ATP-coupled transport of vesicular stomatitis virus G protein: Functional boundaries of secretory compartments. J. Biol. Chem. <u>261</u>: 14690-14696.

38. Bulleid, N.J., and Freedman, R.B. (1988) Defective co-translational formation of disulphide bonds in protein disulphide-isomerase-deficient microsomes. Nature <u>335</u>: 649-651.

39. Kaufman, R.J., Wasley, L.C., Furie, B.C., Furie, B., and Shoemaker, C.B. (1986) Expression, purification, and characterization of recombinant γ-carboxylated factor IX synthesized in Chinese hamster ovary cells. J. Biol. Chem. <u>261</u>: 9622-9628.

40. Pittman, D.D., and Kaufman, R.J. (1989) Structure-function relationships of factor VIII elucidated through recombinant DNA technology. Thrombosis and Haemostasis <u>61</u>: 161-165.

41. Mikkelsen, J., Thomsen, J., Kongerslev, L., Christensen, and Ezban, M. (1989) Heterogeneity in the tyrosine sulfation of Chinese hamster ovary cell produced recombinant FVIII. Thrombosis and Haemostasis <u>62</u>: 197a.

42. Warren, T.G., and Shields, D. (1984) Expression of preprosomatostatin in heterologous cells: Biosynthesis, posttranslational processing, and secretion of mature somatostatin. Cell <u>39</u>: 547-555.

43. Hellerman, J.G., Cone, R.C., Potts, J.T., Rich, A., Mulligan, R.C., and Kronenberg, H.M. (1984) Secretion of human parathyroid hormone from rat pituitary cells infected with a recombinant retrovirus encoding preproparathyroid hormone. Proc. Natl. Acad. Sci. <u>81</u>: 5340-5344.

44. Gentry, L.E., Webb, N.R., Lim, G.J., Brunner, AmM., Ranchalis, J.E., Twardzik, D.R., Lioubin, M.N., Marquardt, H., and Purchio, A.F. (1987) Type 1 transforming growth factor β: Amplified expression and secretion of mature and precursor polypeptides in Chinese hamster ovary cells. Mol. Cell. Biol. <u>7</u>: 3418-3427.

45. Foster, D.C., Sprecher, C.A., Holly, R.D., Gambee, J.E., Walker, K.M., and Kumar, A.A. (1990) Endoproteolytic processing of the dibasic cleavage site in the human protein C precursor in transfected mammalian cells: Effects of sequence alterations on efficiency of cleavage. Biochemistry <u>29</u>: 347-354.

46. Derian, C.K., VanDusen, W., Przysiecki, C.T., Walsh, P.N., Berkner, K.L., Kaufman, R.J., and Friedman, P.A. (1989) Inhibitors of 2-ketoglutarate-dependent dioxygenases block aspartyl β-hydroxylation of recombinant human factor IX in several mammalian expression systems. J. Biol. Chem. <u>264</u>: 6615-6618.

47. Mizuno, K., Nakamura, T., Oshima, T., Tanaka, S., and Matsuo, H. (1989) Yeast <u>KEX2</u> gene encodes an endopeptidase homologous to subtilisin-like serine proteases. Biochem, Biochys Res. Commun. <u>156</u>: 246-254.

48. Fuller, R.S., Brake, A.J., and Thorner, J. (1989) Yeast prohormone processing enzyme (KEX2 gene product) is a Ca^{2+}-dependent serine protease. Proc. Natl. Acad. Sci. <u>86</u>: 1434-1438.

49. Julius, D., Brake, A., Blair, L., Kunisawa, R., and Thorner, J. (1984) Isolation of the putative structural gewne for the lysine-arginine-cleaving endopeptidase required for processing of yeast prepro-alpha-factor. Cell <u>37</u>: 1075-1089.

50. Julius, D., Schekman, R., and Thorner, J. (1984) Glycosylation and processing of prepro-alpha-factor through the yeast secretory pathway. Cell <u>36</u>: 309-318.

51. Thim, L., Hansen, M.T., Norris, K., Hoegh, I., Boel, E., Forstrom, J., Ammerer, G., and Fiil, N.P. (1986) Secretion and processing of insulin precursors in yeast. Proc. Natl. Acad. Sci. <u>83</u>: 6766-6770.

52. Sleep, D., Belfield, G.P., and Goodey, A.R. (1990) The secretion of human serum albumin from the yeast Saccharomyces cerevisiae using five different leader sequences. Bio/Technology 8: 42-48.

53. Thomas, G., Thorne, B.A., Thomas, L., Allen, R.G., Hruby, D.E., Fuller, R., and Thorner, J. (1988) Yeast KEX2 endopeptidase correctly cleaves a neuroendocrine prohormone in mammalian cells. Science 241:226-230.

54. Wise, R.J., Barr, P.J., Wong, P.A., Bathurst, I.C., Brake, A.J., and Kaufman, R.J. (1990) Expression of a cDNA encoding a human proprotein processing enzyme: Correct cleavage of the von Willebrand factor precursor at a paired basic amino acid site. Proc. Natl. Acad. Sci. (in press).

55. Handin R.I., and Wagner, D.D. (1989) in Progress in Hemostssis and Thrombosis 9: B.S. Coller, Ed. W.B. Saunders, Philadelphia, p233-259.

56. Wise, R.J., Pittman, D.D., Handin, R,I., Kaufman, R.J., and Orkin, S.H. (1988) The propeptide of von Willebrand factor independently mediates the assembly of von Willebrand multimers. Cell 52: 229-236.

SYNTHESIS OF BIOLOGICALLY ACTIVE VITAMIN K-DEPENDENT COAGULATION FACTORS

Barbara C. Furie and Bruce Furie

Center for Hemostasis and Thrombosis Research
Division of Hematology/Oncology
New England Medical Center
Departments of Medicine and Biochemistry
Tufts University School of Medicine
Boston, MA

INTRODUCTION

Several of the vitamin K-dependent proteins of blood coagulation have elicited interest as potential therapeutic agents for disorders of hemostasis or thrombosis. In particular, there has been interest in producing recombinant factor IX for treatment of hemophilia B, factor VII for treatment of hemophilia A patients with inhibitors and recombinant protein C for treatment of thrombotic disorders. Since these proteins undergo a unique posttranslational modification, the vitamin K-dependent carboxylation of specific glutamic acid residues in their amino termini to γ-carboxyglutamic acid, their expression in recombinant systems has been more difficult than that of proteins that undergo less extensive processing. A great deal of experience has been accrued on optimal expression systems and methods of growth for the vitamin K-dependent proteins. While some of these proteins are now produced on a scale that makes consideration of therapeutic use feasible, obstacles still remain to the practical production of others. For example, it remains difficult to produce fully carboxylated factor IX at high levels of expression in recombinant systems. Some of these obstacles may be surmountable with an increased understanding of the basic biological processes involved in the synthesis of the vitamin K-dependent proteins. Considerable interest has been focused on two fundamental problems: what are the factors that control the efficient γ-carboxylation of a vitamin K-dependent protein and, what effects efficient cleavage of the propeptide of a vitamin K-dependent protein? Our current understanding of the former, the process of γ-carboxylation, is the subject of this review.

The vitamin K-dependent proteins of blood, factor IX, factor X, factor VII, prothrombin, protein C and protein S, are synthesized in the liver. In the liver vitamin K is reduced to vitamin K hydroquinone by vitamin K reductases. The reduced

Recombinant Technology in Hemostasis and Thrombosis
Edited by L.W. Hoyer and W.N. Drohan, Plenum Press, New York, 1991

form of the vitamin is a cofactor for the vitamin K-dependent carboxylase in the conversion of glutamic acid to γ-carboxyglutamic acid. In addition to the enzyme and the reduced form of vitamin K, the reaction requires molecular oxygen, carbon dioxide and the precursor form of a vitamin K-dependent protein as the substrate. During the course of the reaction γ-carboxyglutamic acid residues are formed in the protein and the vitamin K is converted to vitamin K epoxide. This latter product is recycled to the reduced form of the vitamin to participate in another catalytic event. The question that until recently remained unresolved was the nature of the protein substrate.

SUBSTRATES OF THE VITAMIN K-DEPENDENT CARBOXYLASE

The vitamin K-dependent plasma proteins share a great deal of sequence homology within their γ-carboxyglutamic acid containing regions or Gla-domains. While the uncarboxylated forms of the mature vitamin K-dependent proteins are poor substrates for the vitamin K-dependent carboxylase[1], a slightly larger precursor form of these proteins can be carboxylated[2]. More recently the cDNA sequences for these proteins have been elucidated indicating that they are all synthesized with a short propeptide[3,4,5,6,7,8,9,10,11,12,13]. The propeptides of the plasma vitamin K-dependent proteins are homologous. Furthermore, a homologous propeptide is found on the γ-carboxylated bone Gla protein, a protein that shares no homology with the plasma vitamin K-dependent proteins except for the presence of this propeptide and of γ-carboxyglutamic acid residues[14,15] (Figure 1). Finally, matrix Gla protein, another γ-carboxylated protein of bone, although synthesized without a propeptide, contains a sequence within the mature protein which is homologous to the propeptides of the other γ-carboxyglutamic acid containing proteins[16,17]. The assembly of this data led to the speculation that structures within this conserved sequence were necessary for the recognition of substrates by the vitamin K-dependent carboxylase[18].

Identification of the γ-carboxylation recognition site

To test the hypothesis that the propeptides of the vitamin K-dependent plasma proteins were important for their recognition

	-24	-23	-22	-21	-20	-19	-18	-17	-16	-15	-14	-13	-12	-11	-10	-9	-8	-7	-6	-5	-4	-3	-2	-1	+1
Factor IX							Thr	Val	Phe	Leu	Asp	His	Glu	Asn	Ala	Asn	Lys	Ile	Leu	Asn	Arg	Pro	Lys	Arg	Tyr
Prothrombin	Ser	Leu	Val	His	Ser	Gln	His	Val	Phe	Leu	Ala	Pro	Gln	Gln	Ala	Arg	Ser	Leu	Leu	Gln	Arg	Val	Arg	Arg	Ala
Factor X	Leu	Leu	Leu	Leu	Gly	Glu	Ser	Leu	Phe	Ile	Arg	Arg	Glu	Gln	Ala	Asn	Asn	Ile	Leu	Ala	Arg	Val	Thr	Arg	Ala
Protein C	Thr	Pro	Ala	Pro	Leu	Asp	Ser	Val	Phe	Ser	Ser	Ser	Glu	Arg	Ala	His	Gln	Val	Leu	Arg	Ile	Arg	Lys	Arg	Ala
Factor VII	Trp	Lys	Pro	Gly	Pro	His	Arg	Val	Phe	Val	Thr	Glu	Glu	Glu	Ala	His	Gly	Val	Leu	His	Arg	Arg	Arg	Arg	Ala
Protein S	Val	Leu	Pro	Val	Leu	Glu	Ala	Asn	Phe	Leu	Ser	Arg	Gln	His	Ala	Ser	Gln	Val	Leu	Ile	Arg	Arg	Arg	Arg	Ala
Bone Gla protein	Ser	Gly	Ala	Glu	Ser	Ser	Lys	Ala	Phe	Val	Ser	Lys	Gln	Glu	Gly	Ser	Glu	Val	Val	Lys	Arg	Pro	Arg	Arg	Tyr

Figure 1. The propeptides of some vitamin K-dependent proteins. Residues that demonstrate sequence homology are boxed and shaded; conservative amino acid substitutions are boxed. (from Furie B and Furie BC: Cell 53:505, 1988.)

as substrates for the vitamin K-dependent carboxylase, alterations were made in the factor IX cDNA coding for the propeptide by oligonucleotide directed site specific mutagenesis[19]. The boundary between the signal sequence and the propeptide sequence of factor IX had been established by determining the amino terminal sequence of several mutant factor IX species from patients with hemophilia B whose defects lay in the propeptide cleavage site. These factor IX molecules, factor IX Cambridge (Arg-1 to Ser) and factor IX San Dimas (Arg-4 to Gln) contained 18 residue extensions beyond the amino terminus of mature factor IX[20,21]. The recombinant altered factor IX cDNA species were expressed in Chinese hamster ovary cells and the isolated mutant proteins assayed for γ-carboxyglutamic acid content using conformation-specific antibodies and by direct amino acid analysis for γ-carboxyglutamic acid. Deletion of the entire propeptide (residues -1 to -18) led to the expression of totally uncarboxylated factor IX. Similarly, substitution of alanine for phenylalanine -16 or glutamic acid for alanine -10 almost completely obliterated carboxylation. The factor IX species isolated from cells transfected with cDNA containing mutations in the propeptide had the normal amino terminus of native factor IX, and were identical in molecular weight to the normal protein. These factor IX species also contained the appropriate amount of β-hydroxyaspartic acid indicating that other posttranslational events are unperturbed by the modifications made in the propeptide sequence of factor IX. It thus seems clear that the factor IX propeptide contains a signal important for recognition of the protein as a substrate for the vitamin K-dependent carboxylase. Similar experiments involving deletion of portions of the propeptide of protein C indicated that the propeptide is required for γ-carboxylation of this vitamin K-dependent protein as well[22].

The nature of the recognition site for γ-carboxylation has been further defined by evaluating the effect of mutations in the propeptide of prothrombin[23]. Mutation of histidine -18 to glycine, valine -17 to serine, leucine -15 to either glycine or aspartic acid or of alanine -10 to aspartic acid all resulted in production of prothrombin which was only partially carboxylated. In contrast, mutation of alanine -14 to serine or serine -8 to valine did not inhibit carboxylation. These data suggest that residues -18, -17, -16, -15, and -10 define the carboxylation recognition site (Figure 2). Preliminary data from structural studies of the propeptide suggest that the prothrombin propeptide contains an amphipathic alpha-helical region that is distinct from the γ-carboxylation recognition site.

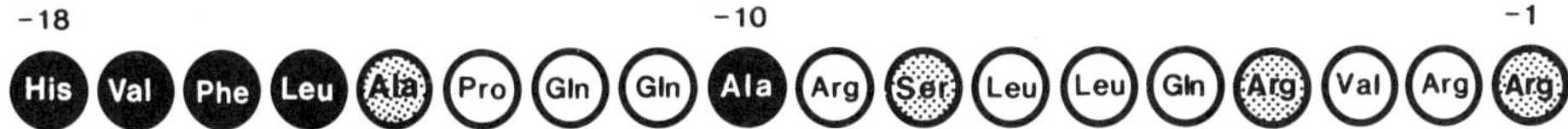

Figure 2. The role of specific amino acids in the propeptide of prothrombin. The amino acids most critical to γ-carboxylation are colored black or gray. Amino acids that do not play a role in carboxylation are shown speckled. The role of amino acids shown in white has not been directly tested. (From Huber P, et al: J. Biol. Chem. 265:12467, 1990).

Peptide substrates of the vitamin K-dependent carboxylase

A great deal of information about the process of γ-carboxylation has been learned from _in vitro_ studies of the vitamin K-dependent carboxylase using small substrates such as the pentapeptide FLEEL (Phe-Leu-Glu-Glu-Leu). However, FLEEL has several disadvantages as a substrate analog. It binds weakly to the enzyme, only the first glutamic acid in the peptide is carboxylated and, only small amounts of substrate are converted to product[24,25]. In order to investigate whether the presence of the propeptide sequence, derived from a plasma vitamin K-dependent protein, could serve to improve small peptides as substrates of the vitamin-K dependent carboxylase, a series of peptides was synthesized incorporating the propeptides of prothrombin or factor IX (Figure 3). A Km of 3.6 μM was determined for the _in vitro_ carboxylation by bovine carboxylase of the peptide ProPT28, which includes the propeptide and the first ten residues of the mature sequence of acarboxyprothrombin. In contrast, FLEEL was carboxylated with a Km of about 2200 μm. A peptide with only part of the

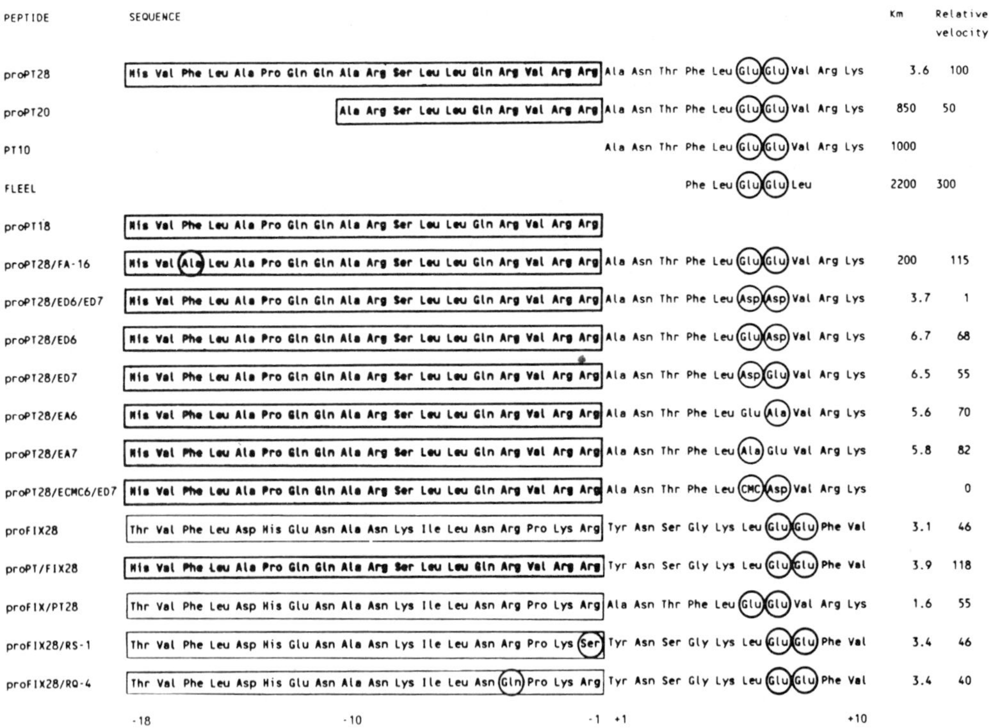

Figure 3. Synthetic peptide substrates for the vitamin K-dependent carboxylase. Kinetic parameters for the various peptides were determined by monitoring the enzyme dependent incorporation of $^{14}CO_2$ into the peptides. The propeptide regions of the substrates are boxed. The Km and Ki values are μM. The velocities of the reactions are given relative to proPT28. (from Furie B and Furie BC: Blood 75:1753-1762, 1990.)

propeptide sequence included, proPT20, residues -10 to +10 of prothrombin and a peptide, PT10, which includes residues +1 to +10 of prothrombin were poor substrates. The carboxylation of proPT28 could be inhibited by proPT18, a peptide which includes residues -18 to -1 of prothrombin, with a Ki of 3.5 μM. This data confirms the importance of an intact carboxylation recognition site and indicates that this site binds tightly to the enzyme[26].

To determine if the _in vitro_ carboxylation of the peptide substrates reflected processes occuring _in vivo_, a peptide substrate equivalent to the phenylalanine -16 to alanine mutation was tested. This peptide, proPT28/FA-16, had a Km of 200 μM, significantly altered from that of proPt28[26]. Further evidence that the _in vitro_ carboxylation system using propeptide containing substrates reflects _in vivo_ events was obtained using the peptide proPT28/ED6/ED7 in which the two glutamic acid residues of proPt28 were replaced by aspartic acids. ProPT28/ED6/ED7 is minimally carboxylated, with a rate of aspartic acid carboxylation about 1% of the rate of glutamic acid carboxylation. The aspartic acid containing peptide is an inhibitor of carboxylation of proPT28 with a Ki of 5 μM. This is consistent with the fact that β-carboxylation of aspartic acid has not been observed in the vitamin K-dependent proteins. Peptides containing Asp6-Glu7 or Glu6-Asp7 were carboxylated at equivalent rates. Neither cysteine nor carboxymethylcysteine was carboxylated when substituted for glutamic acid. This suggests that the vitamin K-dependent carboxylase is fastidious about the distance between the protein backbone and the carboxylatable carbon atom[27].

We have observed that recombinant human prothrombin is more efficiently carboxylated when expressed in Chinese hamster ovary cells then is recombinant human factor IX. In order to determine if this was in some way related to the structures of the propetides of these two proteins the carboxylation of proPT28 and proFIX28 were compared. Both peptides were carboxylated with a Km of about 3 μM. However, the Vmax for carboxylation of proPT28 was about two times greater than that for proFIX28. When peptides were analyzed in which the propeptide of factor IX was substituted for that of prothrombin, proFIX/PT28, and the propeptide of prothrombin was substituted for that of factor IX, proPT/FIX28, it was observed that the prothrombin propeptide, whether linked to the mature amino-terminus of factor IX or prothrombin, participated somewhat more efficiently in carboxylation[27]. More recently we have looked at the carboxylation of a chimeric protein in which the prothrombin propeptide replaced that of factor IX in the factor IX cDNA. At comparable expression levels the presence of the prothrombin propeptide did not appear to significantly enhance the carboxylation of recombinant factor IX _in vivo_. In this instance the results obtained _in vitro_ do not appear to reflect what is observed _in vivo_.

This discrepancy between _in vitro_ and _in vivo_ results has been observed in one other case. Earlier in this review the naturally occuring mutants, factor IX Cambridge (Arg-1 to Ser)[20] and factor IX San Dimas (Arg-4 to Gln)[21], were described as cleavage mutants that were secreted into the circulation with their propeptides still attached. These two hemophiliac factor IXs are also deficient in their γ-carboxyglutamic acid content. To determine whether these proteins have a defect in their carboxylation recognition site or whether the incomplete carboxylation results from steric or conformational factors, we prepared synthetic peptides which mimic these defects and evaluated their ability to be carboxylated. Both proFIX28/RS-1, based upon factor IX Cambridge and proFIX28/RQ-4, based upon factor IX San Dimas, had equivalent Km and Vmax values as those for proFIX28, the peptide based upon the native structure of factor IX[27]. Two naturally occuring mutants of factor IX, factor IX Oxford 3 and factor IX Kawachinagano both with the same mutation as factor IX San Dimas have been reported to be fully carboxylated[28,29]. Finally, expression of a recombinant factor IX in which the Oxford 3 mutation has been created by site specific mutagenesis has been reported to lead to a protein which is only partially carboxylated[30]. Since the details of the posttranslational processing pathway for factor IX are still unknown it is difficult to resolve these discrepancies.

It has been proposed that a concensus sequence identified within the Gla domain of the vitamin K-dependent proteins may play a role in the binding of substrate to the carboxylase[31]. The proposed concensus sequence is Glu-X-X-Glu-X-Cys (Figure 4). To determine whether any components of the carboxylation recognition site are located beyond the carboxy terminal of the propeptide, we have compared the binding of a series of peptides that contain varying amounts of the Gla domain. ProPt18 (residues -18 to -1) is a competitive inhibitor of the carboxylase, with a Ki of 3.5 μM. ProPT28 (residues -18 to +10) is a substrate of the carboxylase, with a Km of 3 μm. ProPT54 (residues -18 to +36), representing the entire Exon II of prothrombin, is a substrate of the carboxylase, with a Km of 2 μM[27]. The similarity of these binding affinities suggests that the carboxylation recognition site is located predominantly in the propeptide. Preliminary experiments determining the efficiency of γ-carboxylation of a recombinant prothrombin in which the cysteine residues at positions 17 and 22 have been mutated to serines indicate carboxylation is normal. The cysteine at position 22 is part of the concensus sequence

proPT18	*HVFLAPQQARSLLQRVRR*	3.5
proPT28	*HVFLAPQQARSLLQRVRR*ANTFLEEVRK	3.6
proPT54	*HVFLAPQQARSLLQRVRR*ANTFLEEVRKGNLER**E**CVE**ET**C**S**YEEAFEALESSTA	2.0

Figure 4. The role of the Gla domain in substrate-vitamin K-dependent carboxylase interaction. The propeptide region is indicated in italics. The concensus sequence in the Gla domain is indicated by underlined bold letters. The binding constant K is an inhibition constant, Ki, for proPT18 and a Km for proPT28 and proPT54 (from Furie B and Furie BC: Annals of the New York Academy of Science, in press, 1990).

suggested to play a role in substrate recognition by the vitamin K-dependent carboxylase. The efficient carboxylation of a mutant recombinant prothrombin missing one of the amino acids of the concensus sequence also suggests that this sequence does not play a role in substrate recognition by the carboxylase.

The vitamin K carboxylase has proved difficult to purify by conventional methods. We have used affinity chromatography employing a synthetic prothrombin propeptide as the ligand to purify the bovine carboxylase. Using this approach we have purified the vitamin K-dependent carboxylase about 10,000-fold over crude liver microsomes[32].

CONCLUSION

A current model of the action of the vitamin K-dependent carboxylase locates the enzyme as an integral membrane protein in the endoplasmic reticulum. The propeptide of the precursor form of a vitamin K-dependent protein interacts with a complementary surface on the vitamin K-dependent carboxylase. The carboxylase then modifies specific glutamic acid residues in close proximity to the mature amino-terminus of the substrate protein, the substrate remaining anchored to the carboxylase. After carboxylation, the propeptide is cleaved. A better preparation of the enzyme will allow the development of tools to further explore the chemical mechanism and the biology of the carboxylase.

REFERENCES

1. Soute BAM, Vermeer C, DeMetz M, Hemker HC, Lijnen HR: In vitro prothrombin synthesis from a purified precursor protein. Biochim. Biophys. Acta 67:101-107, 1981

2. Esmon CT, Grant GA, Suttie JW: Purification of an apparent rat liver prothrombin precursor: characterization and comparison to normal rat prothrombin. Biochemistry 14:1595-1600, 1975

3. Degen SJ, MacGillivray RT, Davie EW: Characterization of the cDNA and gene coding for human prothrombin. Biochemistry 22:2087-2097, 1983

4. Jorgensen MJ, Cantor AB, Furie BC and Furie B: Expression of completely γ-carboxylated recombinant human prothrombin. J. Biol. Chem. 262:6729-6734, 1987

5. Fung MR, Hay CW, MacGillivray, RTA: Characterization of an almost full-length cDNA coding for human factor X. Proc. Natl. Acad. Sci. USA 82:3591-3595, 1985

6. Leytus SP, Chung DW, Kisiel W, Kurachi K, Davie, EW: Characterization of a cDNA coding for human factor X. Proc. Natl. Acad. Sci. USA: 81:3699-3702

7. Hagen FS, Gray CL, O'Hara P, Grant FJ, Saari GC, Woodbury RG, Hart CE, Insley M, Kisiel W, Kurachi K, Davie EW: Characterization of the cDNA coding for human factor VII. Proc. Natl. Acad. Sci. USA: 83:2412-2416, 1986

8.   Kurachi K, Davie EW: Isolation and characterization of a cDNA coding for human factor IX. Proc. Natl., Acad. Sci. USA, 79:6461-6464, 1982.

9.   Choo KH, Gould KG, Rees DJG, Brownlee GG: Molecular cloning of the gene for human anti-haemophilic Factor IX. Nature 299:178-180, 1982

10. Foster D, Davie, EW: Characterization of cDNA coding for human protein C. Proc. Natl. Acad. Sci. USA 81:4766-4770, 1984

11. Long Gl, Belagaje RM, MacGillivray, RTA: Cloning and sequencing of liver cDNA coding for bovine protein C. Proc. Natl. Acad. Sci. USA 81:5653-5656,1984

12. Hoskins J, Norma DK, Beckmann RJ, Long GL: Cloning and characterization of human liver cDNA encoding a protein S precursor. Proc. Natl. Acad. Sci. USA 84:3536, 1987

13. Lundwall A, Dackowski W, Cohen E, Shaffer M, Mahr A, Dahlback B, Stenflo J, Wydro R: Isolation and sequence of the cDNA for human protein S, a regulator of blood coagulation. Proc. Natl. Acad. Sci. USA 83:6716-6720, 1986

14. Price PA, Poser JW, Raman N: Primary structure of the gamma-carboxyglutamic acid-containing protein from bovine bone. Proc. Natl. Acad. Sci. USA 84:8335-8339, 1987

15. Celeste AJ, Buecker JL, Kriz R, Wang EA, Wozney JM: Isolation of the human gene for bone Gla protein utilizing mouse and rat cDNA clones. EMBO J 5:1885-1890, 1986

16. Price PA, Williamson MK: Primary structure of bovine matrix Gla protein, a new vitamin K-dependent bone protein. J. Biol. Chem. 260:14971-14975, 1985

17. Price PA, Fraser JD, Metz-Vira G: Molecular cloning of matrix Gla protein: Implications for substrate recognition by the vitamin K-dependent γ-carboxylase. Proc. Natl. Acad. Sci. USA 84:8335-8339, 1987

18. Pan LC, Price PA: The propeptide of rat bone γ-carboxyglutamic acid protein shares homology with other vitamin K-dependent protein precursors. Proc. Nat. Acad. Sci. USA 82:6109-6113, 1985

19. Jorgensen MJ, Cantor AB, Furie BC, Brown CL, Shoemaker CB, Furie, B. Recognition site directing vitamin K-dependent γ-carboxylation resides on the propeptide of Factor IX. Cell 48:185-191, 1987

20. Diuguid DL, Rabiet M-J, Furie BC, Liebman HA, Furie B: Molecular basis of hemophilia B: A defective enzyme due to an unprocessed propeptide is caused by a point mutation in the factor IX precursor. Proc. Natl. Acad. Sci. USA 83:5803-5807

21. Ware J, Diuguid DL, Liebman HA, Rabiet, M-J, Kasper CK, Furie BC, Furie, B, Stafford, DW: Factor IX San Dimas: Substitution of glutamine for arginine -4 in the propeptide

leads to incomplete γ-carboxylation and altered phospholipid binding properties. J. Biol. Chem. 264:11401-11406, 1989

22. Foster DC, Rudinski MS, Schach BG, Berkner KL, Kumar AA, Hagen FS, Sprecher CA, Insley MY, Davie, EW: Propeptide of human protein C is necessary for γ-carboxylation. Biochemistry 26:7003-7011

23. Huber P, Schmitz T, Furie BC, Furie B: Identification of amino acids in the γ-carboxylation recognition site on the proepetide of prothrombin. J. Biol. Chem. 265:12467-12473, 1990

24. Soute BAM, Ulrich MMW, Vermeer C: Vitamin K-dependent carboxylase. Increased efficiency of the reaction. Thromb. Haemost. 57:5827-5830, 1987

25. Decottignies-Le Marechal P, Rikong-Adie H, Azerad R: Biochem. Biophys. Res. Comm. 90:700-707, 1979

26. Ulrich MMW, Furie, B, Jacobs, M, Vermeer, C, Furie BC: Vitamin K-dependent carboxylation: a synthetic peptide based upon the γ-carboxylation recognition site sequence of prothrombin propeptide is an active substrate for the carboxylase in vitro. J. Biol. Chem. 263: 9697-9702, 1988

27. Hubbard BR, Jacobs M, Ulrich MMW, Walsh C, Furie B, Furie BC: Vitamin K-dependent carboxylation: in vitro modification of synthetic peptides containing the γ-carboxylation recognition site. J. Biol. Chem. 264:14145-14150, 1989

28. Bentley AK, Rees DJG, Rizza C, Brownlee GG: Defective propeptide processing of blood clotting Factor IX caused by mutation of arginine to glutamine at position -4. Cell 45:343-348, 1986

29. Sugimoto M, Miyata T, Kawabata S, Yoshioka A, Fukui G, Iwananga S: Factor IX Kawachinagano: impaired function of the Gla-domain caused by attached propeptide region due to substitution of arginine by glutamine at position -4. British J of Haematology 72:216-221, 1989

30. Galeffi P, Brownlee GG: The propeptide region of clotting factor IX is a signal for a vitamin K dependent carboxylase: evidence from protein engineering of amino acid - 4. Nucleic Acid Research 15:9505-9513, 1987

31. Price PA: Bone Gla protein and matrix Gla protein: Identification of the probable structures involved in substrate recognition by the γ-carboxylase and discovery of tissue differences in vitamin K metabolism. in Suttie J (ed): Current Advances in Vitamin K Research, New York, New York, Elsevier, 1988

32. Hubbard BR, Ulrich MMW, Jacobs M, Vermeer C, Walsh C, Furie B, and Furie BC: Vitamin K-dependent carboxylase: Affinity purification from bovine liver by using a synthetic propeptide containing the γ-carboxylation recognition site. Proc. Natl. Acad. Sci. USA 86:6893-6897, 1989

THE EXPRESSION OF THERAPEUTIC PROTEINS IN TRANSGENIC ANIMALS

Rekha Paleyanda, Janet Young, William Velander[*]
and William Drohan

American Red Cross   Virginia Polytechnic Institute[*]
Rockville, MD 20855   Blacksburg, VA 24061

INTRODUCTION

Human plasma has traditionally been a source of therapeutic proteins.
Hemostatic proteins such as factors VIII, IX, and X, as well as
antithrombin III, albumin, immunoglobulins, fibrinogen and protein C are
routinely isolated from human plasma. Improved protein isolation and
viral inactivation techniques have led to the production of purer and
safer plasma derivatives, although in some cases in quantities
insufficient to meet patient needs.

The expression of therapeutic proteins by recombinant DNA technology
is an attractive alternative to plasma production of proteins, in that it
eliminates the risk of potential contamination with blood-borne viruses
and theoretically provides an unlimited supply of product[1,2,3,4]. However,
the requirement for labor-intensive animal cell production systems, which
utilize expensive culturing media, makes the production of proteins in
these systems rather expensive. More importantly, low expression levels
reported for many proteins may not allow animal cell culture systems to
satisfy the demand for certain proteins[5].

A potential alternative for both plasma and tissue culture sources of
plasma proteins is the transgenic animal carrying additional genetic
information encoding specific human proteins. Although technical
challenges exist in employing this system, the use of transgenic
livestock may provide for the production of large quantities of
therapeutic proteins at reasonable costs. This chapter will review
critical issues and potential benefits involved in the use of transgenic
livestock as protein bioreactors.

TRANSGENIC MICE

Recent advances in gene and embryo manipulation technology have led
to the production of transgenic animals which carry new genetic
information in every tissue, including the germ cells. Various methods
have been employed, especially in the mouse model, to introduce foreign
genes into animals. These include micro-injection of DNA into single cell
embryos, retroviral infection of embryos, and calcium phosphate-mediated
DNA uptake by embryonic stem (ES) cells[6].

The most successful, and consequently most commonly used, technique
is micro-injection of DNA[7,8,9]. This involves isolation of embryos at the
single cell stage, _in vitro_ micro-injection of the DNA encoding the gene
of interest into the isolated embryos, and implantation of manipulated
embryos into pseudo-pregnant females. Transgenic animals can be
identified shortly after birth by analyzing the DNA obtained from a
fragment of the tail with probes specific for the inserted gene.
Integration has been generally found to occur in a head-to-tail,
concatameric fashion at a single genomic site[9]. Incorporation of the
foreign gene at the one-celled stage results in a transgenic animal; at
the multicellular stage, a mosaic. Integration of two different ES cells
may lead to the creation of a chimeric animal. Only microinjection
results in the production of animals that can transmit the genetic
information to their progency in a Mendelian fashion. Germline
integration is essential in order to utilize these transgenic animals as
bioreactors for therapeutic proteins.

Several factors determine the level at which the new protein will be
expressed, as well as its temporal and tissue-specific manner of
expression. Perhaps the most important factors are the promoter and
enhancer employed in controlling the expression of the protein encoded by
the inserted DNA. Regulatory elements which are tissue-specific direct
significant expression to a specific tissue; whereas ubiquitous promoters
permit expression in different tissues within the animal. Promoters
which are temporally controlled would be expressed only at certain times
during embryonic development.

Important cis-acting regulatory elements, other than the 5' upstream
region, may be required for expression at levels equivalent to or higher
than that in the unmodified organ, as indicated in studies with rat beta-
casein promoter/bacterial chloramphenicol acetyltransferase (CAT) fusion
genes[10]. Tissue-specific elements were found in the first noncoding exon
of the beta-casein gene and in the 3' untranslated region of the human
beta-globin gene[11] which contributed to high endogenous expression levels.

The site of DNA integration, apparently a random process, can result
in a complete lack of detectable expression, or of extremely high-level
transgene expression, depending upon the DNA sequences surrounding the
transgene DNA. The presence of intronic sequences within the
transgene[12,13,14], or of dominant control regions surrounding the transgene,
have been shown to eliminate or at least dampen inhibitory effects of the
integration site sequences[15,16,17].

For example, the variable expression pattern reported for whey acidic
protein/tissue plasminogen activator (WAP-tPA) transgenic mice was
probably due to the position effects of transgene insertion, rather than
due to differences in secretion or gene copy[18]. In order to overcome the
position effects of integration upon expression levels, enhancer regions
originally located 10-50 kb upstream[19,20], introns[12] or parts of introns
positioned close to splice junctions[13] have been included in the DNA
constructs.

PROTEIN PRODUCTION IN THE MILK OF TRANSGENIC MICE

Transgenic animal bioreactors for the production of plasma proteins
involve the use of a mammary tissue-specific promoter, directing
expression of the human protein to the mammary gland. The mammary gland
is a naturally self-contained secretory organ that functions to produce
high levels of milk protein without adversely affecting the physiology of
the animal. Once secretion of the protein into milk has been achieved,

it could be readily purified. Several promoters have been successfully
used to target expression of heterologous proteins to the mammary gland
of mice (Table 1). In addition, tissue-specific and hormone-regulated
expression of rat beta-casein has been achieved in mice[21]. It is of
interest that the sheep beta-lactoglobulin promoter, which does not have
a murine analog, has been successfully used to direct the expression of
beta-lactoglobulin (5,000 mg/L)[25] and human alpha-1-antitrypsin (7,000
mg/L)[26] to mouse milk. These levels of expression are remarkably high,
given that beta-lactoglobulin is naturally present at about 4,600 mg/L in
the milk of sheep[25]. In comparison, protein expression levels of between
10-50 mg/L (per 10^6 cells per 24 hrs) in mammalian tissue culture are
considered quite good. These results indicate that regulatory proteins
(e.g. transacting factors) in one species can recognize and functionally
interact with the promoter and regulatory elements of an inserted
transgene from another species, even if the gene is not normally present
in the host species. These results may not be surprising in light of the
observation that the promoter sequences of the milk protein genes of
several species are highly conserved[27,28].

Table 1

PROTEIN EXPRESSION IN MAMMARY GLAND OF TRANSGENIC MICE			
Promoter	Gene Expressed	mg/L[*]	References
Whey Acidic Protein (WAP) (mouse)	WAP (rat)	2,000	22
	tPA (human)	50	18, 37, 38
	CD4	0.5	24
WAP + additional regulatory sequence (mouse)	pS2	1,600	23
Beta-lacto-globulin (sheep)	Beta-lacto-globulin (sheep)	5,000	25
	alpha-1-AT (human)	7,000	26
	factor IX (human)	0.03	38, 49

[*] For comparison: mouse milk contains 2000 mg/L WAP and sheep milk
contains 4600 mg/L beta-lactoglobulin

THE MOUSE WHEY ACIDIC PROTEIN (WAP) PROMOTER

The whey acidic protein (WAP) is the major whey protein in rodent
milk[29] and comprises about 2.4% of the total milk protein. The WAP cDNA
has been cloned[30,31] and the WAP gene has been shown to consist of 4 exons
and three introns. In lactating mammary gland tissue, WAP mRNA accounts
for 15% of the total poly (A) mRNA[30,31,32,33]. The levels of WAP mRNA increase
340-fold from the virgin to the lactating state. Hormonal regulation of
milk protein secretion was shown to be mediated by an increase in

transcriptional activity, as well as an increase in mRNA stability in the mammary gland. These observations with the endogenous WAP RNA and protein suggest that this promoter might be good candidates for promoting mammary-specific expression of heterologous proteins in transgenic mice.

The mouse WAP promoter (comprising 2.6 kb of the 5' WAP sequence) has been shown capable of promoting the transcription of the human <u>Ha-ras</u> gene[35], the human <u>c-myc</u> gene[36], and the human <u>tPA</u> cDNA[18,37] in the mammary gland of transgenic mice. Expression of the <u>Ha-ras</u> transgene was 1-2% of endogenous WAP gene expression (20-40 mg/L), while <u>c-myc</u> was 10% (200 mg/L) and <u>tPA</u> 2.5-5% (50-100 mg/L). The promoter conferred tissue specificity of transgene expression to the mammary gland and this expression was inducible by hormones upon lactation. These experiments indicate that by manipulating the sequences regulating gene expression, it may be possible to express or overexpress foreign genes in the milk of transgenic animals.

Secretion of human tissue plasminogen activator (tPA) from mammary epithelium into the milk of transgenic mice was achieved by Gordon et al.[38] and deserves special mention as the protein product has been extensively characterized. The promoter region of the murine WAP was fused to a tPA cDNA containing its natural signal sequence, and tPA was found secreted in the mouse milk. Consequently, no mammary-specific signal sequence appears to be required for proper protein secretion. In fact, mammary cells have been shown to secrete, with equal efficiency, proteins with different signal peptides[39]. WAP-tPA RNA was detected predominantly in the lactating mammary gland, indicating that the 2.6 Kb of 5' WAP DNA used to regulate tPA gene expression was sufficient for targeting expression to the mammary gland[18].

Experiments characterizing the tPA product demonstrated that the specific activity of the t-PA produced in transgenic mouse milk was equivalent to that of tissue-culture or plasma-derived t-PA[38], but the level of expression (50-100 mg/L) exceeded the highest levels reported in cell culture production systems by up to ten fold. These results also suggest that post-translational modifications required for biological activity can be properly performed in the mammary epithelium for proteins naturally expressed in other differentiated cell types.

PRODUCTION OF PLASMA PROTEINS IN THE MILK OF TRANSGENIC LIVESTOCK

There is relatively little available data describing the expression of heterologous proteins in the milk of transgenic livestock (sheep, goats, pigs or cattle)[40,41,42,43]. The animals' long gestation period (7 to 9 months), lack of information concerning the manipulation of large animal oocytes, and the relatively small number of investigators in the field have limited the amount of information. Of course, large amounts of naturally occuring proteins are present in milk, e.g. bovine caseins are present at levels of 24-28 g/L, whey proteins at 5-7 g/L[44,45]. Two recent reports, however, have provided encouraging results. Scientists at the Institute of Animal Physiology and Genetics Research, Edinburgh, Scotland have used the sheep beta-lactoglobulin (BLG) promoter to regulate the expression of human coagulation factor IX (CFIX) in the milk of sheep[27]. Although expression levels are quite low (0.03 mg/L), the authors can detect CFIX by both sensitive ELISA assays and Western blots[27]. In separate experiments, scientists from the National Institutes of Health and the U.S. Department of Agriculture have used the murine WAP gene, both regulatory and structural sequences, to construct transgenic pigs (Pursel VG, Wall RJ, Hennighausen L, Pittius CW, King K: personal communication). Although the porcine genome does not contain the WAP gene, the levels of WAP expressed in the milk of transgenic pigs

approached the levels of 2000 mg/L present in the milk of lactating mice
(R. Wall and L. Hennighausen, personal communication). However, the long
term effects on the animal of the expression of WAP in porcine milk have
yet to be evaluated. These experiments are the first establishing that
heterologous genes can be expressed at high levels in the milk of
livestock animals (Table 2). In general, widely variable expression
levels of heterologous proteins indicate a need for better understanding
of the regulatory sequences required to achieve high levels of expression
after random insertion of the transgene into the genome of the host
animal.

Table 2

PROTEINS EXPRESSED IN THE MAMMARY GLAND OF TRANSGENIC LIVESTOCK				
	Promoter	Gene Expressed	mg/L[*]	References
PIGS	Whey Acidic Protein (mouse)	WAP (mouse)	2000	Pursel VG et. al, unpubl.
		RSV CAT	<.001	Wall RD et. al, unpubl.
SHEEP	Beta-lacto-globulin (sheep)	factor IX (human)	0.03	27

[*] For comparison: mouse milk contains 2000 mg/L of WAP and sheep milk
contains 4600 mg/L of beta-lactoglobulin.

MARKET ANALYSIS OF THE USE OF TRANSGENIC LIVESTOCK FOR PRODUCTION OF
PLASMA PROTEINS

A number of academic, government and commercial laboratories are
presently considering the use of large farm animals as protein
bioreactors for a variety of therapeutic proteins, including plasma
proteins (Table 3). An economic analysis of the advantages of producing
plasma proteins in the milk of large animal transgenics is fairly
straightforward, in that the commercial value and market for many of
these plasma proteins is known, and the quantity of milk produced per
year by most large farm animals is well established. For illustrative
purposes, we have chosen to set the levels of expression of plasma
proteins in transgenic livestock milk at those levels reported in mouse
milk for tPA, and assume that these levels can eventually be duplicated
in large animals. If lower levels are achieved, the herd required to
produce these proteins would necessarily be larger.

CURRENT RESEARCH ON TRANSGENIC FARM ANIMALS

ACADEMIC	GOVERNMENT
Tufts Univ. - Goats, Pigs	Agriculture & Food Research Council, U.K. - Sheep
Univ. Adelaide - Pigs	Agricultural Research Service- U.S.D.A., Sheep, Pigs, Cattle
Univ. Munich - Pigs	Council of Scientific and Industrial Research Research Organizations, Australia- Sheep
Univ. of Calif. at Davis - Cattle	National Institutes of Health- Pigs
Virginia Tech - Pigs, Cattle	

PRIVATE

ABC - Pigs
American Red Cross - Pigs
DNX (Embryogen) - Pigs
Gene Pharming International - Cattle
Genzyme (Integrated Genetics) - Goats
Granada - Cattle
Pharmaceutical Proteins Limited - Goats
Transgenic Science - Pigs

For example, the current U.S. market for Coagulation Factor VIII (CFVIII) is currently 120 grams (600,000,000 units). Assuming that expression levels of 0.05 g/L can be achieved in bovine milk (equal to the level of t-PA produced in mouse milk), a single cow could produce the entire U.S. supply of FVIII (Table 4). Assuming similar levels of expression, 5,333 rabbits (900 mL milk/yr) or ten goats (500L milk/year) could also produce the 120 g FVIII annual requirement. These estimates take into consideration a 50% efficiency of FVIII protein recovery upon purification from milk.

Table 4

U.S. REQUIREMENT FOR COAGULATION FACTOR VIII

US Market = 600,000,000 U

Specific Activity of Factor VIII = 5,000 U/mg

$$\frac{600,000,000 \text{ U/year}}{5,000 \text{ U/mg}} = 120 \text{ g/year}$$

Assume: 0.05 g/L is an achievable level of expression:

(0.05 g/L) (20 L/day) = 1 g of protein per day

(350 day/year) (1 g/day) = 350 g/year/cow

With 50% efficiency in protein purification:

$$\frac{120 \text{ g FVIII/year}}{350 \text{ g FVIII/year/cow} \times 0.5} = 0.7 \text{ cows}$$

Human coagulation factor IX has already been produced in the milk of transgenic sheep[27], although at low levels (0.03 mg/L). Considering that the U.S market for CFIX is estimated to be 750 grams per year, four transgenic cows could produce the entire U.S. supply, if expression levels can be maintained at 0.05 g/L and losses during purification can be limited to no more than 50% (Table 5).

Table 5

US REQUIREMENT FOR COAGULATION FACTOR IX

US Market = 150,000,000 units

Specific Activity of Factor IX is 200 U/mg

$$\frac{150,000,000 \ U}{200 \ U/mg} = 750 \ g \ per \ year$$

Assume: 0.05 g/L is an achievable level of expression:

(0.05 g/L) (20 L/day) = 1 g of protein a day

(350 day/year) (1 g/day) = 350 g/year/cow

With 50% efficiency in protein purification

$$\frac{750 \ g \ CFIX/year}{350 \ g \ CFIX/year/cow \ x \ 0.5} = 4.3 \ cows$$

Our laboratory has developed methods for immunoaffinity purification of Protein C (and Activated Protein C) from plasma. These methods for isolation of Protein C from plasma could be modified to isolate Protein C at similar yields from the milk of transgenic animals. Due to the fact that clinical data are not yet available, neither dose ranges nor clearly defined clinical indications have been established. Nevertheless, if protein C proves useful for even some of the postulated indications (Table 6), we estimate that plasma-derived supplies of activated protein C will be insufficient to meet clinical requirements. If it were possible to use all plasma collected in the U.S. for the recovery of Protein C, with normal losses during purification, only about 12 Kg of protein C could be produced. In our estimate, this may only represent about 10% of the U.S. demand. A shortfall of up to 94 kg may occur if effective use of activated protein C requires the highest doses estimated for only 3 indications: septic shock, cardiovascular thrombolytic therapy and prevention of deep vein thrombosis as a consequence of total hip replacement. Since cell culture systems have so far been unsatisfactory for the production of fully active Protein C, the production of protein C by transgenic animals is an attractive alternative. For protein C produced in the milk of transgenic animals at 0.05 g/L, 537 cows would suffice to produce the U.S. supply.

Table 6

PARTIAL ESTIMATE OF U.S. CLINICAL REQUIREMENTS FOR
PROTEIN C AND ACTIVATED PROTEIN C

Indication	Estimated Dose (mg) Per Treatment	# Treatments Per Year	Total U.S. Req. (Kg)
Septic Shock	5-50	120,000	0.6-6.0
Thrombolytic Therapy**	10-100	800,000	8-80
Hip Replacement	10-100	200,000	2-20
Homozygous Deficient	3	100 x 365*	0.10
Heterozygous Deficient	50	1,000	0.05
Total			10.8-106.2

* 100 individuals in U.S. x 365 treatments/year
** Refers to the use of APC, following thrombolytic therapy, to prevent the reformation of blood clots.

Production of other plasma derivatives in transgenic livestock at levels designed to meet market demands may prove more problematic than the proteins reviewed above. For example, in 1987 the U.S. utilization of human serum albumin was 162,000 Kg. Even assuming an increased expression level of 0.1g/L in bovine milk and 75% recovery during processing, 308,571 cows would be required to produce the protein. Losses during what is anticipated to be a difficult purification process may be greater than those assumed for this calculation and would further increase the number of animals needed. Obviously, at this level of expression, production and purification of human serum albumin from the milk of transgenic farm animals is a formidable task.

A summary of these calculations is presented in Table 7.

Table 7

PRODUCTION OF HUMAN THERAPEUTIC PROTEINS IN TRANSGENIC ANIMALS:

Plasma Protein	Total U.S. Requirement (kg/year)	Expression level assumed (g/L) *	Transgenic animals**
Factor VIII	0.12	0.05	0.7 cows
		0.05	9.6 goats,sheep,pigs
		0.05	5,333 rabbits
Factor IX	0.75	0.05	4.3 cows
Protein C	90	0.05	537 cows
Human Serum Albumin	162,000	0.1	308,571 cows
		1.0	30,857 cows
		10.0	3,086 cows

* Calculations assume that the total volume of milk produced by a cow is 6,000 L; by a goat, sheep or pig is 500 L/yr and by a rabbit is 900 mL/yr.
** Estimates include 50% loss occuring during protein purification from milk except for human serum albumin, where a 25% purification loss is assumed.

CHALLENGES IN PRODUCING LARGE ANIMAL TRANSGENICS

A variety of scientific and economic issues need to be successfully resolved before large animal transgenics are employed as protein bioreactors. Most importantly, we need to understand more about the regulatory elements controlling mammary-specific expression[46,47] and we need to identify the dominant control sequences which may eliminate dependence of high level expression upon integration site. Such information should enable us to design and introduce DNA constructs which can predictably result in overexpression of proteins in the milk of transgenic livestock.

Additional information is also required regarding the stability of proteins present in the milk of livestock. A number of serine proteases, with broad specificities and pH tolerance, are found in milk[48,49] and may proteolytically activate the transgene proteins which are to be expressed in the form of zymogens, or degrade these proteins entirely. We require further knowledge of the types of proteases present in animal milk and need to develop techniques to inhibit the action of specific proteases which may degrade transgene protein products. In fact, tPA expressed in mouse milk was in a predominantly proteolyzed form[18]. We shall also have to determine if disruption of cellular processes occurs and whether prolonged lactation and collection of milk from overexpressing transgenic animals is deleterious to the animal or to the mammary gland itself. The possibility that transgenic proteins produced in the mammary gland can be found in the plasma also needs to be assessed.

Likewise, little is known about viruses and other pathogens that might naturally occur in the milk of large animals. If present, they pose a threat as contaminants for parenteral administration of therapeutic protein concentrates. Consequently, a better understanding of viruses present in animal milk is needed, as well as an evaluation of methods that may be required to remove or inactivate these viruses.

The length of time required to establish a transgenic herd of animals is considerable. Depending upon the species and the probability of obtaining male or female offspring at critical times, it could take a minimum of 5-10 years to generate a producing herd. It would be helpful if technology could be developed to reduce this time period. Methods currently being developed should permit determination of the integration of heterologous DNA into embryo DNA, prior to implantation into recipient animals, and, in addition, of the expression potential of a transgenic animal prior to lactation by _in vitro_ cultures and hormonal stimulation of mammary tissue biopsies[7,50,51,52].

The advantages of one species over another must be carefully evaluated. Rabbits, for example, have 3 lactation periods a year and may produce a total of 900 mL of milk a year. Goats produce as much as 500 liters of milk per year. Cows, on the other hand, produce from 6,000 to 20,000 liters a year. The porcine system has the advantage of a more rapid rate of reproduction and a larger number of offspring than sheep, goats or cows. However, the quantity of milk produced is lower and milking may be more difficult. The duration of the gestation period also plays a role in these considerations, as it ranges from 19-21 days in a mouse to nine months in a cow.

An important economic issue to be considered is the high cost of producing a large transgenic animal. The cost of generating just one expressing transgenic pig is estimated to be around 25,000 dollars by Dr. Robert Wall of the U.S. Department of Agriculture, while the cost of producing an expressing cow could well exceed five hundred thousand

dollars. Regulatory issues raised by this new technology will have to be
addressed and new guidelines written.

Finally, it is unclear what part, if any, of the process of
generating transgenic animals will be patentable. How will companies
that commit significant resources over a long period of time, protect and
share such intellectual property? These issues are important and must be
addressed. However, progress in transgenic technology is beginning to
advance more rapidly. More and more proteins are being produced at
higher and higher levels in animal milk. The objective of producing
large quantities of proteins economically appears itself to be adequate
motivation for the support of significant research in this exciting area.

ACKNOWLEDGEMENT

In partial fulfillment of the doctoral thesis (RP) at The George
Washington University, Department of Genetics, Washington, DC 20052, USA

REFERENCES

1.   Anson DS, Austen DEG, Brownlee GG: Expression of active human
     clotting factor IX from recombinant DNA clones in mammalian cells.
     Nature 315:683, 1985

2.   Kaufman RJ, Wasley LC, Furie BC, Furie B, Shoemaker CB: Expression,
     purification and characterization of recombinant gamma-carboxylated
     Factor IX synthesis in Chinese hamster ovary cells. J Biol Chem 261:
     9622, 1986

3.   Kaufman RJ, Wasley LC, Dorner AJ: Synthesis, processing, and
     secretion of recombinant human factor VIII expressed in mammalian
     cells. J Biol Chem 263:6352, 1988

4.   Sarver N, Ricca GA, Link J, Nathan MH, Newman J, Drohan WN: Stable
     expression of recombinant Factor VIII molecules using a bovine
     papillomavirus vector. DNA 6:553, 1987

5.   Clark AJ, Simons P, Wilmut I, Lathe R: Pharmaceuticals from
     transgenic livestock. Tibtech 5:20, 1987

6.   Jaenish R: Transgenic animals. Science 240:1468, 1988

7.   Hammer RE, Pursel VG, Rexroad CF, Wall RJ, Bolt DJ, Palmiter RD,
     Brinster RL: Genetic engineering of mammalian embryos. J Anim Sci
     63:269, 1986

8.   Gordon JW, Ruddle FH: Integration and stable germ line transmission
     of genes injected into mouse pronuclei. Science 214:1244, 1981

9.   Brinster RL, Chen HY, Trumbauer ME, Yagle MK, Palmiter RD: Factors
     affecting the efficiency of introducing foreign DNA into mice by
     microinjecting eggs. Proc Natl Acad Sci 82:4438, 1985

10. Lee K-F, Atiee SH, Rosen JM: Differential regulation of rat beta-
     casein chloramphenicol acetyltransferase fusion gene expression in
     transgenic mice. Mol Cell Biol 9:560, 1989

11. Behringer RR, Hammer RE, Brinster, RL, Palmiter RD, Townes TM: Two
    3'sequences direct adult erythroid-specific expression of human
    beta-globin genes in transgenic mice. Proc Natl Acad Sci USA
    84:7056, 1987

12. Brinster RL, Allen JM, Behringer, RR, Gelinas RE, Palmiter RD:
    Introns increse transcriptional efficiency in transgenic mice. Proc
    Natl Acad Sci USA 85:836, 1988

13. Buchman AR, Berg P: Comparison of Intron-Dependent and Intron-
    Independent Gene Expression. Mol Cell Biol 8:4395, 1988

14. Carlberg K, Ryden TA, Beemon K: Localization and footprinting of an
    enhancer within the Avian Sarcoma virus gag gene. J. Virol 62:1617,
    1988

15. van Assendelft GB, Hanscombe O, Grosveld F, Greaves DR: The beta-
    globin dominant control region activates homologous and heterologous
    promoters in a tissue-specific manner. Cell 56:969, 1989

16. Behringer RR, Ryan TM, Reilly MP, Asakura T, Palmiter RD, Brinster
    RL, Townes TM: Synthesis of functional human hemoglobin in
    transgenic mice. Science 245:971, 1989

17. Lang G, Wotton D, Owen MJ, Sewell WA, Brown MH, Mason DY, Crumpton
    MJ, Kioussis D: The structure of the human CD2 gene and its
    expression in transgenic mice. EMBO J 7:1675, 1988

18. Pittius CW, Hennighausen L, Lee E, Westphal H, Nicols E, Vitale J,
    Gordon K: A milk protein gene promoter directs the expression of
    human tissue plasminogen activator cDNA to the mammary gland in
    transgenic mice. Proc Natl Acad Sci 85:5874, 1989

19. Pinkert CA, Ornitz DM, Brinster RL, Palmiter RD: An albumin enhancer
    located 10 kb upstream functions along with its promoter to direct
    efficient, liver-specific expression in transgenic mice. Genes Devel
    1:268, 1987

20. Grosveld F, van Assendelft GB, Greaves DR, Kollias G: Position-
    independent, high-level expression of the human beta-globin gene in
    transgenic mice. Cell 51:975, 1987

21. Lee K-F, DeMayo, FJ, Atiee SH, Rosen JM: Tissue-specific expression
    of the rat beta-casein gene in transgenic mice. Nuc Ac Res 16:1027,
    1988

22. Bayna EM, Rosen JM: Tissue specific, high level expression of rat
    whey acidic protein gene in transgenic mice. Nuc Ac Res 18:2977,
    1990

23. Tomasetto C, Wolf C, Rio MC, Mehtali M, LeMeur M, Gerhinger P,
    Chambon P, Lathe R: Breast cancer protein PS2 synthesis in mammary
    gland of transgenic mice and secretion into milk. Mol Endocrinol
    3:1579, 1989

24. Yu SH, Deen KC, Lee E, Hennighausen L, Sweet RW, Rosenberg M,
    Westphal, H: Functional human CD4 protein produced in milk of
    transgenic mice. Mol Biol Med 6:255, 1989

25. Simons JP, McClenaghan M, Clark AJ: Alteration of the quality of milk by expression of sheep beta-lactoglobulin in transgenic mice. Nature 328:530, 1987

26. Archibald AL, McClenaghan M, Homsey V, Simons JP, Clark AJ: High level expression of biologically active human alpha-1-antitrypsin in the milk of transgenic mice. Proc Natl Acad Sci USA 87:5178, 1990

27. Clark AJ, Bessos H, Bishop JO, Brown P, Harris S, Lathe R, McClenaghan M, Prowse C, Simons JP, Whitelaw CBA, and Wilmut I: Expression of human anti-hemophilic factor IX in the milk of transgenic sheep. Biotechnology 7, 1989

28. Hall L, Emery DC, Davies MS, Parker D, Craig RK: Organization and sequence of the human alpha-lactalbumin gene. Biochem J 242:735, 1987

29. Lubon H, Hennighausen L: Conserved region of the rat alpha-lactalbumin promoter is a target for protein binding in vitro. Biochem J 256:391, 1988

30. Piletz JE, Heinleu M, Ganshow RE: Biochemical characterization of a novel whey protein from murine milk. J Biol Chem 256:11509, 1981

31. Hennighausen LG Sippel AE: Comparitive sequence analysis of the mRNAs coding for mouse and rat whey proteins. Nuc Ac Res 10:3733, 1982

32. Campbell SM, Rosen JM, Hennighausen LG, Strech-Jurk U, Sippel AE: Comparison of the whey acidic protein genes of the rat and the mouse. Nuc Ac Res 12:8685, 1984

33. Hobbs AA, Richards DA, Kessler DJ Rosen JM: Complex hormonal regulation of rat casein gene expression. J Biol Chem 257:3598, 1982

34. Richards DA, Rodgers JR, Supowit SC, Rosen JM: Construction and preliminary characterization of the rat casein and lactalbumin cDNA clones. J Biol Chem 256:526, 1981

35. Andres A-C, Schoenenberger C-A, Groner B, Hennighausen L, LeMeur M, Gerlinger P: Ha-ras oncogene expression by a milk protein gene promoter: tissue specificity, hormonal regulation and tumor induction in transgenic mice. Proc Natl Acad Sci USA 84:1299, 1987

36. Schoenenberger C-A, Andres A-C, Groner B, van der Valk M, LeMeur M, Gerlinger P: Targeted c-myc expression in mammary glands of transgenic mice induces mammary tumors with constitutive milk protein gene transcription. EMBO J 7:169, 1988

37. Pittius CW, Sankaran L, Topper YL, Hennighausen L: Comparison of the regulation of the whey acidic protein gene with that of a hybrid gene containing the whey acidic gene promoter in transgenic mice. Mol Endocrinol 2:1027, 1988b

38. Gordon K, Lee E, Vitale JA, Smith AE, Westphal H, Hennighausen L: Production of human tissue plasminogen activator in transgenic mouse milk. Biotech 5:1183, 1987

39. Craig RK, Perera PAJ, Mellor A, Smith A: Initiation and processing <u>in vitro</u> of the primary translational product of guinea-pig caseins. Biochem J 184:261, 1979

40. Church RB: Can. Pacific Biotech Symp Agric and Forest Bull 8:22-23, 1986

41. Simons JP, Land RB: Transgenic Livestock. J Reprod Fert Suppl 34:237, 1987

42. Church RB: Embryo manipulation and gene transfer in domestic animals. Tibtech 5:13-19, 1987

43. Hammer RE, Pursel VG, Rexroad Jr CE, Wall RJ, Bolt DJ, Ebert KM, Palmiter RD, Brinster RL: Production of transgenic rabbits, sheep and pigs by microinjection. Nature 315:680, 1985

44. Wong NP, ed.: Fundamentals of Dairy Chemistry, 3rd Edition. New York, NY, Van Nostrand Reinhold Co., 1988

45. Webb B II, ed.: Fundamentals of Dairy Chemistry, 2nd Edition. Westport, CT, The Ari Publishing Co, Inc., 1973

46. Harris S, McClenaghan M, Simons JP, Ali S, and Clark AJ: Gene expression in the mammary gland. J. Reprod. Fert 88:707-715, 1990

47. Clark AJ, Ali S, Archibald AL, Brown P, Harris S, McClenaghan M, Prowse C, Simons JP, Whitelaw CBA, and Wilmut I: The molecular manipulation of milk composition. Genome 31:950-955, 1989

48. Reimerdes EH: Natural protease activity in milk. J Dairy Sci 66:1591, 1983

49. Korycka-Dahl M, Ribadeau-Dumas B, Chene N, Martal J: Plasmin activity in milk. J Dairy Sci 66:704, 1983

50. Ebert KM, Papaioannou VE: In vivo culture of embryos in the mature mouse oviduct. Theriogenology 31:299, 1989

51. Wall RJ, Hawk HW: Development of centrifuged cow zygotes cultured in rabbit oviducts. J Reprod Fert 82:673, 1988

52. Wall RJ, Pursel VG, Hammer RE, Brinster RL: Development of porcine ova that were centrifuged to permit visualization of pronuclei and nuclei. Biol Reprod 32:645, 1985

CLINICAL USE OF PROTEINS PRODUCED BY RECOMBINANT TECHNOLOGY

THE USE OF ANIMAL MODELS TO EVALUATE PROTEINS PRODUCED BY RECOMBINANT

TECHNOLOGY

Alan R. Giles

Department of Pathology and Medicine
Queen's University
Kingston, Ontario, Canada, K7L 3N6

Although the ultimate test of a recombinant protein's safety and
efficacy for its intended use rests with rigorous evaluations in humans,
pre-clinical testing of new therapies in animals is required by many
regulatory jurisdictions to provide both the justification for human study
and information upon which the design of these studies will be based.
Moreover, in more traditional pharmaceuticals, _in vivo_ evaluations in
appropriate animal models have played an essential role in primary
development in many cases. Although frequently modelled on "natural"
products, recombinant proteins are no exceptions. Given the relatively
contrived nature of the expression systems used, minor variations in their
molecular biology, eg. carbohydrate composition, may result in functional
consequences that cannot necessarily be excluded by _in vitro_ study alone.
Moreover, the very technology permits the deliberate manipulation of the
protein structure designed to improve functional performance such as
increasing the t1/2 _in vivo_. The success or failure of such endeavours may
be more safely and, in many cases, effectively evaluated in appropriate
animal models prior to human study. The questions that may be addressed by
such an approach fall into two general areas.

1. Efficacy
    i. Are effective blood levels attained and maintained?
    ii. Is the anticipated biological response observed?

2. Safety
    i. Is there evidence of acute toxicity and, if there is, is the
        margin between the effective and toxic doses acceptable?
    ii. Is there evidence of medium to long-term side effects?

In order to provide acceptable information as to the efficacy of a
recombinant protein in the management of any given diseases state, the
model itself must be validated as being appropriate and, wherever
possible, exhibit a corrective response to an established therapy.
Consequently, the condition of the animal should mimic that to be studied
in the human as closely as possible and, generally speaking, a congenital
rather than acquired condition is preferred in order to minimize the
biological variables involved. Having achieved this, appropriate
methodology must be available to measure objectively the studied protein's
activity, both _in vivo_ and _in vitro_.

Recombinant Technology in Hemostasis and Thrombosis
Edited by L.W. Hoyer and W.N. Drohan, Plenum Press, New York, 1991

In the study of recombinant proteins in hemostasis, we are fortunate
that a number of well-described animal models of human deficiency states
affecting hemostasis exist.[1] Many of these appear to be totally analogous
to the human condition both in their pathophysiology and their functional
consequences. However, this has not eased the requirement for the
development of appropriate methodology for their study. This includes the
accurate determination of endpoints in documenting dose response, etc, in
replacement therapy and taking account of the known inter-species
differences with regard to the interaction of the different components of
the hemostatic mechanism. Such methodology has been developed for the
study of factor VIII deficiency in the dog and this has now been
successfully applied to the study of recombinant factor VIII prior to
clinical evaluation that is now underway. This experience will be used to
illustrate the development of the methodology and underline the utility of
animal models for both current and future applications in the evaluation of
recombinant proteins in hemostasis.

<u>Canine Factor VIII Deficiency</u>

A number of mammals have been found to suffer from congenital factor
VIII deficiency.[2] As in man, it is a sex-linked recessive disorder and
has an identical clinical phenotype in that severe deficiency is associated
with the frequent occurrence of hemarthroses, soft tissue hematomas, etc.
The dog has been most exhaustively studied and important contributions have
been made to the overall understanding of factor VIII deficiency in both
man and animals.[3] A major colony has been developed at Queen's University
specifically for the purpose of developing a model to evaluate therapeutic
interventions that either bypass or replace factor VIII <u>in vivo</u>. Figure 1
shows the current family tree. The condition occurs in all breeds of dogs
but in this case the defect was first demonstrated in a family of purebred
miniature Schnauzers. For reasons to be discussed later, the affected male
miniature Schnauzer (II.2) was bred to a purebred Brittany spaniel (II.1)
resulting in a litter of four obligate carrier females (III.2-5). II.2 was
bred to his litter mate, II.3, and his male litter mate, II.4, was bred as
illustrated. Of particular significance was the observation that the

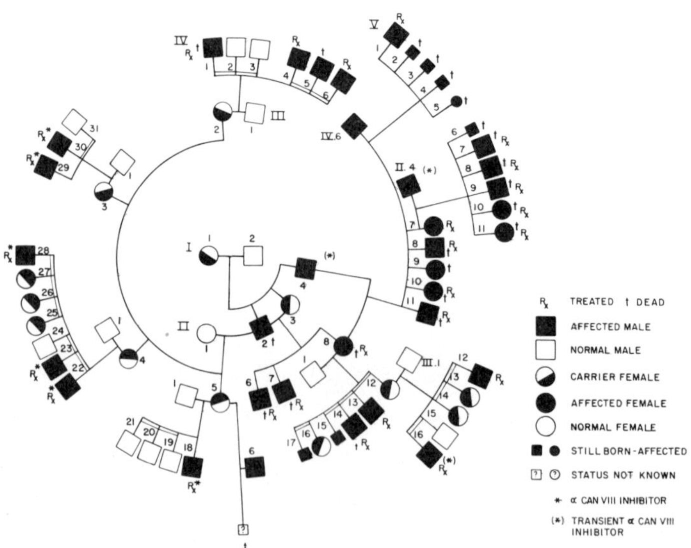

Fig. 1. Family tree of Queen's University hemophilic
(factor VIII deficiency) dog colony

214

offspring of the four obligate carriers related to the Brittany spaniel
have exhibited the propensity for developing canine/anti-canine factor VIII
antibodies following replacement therapy with canine cryoprecipitate.[4]
This further extends the analogy between canine and human factor VIII
deficiency where approximately 10% of treated hemophiliacs demonstrate a
similar propensity.[5] It is of great interest that this does not appear to
be directly related to the ancestral hemophilia gene, given that it is only
the relatives of the Brittany spaniel that demonstrate this apparently
genetically pre-determined condition. Again, this is in line with our
current view of the human condition where it appears that genetically
determined characteristics of the individual immune response appear to play
a vital role.[6]

The animals described are severely affected from a clinical standpoint
although their factor VIII levels may measure variably from 1 - 4%. With
the exception of the inhibitor animals, they show an effective response to
factor VIII replacement given in the form of canine cryoprecipitate
prepared by standard blood transfusion techniques. Nonetheless, the
familiar difficulty in the human situation of evaluating response in
bleeding that is for the most part concealed[7] is compounded by the
additional difficulty of having little if any equivalent means of
establishing such criteria as pain. In order to circumvent this, a simple
and reproducible method of observing bleeding and its response to
replacement therapy has been developed.[8]

The Cuticle Bleeding Time (CBT)[8]

The canine cuticle (Figure 2) is richly vascularized and bleeds
profusely when injured. However, in the hemostatically competent animal
bleeding ceases within a short time and a permanent and effective
hemostatic plug develops. This is not the case in hemostatically

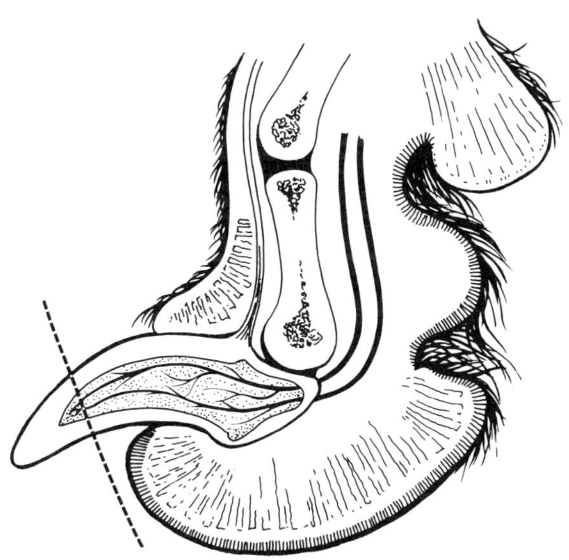

Fig. 2. The saggital section of the paw of a dog
showing the vascularized nail cuticle. The
dotted line indicates the plane of
transection of the nail cuticle. Reproduced
from Reference 8 with permission of Grune and
Stratton Inc.

incompetent animals. An approach has been developed where, under light anesthetic, the cuticle is minimally injured and direct observations of bleeding may be made including a precise notation on when the cessation of bleeding occurs. As shown in Figure 3, clear cut differences can be observed between normal and factor VIII deficient animals. Of importance is that, in the case of the latter, replacement with homologous factor VIII corrects the abnormality. The model is particularly useful in that, given the availability of 16 nails, an appropriate number of baseline observations may be made before studying the effect of intervention at treatment intervals over a sustained period of time. The only limit on this is the period of anesthesia permissible which, in our experience, should not exceed three hours. Providing the injury to the cuticle is minimized both in terms of degree and frequency of usage, it regrows without residual abnormality. The introduction of the Brittany spaniel in the bleeding program previously described was directed at developing animals with colourless rather than black (miniature Schnauzer) nails in order that the cuticle tip could be directly visualized and thus aid in minimizing the injury induced. The introduction of the Beagle (III.1 – Figure 1) was similarly directed.

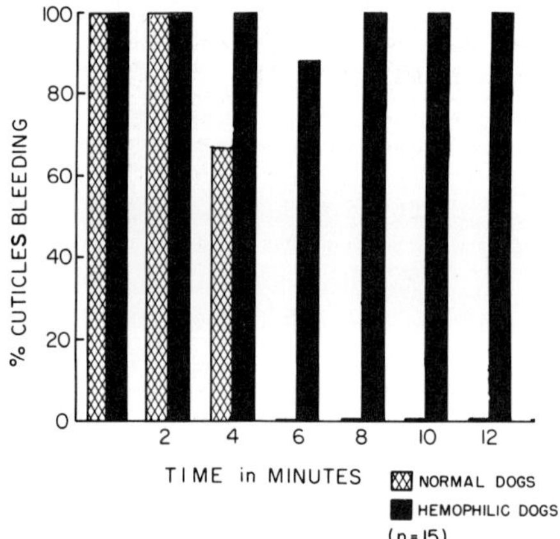

Fig. 3. The duration of bleeding following injury to the nail cuticle of normal and hemophilic (factor VIII deficient) animals is shown. The verticle bars indicate the percentage of the total number of cuticles injured at times "0" observed to be bleeding at successive time intervals over a 12 minute period. In the case of the hemophilic animals, at the completion of the observation period (12 minutes), bleeding is stopped by local cautery with silver nitrate application. Reproduced from Reference 8 with permission of Grune and Stratton Inc.

The animals developing canine/anti-canine factor VIII antibodies present particular management difficulties. However, as in the human condition, these antibodies have been demonstrated to show species specificity in that, although recognizing both canine and human factor VIII, in _in vitro_ assay, porcine factor VIII is not inactivated.[4] Figure 4

demonstrates the efficacy of porcine factor VIII _in vivo_ in one of
these animals. These data further demonstrate the analogy between the
canine and human conditions and also that, although there are clearly
species differences in factor VIII at the molecular level, there is
functional equivalence when given as replacement therapy.[9]

EVALUATION OF RECOMBINANT FACTOR VIII

The methodology described above has now been applied to the evaluation
of recombinant factor VIII in terms of (1) its recovery and survival
following infusion; (2) its ability to form a bimolecular complex with
circulating canine von Willebrand factor; and (3) its ability to correct
the hemostatic defect.

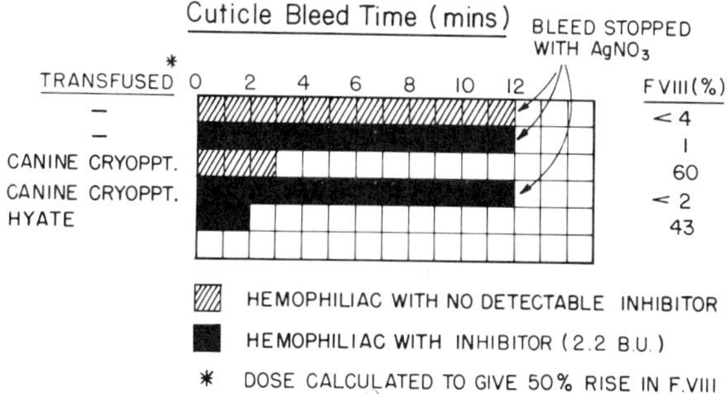

Fig. 4. The effective treatment with either canine or
porcine factor VIII:C on the cuticle bleeding time
in hemophilic dogs with and without inhibitors to
canine factor VIII:C. Reproduced from Reference 4
with permission of Grune and Stratton Inc.

Wild Type Factor VIII

Wild type factor VIII was shown to demonstrate the anticipated
recovery and survival characteristics anticipated from those noted in
studies of infusion of the homologous protein into either human or canine
subjects (Figure 5).[2,10,11]

Moreover, correction of the hemostatic defect observed using the
cuticle bleeding time was observed. Of particular importance was the
observation that factor VIII, in the absence of co-infused von Willebrand
factor, formed a bimolecular complex with the endogenous circulating protein
(Figure 6)[11]. Thus, these studies confirmed the previous expectations
based on _in vitro_ study that wild type recombinant factor VIII had the
potential to provide an effective and safe means of replacing factor VIII
in deficient humans. The clinical studies, which are now well advanced,
are entirely compatible with those performed in dogs with regard to
observations on _in vivo_ kinetics and functional potential.[12]

<u>Factor VIII Mutants</u>

In many regards, the results obtained with the wild type form could be anticipated. However, one of the real strengths of recombinant technology lies in the ability to induce specific changes in the protein that may confer advantages either in its large scale production or in its efficacy in vivo. The domain structure of factor VIII has been described in detail and is covered at length in other chapters in this book.[13] In the early studies by the group at Genetics Institute, the lack of homology of the B domain between human and porcine factor VIII was noted and was in striking contrast to the high degree of homology observed in the A and C domains. As porcine factor VIII had been shown to be an effective replacement therapy in other species,[4,9] it was speculated that the B domain may not be necessary for the cofactor function of factor VIII. This was subsequently confirmed by <u>in vitro</u> study of various B domain deletion mutants. Moreover, Toole and co-workers demonstrated that expression of these mutants appear to be significantly augmented[14]. This observation may have major importance in optimizing the yields of recombinant protein that may be obtained in satisfying the total clinical demand. It was therefore essential to establish that the lack of functional consequences of this manoeuvre observed <u>in vitro</u> could be duplicated <u>in vivo</u>. Using the methodology described, we have now demonstrated equivalence between a number of B domain deletion mutants and wild type recombinant factor VIII both in terms of their recovery and survival, von Willebrand factor binding and functional activity in correcting the defect in the cuticle bleeding time.[15]

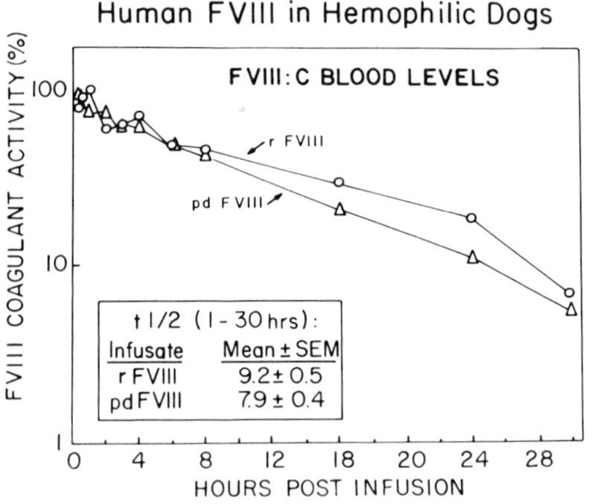

Fig. 5.   Post-transfusion survival of factor VIII:C functional activity. Sequential plasma sampling was performed over a 30 hour period following the infusion of recombinant factor VIII (0-------0) and plasma derived factor VIII (△ ----- △). Factor VIII:C levels were determined by a one stage assay using a 20 normal dog pool plasma as the standard reference plasma. Each data point is the mean of two observations. The t1/2 values were determined by linear regression analysis of the data plotted. Reproduced from Reference 11 with permission of Grune and Stratton Inc.

Protein C is now recognized as a major physiological anticoagulant which exerts its effect _in vivo_ by inactivating the essential cofactors, Va and VIIIa.[16] The putative activated protein C cleavage site is thought to reside at Arginine 336 in the heavy chain of the activated protein.[17] Theoretically, replacement of the Arginine at this position with Isoleucine could impair the APC induced inactivation of the activated protein.[17] We have explored this possibility by studying the fate of the recombinant protein (RI-rVIII) where such a change has been engineered by site directed mutagenesis[18] following its infusion into a normal dog and then triggering protein C activation. We have previously demonstrated that protein C may be activated _in vivo_ by the infusion of purified factor Xa in combination with coagulant active phospholipid in the form of phosphatidylcholine / phosphatidlyserine lipid vesicles (PCPS).[19,20] [125]I radiolabelled RI or wild type recombinant factor VIII were infused into several normal dogs and their survival studied by measuring the level of radioactivity in plasma samples under basal conditions for 30 minutes prior to inducing protein C activation . The animals' own factor VIII levels were used as internal controls and monitored by one stage factor VIII assays made on the same samples. Following protein C activation as described above, significant falls occurred in both radioactive counts and factor VIII clotting activity indicating that both the radiolabelled recombinant factor VIII preparations and the endogenous factor VIII of the dog had been inactivated. However, in the case of the RI mutant there was a significant delay in this as compared with the internal control, ie. the dogs own endogenous factor VIII.

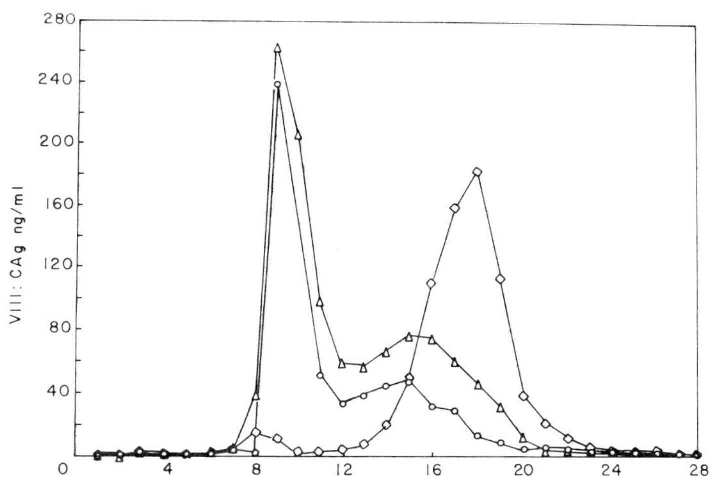

Fig. 6. Elution profile on sepharose 4b of factor VIII:Ag and vWf:Ag or rF.VIII and post-transfusion plasma samples containing rF.VIII or pdF.VIII. Plasma samples obtained one hour after the infusion of either recombinant (0------0) or plasma-derived (△ ------ △) factor VIII were subjected to chromatography on Sepharose 4B. Eluting plasma fractions were assayed for factor VIII:Ag and vWf:Ag. Recombinant factor VIII (◊ ----- ◊) were processed in an identical fashion. vWf:Ag was detectable in fractions 7 - 12 (data not shown in figure). Reproduced from Reference 11 with permission of Grune and Stratton Inc.

These studies demonstrated that the point mutation had indeed conferred
some degree of protection from protein C induced inactivation. Further
studies are required to establish the mechanism by which subsequent
inactivation, although delayed, subsequently did occur. In parallel
studies, in hemophilic animals, full functional replacement activity was
observed with the mutant form and no evidence of toxicity was documented.[18]

SUMMARY AND CONCLUSION

In summary, appropriate animals models for evaluating hemostatically
active recombinant proteins do exist. The model of factor VIII deficiency
is particularly well developed but the methodology described is
transferrable to other deficiency states such as factor IX and von
Willebrand factor deficiency, etc. Initial studies of wild type factor
VIII have confirmed the utility of this approach and contributed
significantly to the overall development of this vital new form of therapy.
The further application to the development of even more novel forms of
recombinant proteins are now under investigation and have again
demonstrated the usefulness of this approach in documenting the validity of
in vitro evaluations with regard to in vivo performance.

ACKNOWLEDGMENTS

The author is a Distinguished Research Professor of the Heart & Stroke
Foundation of Ontario. The work describes involved the participation of
many collaborators and agencies. The author would particularly like to
acknowledge the generous support of Dr. Walter Bowie in establishing the
hemophilic dog colony by providing the initial hemophilic animals. The
technical support of my research staff, particularly Mr. Shawn Tinlin
and Mr. Hugh Hoogendoorn, is gratefully acknowledged. Financial support for
the development of a hemophilic colony and the methodology described came
from the Medical Research Council of Canada (MA-7667) and Alpha
Therapeutic Corporation. The studies on recombinant factor VIII were
performed in collaboration with Genetics Institute and Cutter Biologicals.
I am grateful to Ms Barbara Saunders for the preparation of the manuscript
and to Stan Morton for most of the illustrations. Permission to republish
Figures 2 - 6 from Grune & Stratton Inc. is gratefully acknowledged.

Finally, I should like to dedicate this paper as a small token of a
very large respect and debt to the late Ted Zimmerman, who from the very
beginning supported my view of the essential need for methodologically
appropriate animal models to address issues such as these.

REFERENCES

1.   Dodds WJ: Second international registry of animal models of thrombosis
and hemorrhagic diseases. ILAR News 24:3, 1981
2.   Bowie EJW, Owen CA, Giles AR: Animal models for the study of factor
VIII and von Willebrand factor, in Zimmerman TS, Ruggeri ZM (eds):
Coagulation and Bleeding Disorders - The Role of Factor VIII and von
Willebrand Factor. New York & Basel, Marcel Dekker Inc, 1989, pages 305-324
3.   Dodds WJ: Learning from animal models of bleeding disorders. Ann NY
Acad Sci 406:59, 1983
4.   Giles AR, Tinlin S, Hoogendoorn H, Greenwood P, Greenwood R:
Development of factor VIII:C antibodies in dogs with hemophilia A (Factor
VIII:C deficiency). Blood 663:451, 1984

5.   McMillan CW, Shapiro SS, Whitehurst D, Hoyer LW, Rao AV, Lazerson J: The natural history of factor VIII:C inhibitors in patients with hemophilia A: A national cooperative study. II. Observations on the initial development of factor VIII:C inhibitors. Blood 71:344, 1988

6.   Shapiro SS: Genetic predisposition to inhibitor formation. Prog Clin Biol Res 150:45, 1984

7.   Lusher JM, Blatt PM, Penner JA, Aledort LM, Levine PH, White GC, Warrier AI, Whitehurst DA: Autoplex versus proplex: A controlled, double-blind study of effectiveness in acute hemarthroses in hemophiliacs with inhibitors to factor VIII. Blood 62:1135, 1983

8.   Giles AR, Tinlin A, Greenwood R: A canine model of hemophilic (factor VIII:C deficiency) bleeding. Blood 60:727, 1982

9.   Kernoff PBA: Porcine factor VIII: Preparation and use in treatment in inhibitor patients. Prog Clin Biol Res 150:207, 1984

10. Nilsson IM, Hedner U: Characteristics of various factor VIII concentrates used in the treatment of haemophilia A. Brit J Haematol 37:543, 1977

11. Giles AR, Tinlin S, Hoogendoorn H, Fournel MA, Ng P, Pancham N: The in vivo characterisation of recombinant factor VIII in a canine model of hemophilia A (factor VIII deficiency). Blood 72:335, 1988

12. White II GC, McMillan CW, Kingdon HS, Shoemaker CB: Use of recombinant antihemophilic factor in the treatment of two patients with classic hemophilia. New Engl J Med 320:166, 1989

13. Vehar GA, Keyt B, Eaton D, Rodriguez H, O'Brien DP, Rotblat F, Oppermann H, Keck R, Wood WI, Harkins RN, Tuddenham EGD, Lawn RM, Capon DJ: Structure of human factor VIII. Nature 312:337, 1984

14. Toole JJ, Pittman DD, Orr EC, Murtha P, Wasley LC, Kaufman RJ: A large region (~95 kDa) of human factor VIII is dispensable for in vitro activity. Proc Natl Acad Sci USA 83:5939, 1986

15. Giles AR, Pittman DD, Kaufman RJ: Various deletions of the B domain of factor VIII do not compromise its cofactor function in vivo. Manuscript in Preparation

16. Esmon CT, Esmon NL: Protein C activation. Seminars in Thrombosis and Hemostasis 10:122, 1984

17. Pittman DD, Kaufman RJ: Proteolytic requirements for thrombin activation of anti-hemophilic factor (factor VIII). Proc Natl Acad Sci USA, 85:2429, 1988

18. Kaufman RJ, Pittman DD, Wasley LC, Wang JH, Israel DI, Giles AR: Expression and characterisation of F.VIII produced by recombinant technology, in Roberts HR (eds): Pure Factor VIII and the Promise of Biotechnology. Baxter Health Care Publications 1989, pages 119-145

19. Giles AR, Nesheim ME, Mann KG: Studies of factors V and VIII:C in an animal model of disseminated intravascular coagulation. J Clin Invest 74:2219, 1984

20. Hoogendoorn H, Nesheim ME, Giles AR; A qualitative and quantitative analysis of the activation and inactivation of protein C in vivo in a primate model. Blood 75:2164, 1990

EXPERIENCES WITH RECOMBINANT FACTOR VIIa IN HEMOPHILIACS

Ulla Hedner

Biopharmaceuticals Division
Novo Nordisk
DK-2880 Bagsvaerd, Denmark

INTRODUCTION

Treatment of hemophiliacs with antibodies against factor VIII and factor IX is a matter of major concern in hemophilia care. Although a number of sophisticated methods for removal of antibodies have been designed, such procedures are suitable only in life-threatening situations since they are complicated and expensive.[1,2] Much effort has been focused on finding an agent capable of activating the final common pathway of the coagulation cascade, inducing hemostasis independent of the presence of factor VIII and factor IX. Such an agent should be hemostatically active, but also safe with regard to thromboembolic complications. Currently used prothrombin complex concentrates (PCC) or activated PCC (APCC) seem only to be active in about 50% of the bleeds.[3,4] Furthermore, thromboembolic complications are being reported following the use of PCC or APCC.[5,6] Activated factor VII (FVIIa) should be an attractive candidate since it is not proteolytically active until after having formed a complex with tissue factor or other phospholipids, thereby minimizing the risk of inducing a systemic activation of the coagulation system.

PLASMA-DERIVED FACTOR VIIa (pFVIIa)

Factor VII was purified from human plasma and given to 2 hemophilia A patients with antibodies against factor VIII in association with muscle or joint bleeds, as well as extraction of primary molars. These patients were given pFVIIa in doses of 2-5 μg/kg, resulting in plasma levels of FVII of 1.4-2.0 U/ml, no clear shortening of the APTT, and a shortening of the prothrombin time of no more than 2 seconds. A hemostatic effect was achieved.[7] Later it was shown that higher doses (up to 10-15 μg/kg) of pFVIIa given at shorter intervals were also effective in major joint bleeds.[8] Following those doses the plasma levels of factor VII rose to 6-7 U/ml. In parallel, a transient shortening of the APTT was seen, as well as a prothrombin time shortening. As encouraging as these results were, it seemed obvious that a large scale production of a pure plasma-derived factor VIIa would meet with substantial difficulties since the plasma concentration of factor VII is low (about 0.5 μg/ml). Furthermore, the risk of transmitting blood-borne viruses makes alternative ways of production such as gene technology preferable.

RECOMBINANT FACTOR VIIa

Human FVIIa was cloned and expressed in BHK cells.[9] These cells were capable of γ-carboxylation, and the factor VII was secreted in the single-chain form. A spontaneous conversion of factor VII into FVIIa occurred during the purification procedure.[10] The recombinant FVIIa (rFVIIa) purified from the culture medium was characterized with regard to amino acid sequence, carbohydrate composition, and γ-carboxylation.[11] The amino acid sequence was identical with that found in pFVIIa, and the carbohydrate structure was very similar. Out of the 10 γ-carboxylated glutamic acid residues of pFVIIa, 9 were fully and 1 partially γ-carboxylated in the rFVIIa.[11]

The rFVIIa normalized the prolonged APTT of hemophilic plasmas (both A and B) *in vitro*, provided it was added in amounts of 3.8 μg/ml plasma. This roughly corresponds to a dose of 150 μg/kg.[12] Furthermore, the cuticle bleeding time was normalized in hemophilia A and hemophilia B dogs.[13]

CLINICAL EXPERIENCE WITH rFVIIa

Purified rFVIIa (Novostase™, NovoSeven™, NOVO NORDISK, Denmark) has been given to 22 patients in total, 20 of these being hemophiliacs (1 hemophilia B and 19 hemophilia A) with inhibitors. One patient had an acquired inhibitor against factor VIII and one had a factor XI deficiency further complicated by an inhibitor against factor XI. Out of the 22 patients, 13 were treated for life-threatening bleeding episode and 9 underwent surgery. The patients were treated on a compassionate use basis after the approval by the FDA or corresponding authorization from Health Authorities in countries other than the United States.

Table 1 Patients Given rFVIIa (Novostase, Novoseven)

---

13 Patients Treated for Bleeding Episodes
   11 Hemophilia A + Inhibitor
   1 Hemophilia B + Inhibitor
   1 Acquired Inhibitor Against FVIII:C

9 Patients Subjected to Surgery
   8 Hemophilia A + Inhibitor
   1 FXI Deficiency + Inhibitor Against FXI

Surgery

Nine patients have been subjected to surgery under the cover of rFVIIa as the sole coagulation factor. The first patient given rFVIIa was subjected to synovectomy.[14] He was given a dose of 54 μg/kg every 4 h for the first 2 days and then at every 6 h for 10 days. According to the routine in hemophilia surgery in Sweden, this patient was also given antifibrinolytic therapy (tranexamic acid). No abnormal peri- or postoperative bleeding occurred.

Three extensive tooth extractions (dental surgery) were uneventfully performed and one patient, 28 months old, was treated in association with a posttraumatic mouth bleed that was not controlled by repeated factor VIII infusions and tranexamic acid. He had a loose front tooth extracted, was bleeding from several venopunctures including one in the jugular vein and had a hemoglobin of 49 g/l when a factor VIII inhibitor

·was diagnosed. The initial Novostase (NovoSeven) dose of 60 μg/kg was
increased to 90 μg/kg every 3 h for 12 h, after which the dose intervals
were prolonged and the dose reduced. His bleeding was immediately
controlled.

One patient was subjected to an extensive herniotomy on a rFVIIa dose
of 75 μg/kg. No perioperative bleeding occurred. A second dose was
given 2 h later with the intention that he be given the same dose every 4
h. However, he started oozing from the wound just before the third dose
so the interval between doses was shortened to 3 h. Overt bleeding from
the wound then started at 26 h after surgery but it stopped after the
administration of extra rFVIIa. It began again on day 3 (45 h postop).
In spite of addition of tranexamic acid at this point, and extra rFVIIa
given by continuous infusion (38 μg/h), the bleeding was not fully
controlled and porcine FVIII had to be given.

Another patient, 21 years old, had a spontaneously ruptured <u>Salmonella</u>
abscess of the left thigh that made drainage of the area imperative. He
was subjected to explorative surgery under the cover of rFVIIa (90 μg/kg)
and extensive myonecrotic tissue was removed. No bleeding occurred
initially, but towards the latter part of surgery an episode of profuse
bleeding developed. Low levels of fibrinogen (0.8 g/L), α_2-antiplasmin
(46%) as well as AT III (24%) were detected, as well as high levels of
fibrin D-dimer. The bleeding slowed on addition of cryoprecipitate,
fresh frozen plasma, and red cells. He was initially kept on rFVIIa at
70 μg/kg every 3 h, and the dose was later increased to 90 μg/kg. Two
days later another surgical exploration was performed when the patient
was in a stable condition. He was kept on rFVIIa at a dose of 90 ug/kg
and there was no bleeding until several hours after the surgery when
bleeding recurred. Abnormal laboratory values were found. At this time
the bleeding again slowed with the same therapy and the patient was
continuously kept on rFVIIa at a dose of 90 μg/kg every 3 h.
Unfortunately, he then started to bleed excessively from the surgical
wound, became hypovolemic, developed bradycardia, and died. The
physician in charge of the treatment believed that the rFVIIa
successfully allowed the surgery to be performed and that it supported
haemostasis. He did not believe that the rFVIIa caused coagulation
factor consumption in this patient, but that this was related to the
bacteriaemia during the surgical manipulation and the extensive tissue
necrosis.

One patient, 66 years old, had <u>factor XI deficiency</u>, with substantial
bleeding problems in the past and an inhibitor against factor XI. He was
subjected to orchiectomy because of a massive bleeding from his left
testis, under the cover of rFVIIa (60 μg/kg) every 3 h. No bleeding
occurred. At 16 h after surgery the patient developed substernal chest
pain similar to that he had experienced once before, several years
previously. The EKG showed no ischaemic changes and the pain was
relieved by nitroglycerine.

<u>Life Threatening Bleeding</u>

Thirteen patients were given rFVIIa for life-threatening bleeds for
which there was considered to be no alternative treatment. Prothrombin
complex concentrate (PCC)/activated prothrombin complex concentrate
(APCC) administration and, in most cases, porcine FVIII had been proved
to be ineffective. One patient with massive laryngeal bleed that was
threatening to suffocate him was successfully treated with rFVIIa in
doses of 60 μg/kg[15]; another was successfully treated in association with
a cerebellar bleed; one with a post-traumatic epidural bleed with severe

neurological symptoms; one with a bleeding in the soft tissue of the neck-
following a failed venipuncture; one with a compartment syndrome in the
forearm; one with a joint bleed; and one with a severe muscle bleed in
the thigh. One of the patients was treated in association with severe
mesenteric-retroperitoneal bleeding. The rFVIIa controlled this bleed
within 24 hours and the resolution of the massive intra-abdominal
hematoma was followed on CT scans. All these bleeding episodes were
effectively controlled by rFVIIa given in doses of 60-90 μg/kg every 3-4
hours.

One patient was a boy with a profuse epistaxis whose nasal septum was
destroyed as a result of repeated episodes of severe epistaxis in the
past. Novostase (NovoSeven) in a dose of 52-75 μg/kg every 3-4 hours for
7 days controlled the bleeding initially, but the patient started to
rebleed. In order to control this heavy bleeding, FEIBA and PCC were
added. No side-effects were observed during the 11 days of rFVIIa
treatment.[16]

The one hemophilia B patient who was treated was 11 years old. He was
given rFVIIa because of profuse bleeding with hypotension and falling
haemoglobin following surgical fasciotomy for a compartment syndrome in
his left forearm. Serious anaphylactic reactions had previously been
recognized after the administration of factor IX concentrates. Novostase
(NovoSeven) was given in doses of 60-90 μg/kg every 3 h for 3 weeks,
whereafter the dose intervals were increased to every 4 h.
Antifibrinolytic treatment was instituted 4 days after the start of the
treatment. The bleeding was controlled during the first week of rFVIIa
treatment. Three weeks after the start of treatment, skin grafting was
performed under the cover of rFVIIa and tranexamic acid. The estimated
blood loss during the operation was 200 ml. During the treatment,
resistant Staph aureus was present in the wounds, and E.Coli was cultured
from the central line. No complications occurred. This patient received
as much as 0.7 mg/kg/24 h of rFVIIa without any side effects. He has
later been treated with rFVIIa for joint bleeds with good effect.[17]

Two patients treated with rVIIa died. In one, a retroperitoneal bleed
was complicated by severe renal insufficiency. While hemostasis was
achieved and the effect of rFVIIa treatment was described as very good,
insufficient amounts of rFVII were available at the time, and the
treatment could not be continued. The other patient who died was an
elderly man (68 years old) with an acquired factor VIII inhibitor. He
developed a bleed in the tongue that extended into the nasopharynx and
threatened to suffocate him. He was subjected to an emergency
tracheotomy and then bled profusely. Bleeding was controlled by rFVIIa
given in doses of 75-90 μg/kg every 3-4 h. Unfortunately, he died 2 days
later from profuse gastrointestinal bleeding. The steroid therapy he was
receiving may have contributed to the development of a hemorrhagic
gastritis and the fatal bleeding.

In summary, the experience so far supports the conclusion that rFVIIa
is hemostatically active in major surgery as well as in life-threatening
bleeds provided it is given frequently (every 3-4 hours) in doses of 70-
90 μg/kg. There is also some evidence that adequate doses of
antifibrinolytic therapy may permit hemostasis at a lower dose of rFVIIa.
It should be stressed that antifibrinolytic therapy has been recommended
for routine use in hemophilia surgery, together with factor VIII or
factor IX concentrates.[18,19]

MONITORING FVIIa ADMINISTRATION

Peak plasma FVII levels of 30-40 U/ml seem to be adequate for
hemostasis in serious bleeding episodes and major surgery. Trough levels
of ≥10 U/ml seem to be advisable in these conditions. The prothrombin
time shortens and is approximately 9-10 s during hemostatically effective
treatment. The APTT also shortens. It is, however, hard to establish a
recommended APTT effect since the results vary with the reagents used.

SAFETY

As pointed out, a factor VIII by-passing agent should be not only
hemostatically effective, but it must also be safe with regard to the
development of thromboembolic complications. The safety of FVIIa was
addressed in a number of ways. Plasma-derived FVIIa was given to normal
dog with monitoring of coagulation factors that are affected by
administration of PCC/APCC[20]. No signs of a generalized consumption
coagulopathy were observed[7]. The safety, especially with regard to the
induction of a generalized coagulopathy, was also evaluated in rabbits
for their tissue factor is known to form proteolytically active complexes
with human FVIIa. In doses of rFVIIa up to 5000 u/kg (roughly
corresponding to a dose of 150 μg/kg), no coagulation changes were
observed that indicate generalized proteolytic activity in the
circulating blood[12]. The possible enhancing activity of FVIIa on the
development of endotoxin-induced systemic coagulopathy was also tested in
a rabbit model. The rabbits received two different doses of endotoxin
(0.03 μg/kg/h or 0.30 μg/kg/h). Each group of animals was divided into
two subgroups, receiving either 100 μg/kg or 300 μg/kg of rFVIIa in
addition to the endotoxin infusion. A negative control group was given
only saline and positive control groups received only endotoxin. No
rFVIIa effect was found, as measured by the leucocyte and platelet
counts, the fibrinogen level and the ethanol gelation test as a measure
of soluble fibrin--and indirectly of free thrombin activity (Diness et
al. To be published).

Thus rFVIIa seems to be safe in regard to the development of
thromboembolic complications, as well as when given with endotoxin so
that tissue factor exposure is induced, e.g. on monocytes[21]. This was
also observed in patients given rFVIIa, one of whom had <u>Staphylococci</u>
grown from wound fluid as well as <u>E.Coli</u> from his central venous line.
Furthermore, the coagulopathy that developed during surgery in the
patient with a ruptured abscess and <u>Salmonella</u> bacteriaemia was reversed
during treatment with rFVIIa.

REFERENCES

1. Nilsson IM, Berntorp E, Zettervall O: Induction of immune tolerance
    in patients with hemophilia and antibodies to factor VIII by
    combined treatment with intravenous IgG, cyclophosphamide, and
    factor VIII. New Engl J Med 318:947, 1988
2. White GC, Taylor RE, Blatt PM, Roberts HR: Treatment of a high titre
    anti-factor-VIII antibody by continous factor VIII administration:
    Report of a case. Blood 58:141, 1983
3. Sjamsoedin LJN, Heijnen L, Mauser-Bunschoten EP, van Geijlswijk JL,
    van Houwelingen H, van Asten P, Sixma PJ: The effect of activated
    prothrombin-complex concentrate (FEIBA) on joint and muscle bleeding
    in patients with hemophilia A and antibodies to factor VIII. N Engl
    J Med 305:717, 1981

4. Lusher JM, Shapiro SS, Palascak JR, A. Rao AV, Levine PH, Blatt PM, and the Hemophilia Study Group: Efficacy of prothrombin-complex concentrates in hemophiliacs with antibodies to factor VIII. N Engl J Med 303:421, 1980

5. Sullivan DW, Purdy LJ, Billingham M, Glader BE: Fatal myocardial infarction following therapy with prothrombin complex centrates in a young man with hemophilia A. Pediatrics 74:279, 1984

6. Schimpf KL, Zeltsch C, Zeltsch P: Myocardial infarction complicating activated prothrombin complex concentrate substitution in patient with haemophilia A. Lancet ii:1043, 1982

7. Hedner U, Kisiel W: The use of human factor VIIa in the treatment of two hemophilia patients with high-titer inhibitors. J Clin Invest 71:1836, 1983

8. Hedner U, Bjoern S, Bernvil SS, Tengborn L, Stigendahl L: Clinical experience with human plasma-derived factor VIIa in patients with hemophilia A and high-titer inhibitors. Haemostasis 19:335, 1989

9. Hagen FS, Gray CL, O'Hara P, F. Grand FJ, Saari GC, Woolbury RG, Hart CE, Insely M, Kisiel W, Kurachi K, E. Davie EW: Characterization of a cDNA coding for human factor VII. Proc Natl Acad Sci USA 83:2412, 1986

10. Bjoern S, Thim L: Activation of coagulation factor VII to VIIa. Res Discl 269, 1986

11. Thim L, Bjoern S, Christensen M, Nicolaisen EM, Lund-Hansen T, Pedersen AH, Hedner U: Amino acid sequence and posttranslational modifications of human factor VIIa from plasma and transfected baby hamster kidney cells. Biochemistry 27:7785, 1988

12. Hedner U, Ljungberg, Lund-Hansen T: Comparison of the effect of plasma-derived and recombinant human FVIIa *in vitro* and in a rabbit model. Blood Coagulation and Fibrnolysis. Accepted for publication 1990.

13. Brinkhous KM, Hedner U, Garris JB, Diness V, Read MS: Effect of recombinant factor VIIa on the hemostatic defect in dogs with hemophilia, hemophilia B, and von Willebrand disease. Proc Natl Acad Sci USA 86:1382, 1989

14. Hedner U, Glazer S, Pingel K, Alberts KA, Blombäck M, Schulman S, Johnsson H: Successful use of recombinant factor VIIa in patient with severe hemophilia A during synovectomy. Lancet ii:1193, 1988

15. Macik BG, Hohneker J, Roberts HR, Griffin AM: The use of recombinant activated factor VII for treatment of a retropharyngeal hemorrhage in a hemophilic patient with a high titer inhibitor. Am j Med 32:232, 1989

16. Grupp SA, Glazer S, Williams DA: Recalcitrant epistaxis in a hemophiliac with inhibitors: Experience with recombinant factor VII. Blood 74 (Suppl. 1):390a, 1989

17. Schmidt ML, Smith HE, Gamerman S, Glazer S, Scott JP: Prolonged recombinant activated factor VII (rFVIIa) treatment for severe bleeding and surgical hemostasis in a factor IX-deficient patient with an inhibitor. Abstracts, Soc Ped Res, May, 1990

18. Forbes CD, Barr RD, Reid G: Tranexamic acid in control of haemorrhage after dental extraction in haemophilia and Christmas disease. Br Med J 2:311, 1972

19. Nilsson IM, Hedner U, Ahlberg A, Larsson SA, Bergentz SE: Surgery of hemophiliacs - 20 years' experience. World J Surg 1:55, 1977

20. Hedner U, Nilsson IM, Bergentz SE: Various prothrombin complex concentrates and their effect on coagulation and fibrinolysis *in vivo*. Thromb Haemostas 35:386, 1976

21. Noguchi M, Sakai T, Kisiel W: Correlation between antigenic and functional expression of tissue factor on the surface of cultured human endothelial cells following stimulation by lipopolysaccharide endotoxin. Thrombo Res 55:87, 1989

CLINICAL TRIALS OF FACTOR VIII PRODUCED BY RECOMBINANT TECHNOLOGY

Richard S. Schwartz

Clinical Research Department
Cutter Biological, Miles Inc.
Berkeley, CA USA

Hemophilia is a very old disease; references to a bleeding disorder appearing in males are found as early as the 2nd century AD in the Talmud. However, treatment has only been available since the 1950's when treatment with fresh frozen plasma became available. The major therapeutic advance for hemophilia A occurred in the mid-1960s when Judith Pool of Stanford University described a methodology to isolate an anti-hemophilic factor from cryoprecipitate[1]. This observation led to the rapid introduction of commercial scale production of factor VIII concentrates for treatment of hemophilia A. Factor IX concentrates for treatment of hemophilia B were also developed shortly thereafter.

Development of factor VIII concentrate for treatment of hemophilia A has unquestionably resulted in a dramatic increase in both the quality of life and life expectancy of individuals with this disorder[2,3]. However, treatment with plasma-derived clotting concentrates has been associated with some disadvantages. Reliance on plasma as a source of factor VIII concentrate has resulted in exposure of treated hemophiliacs to alloantigens as well as transfusion-related viral disorders including hepatitis B, non-A, non-B hepatitis, and human immunodeficiency virus (HIV)[4-6]. Fortunately, a number of methodologies have been described in the last several years, such that all the factor VIII concentrates currently on the market are now safe from transmission of HIV[4-6]. However, these new technologies, developed to inactivate potential viruses in human plasma, have decreased yields, initially exacerbating a world-wide shortage of clotting factor concentrates for treatment of hemophilia[6].

There has therefore been great interest in developing a method to manufacture factor VIII for treatment of hemophilia A using recombinant technology, obviating the need for reliance on plasma. This review will summarize material that will be reported in reference 7, documenting the rapid progress made in the past 6 years towards this goal.

As a result of major advances in molecular biology and genetic engineering, we have learned a tremendous amount about the structure and biology of factor VIII[8-12]. We now know that the factor VIII protein consists of 2332 amino acids. It is synthesized as a 2351 amino acid single chain precursor in the cell but is secreted as a 2332 amino acid peptide following N-terminal processing[9,11].

Recombinant Technology in Hemostasis and Thrombosis
Edited by L.W. Hoyer and W.N. Drohan, Plenum Press, New York, 1991

Factor VIII appears to circulate in the plasma as a two-chain metal-ion stabilized complex consisting of a heavy chain, typically with a molecular weight ≥90 kDa, and a light chain of 80 kDa. The factor VIII protein has several repeating regions (A1, A2, and A3) that share a high degree of homology with each other, as well as with similar domains in ceruloplasmin. There is also an intervening region, the B region, separating the heavy and light chains in the single peptide precursor which appears to be non-essential for factor VIII coagulant activity. The molecule is heavily glycosylated, mostly in the B region. Both the heavy and light chains are cleaved into lower molecular weight components upon activation by thrombin[13]. Factor VIII circulates in the plasma complexed with von Willebrand's factor.

The gene encoding factor VIII is 186,000 base-pairs in length, consisting of 26 exons and 25 introns[8]. The messenger RNA is some 9000 base-pairs in length, coding for a protein described previously with a total molecular weight after post-translational modifications estimated at 330,000 kDa.

The concentration of factor VIII in plasma is very low. Rotblat et al[14] and Fass et al[15] succeeded in purifying human and porcine factor VIII, respectively, to an extremely high degree. This permitted two different groups, one at Genentech and the other at Genetics Institute, to create synthetic DNA probes which were then used to probe the DNA genome of human cells to try to identify the gene for factor VIII[8-11]. The creation of synthetic DNA probes was an important initial step to be able to succeed in this endeavor. By utilizing a number of different overlapping probes, it was ultimately possible to identify the entire DNA sequence of the factor VIII gene. The DNA fragments were then used to probe the messenger RNA of the human cells to identify and sequence the messenger RNA for factor VIII. By this means, it was possible to prepare a cDNA encoding only the coding sequences (exons) required.

A very simplified overview of the process by which recombinant factor VIII concentrate is prepared by Cutter Biological is summarized in Fig. 1. In work done by Genentech, Inc., the factor VIII gene was expressed and copied, as just discussed, and genetically engineered factor VIII was produced using mammalian cell culture. Next, scientists at Cutter Biological, working with their colleagues at Bayer AG, developed a manufacturing scale process to produce recombinant factor

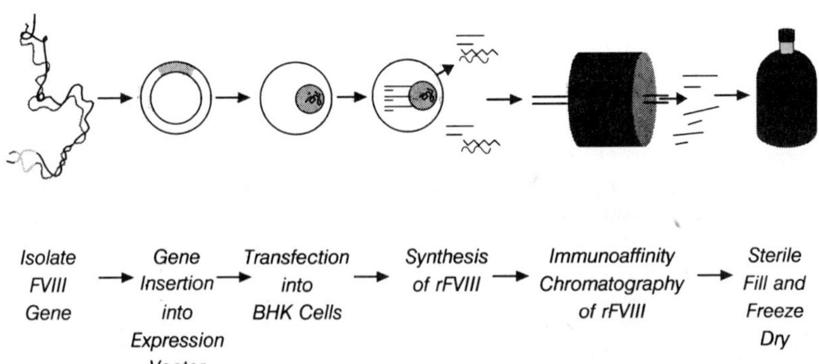

Fig. 1.   Schematic representation of developmental process for manufacture of factor VIII (FVIII) by recombinant DNA technology (Reprinted with permission from reference 7)

VIII in sufficient quantities. The factor VIII molecule is secreted into medium along with other proteins from the cell substrate. A high degree of purification of the final product is accomplished by immunoaffinity chromatography using a murine monoclonal antibody specific for factor VIII after preliminary preparative adsorption on and elution from an ion exchange matrix. The molecule is then stabilized with albumin, sterile filled, and lyophilized.

About 40% of the recombinant factor VIII (rFVIII) has been sequenced to date and is in 100% agreement with the consensus structure sequence predicted from the cDNA. On sodium dodecyl sulfate-polyacrylamide gel electrophoresis, recombinant factor VIII demonstrates the same banding pattern as plasma-derived human factor VIII, with the exception of a 210 kDa band sometimes seen with plasma-derived factor VIII corresponding to von Willebrand's factor which is not present in the recombinant preparation.

Laboratory studies have been conducted to determine the equivalence of recombinant factor VIII with plasma-derived factor VIII. The one-stage clotting assay gives parallel curves for both preparations, and the initial rates of factor Xa generation by plasma-derived factor VIII and recombinant factor VIII are similar[16]. Activated protein C inactivates both factor VIII preparations in the same manner. A series of studies in haemophilic dogs have been conducted in collaboration with Alan Giles in Canada, demonstrating that plasma-derived factor VIII and recombinant factor VIII yield similar *in vivo* pharmacokinetics and correction of bleeding times in an animal model of hemophilia A[17].

Clinical studies with recombinant factor VIII have been an international effort, with studies conducted in multiple centers. To date, 96 subjects have been enrolled in the clinical investigation of recombinant factor VIII. The study has been conducted in three stages:

Stage   I: safety and pharmacokinetics.
Stage II: safety and efficacy.
Stage III: treatment for surgical procedures and acute bleeding
             episodes.

Previously untreated patients (PUPs) are also being treated.

In Stage I of the investigation, comparative studies of half-life determination and pharmacokinetics using plasma-derived factor VIII and recombinant factor VIII were conducted in 17 asymptomatic hemophiliacs. Plasma factor VIII levels were increased equivalently with both preparations, and *in vivo* falloff curves were comparable. The *in vivo* half-life of recombinant factor VIII, 15-19 hours, equalled or exceeded the half-life of 15 hours measured with plasma-derived factor VIII.

In Stage II, 44 hemophilic subjects are receiving exclusively recombinant factor VIII for standard home-treatment and prophylaxis of bleeding episodes. Many of the subjects have now been on study more than one year. The data indicate a high degree of success as evidenced by the assessment of the subjects participating in the investigation, and the number of treatments being required to treat bleeding episodes.

As part of Stage III of the study, subjects with hemophilia A are undergoing surgical procedures or are being treated for major hemorrhagic episodes in the hospital under exclusive recombinant factor VIII coverage. A series of bleeding episodes or surgical procedures have now been treated with recombinant factor VIII exclusively, and hemostasis has been achieved in all the cases.

Infusions of recombinant factor VIII have been well tolerated, with only infrequent and relatively minor reactions. No individual participating in these studies has had to terminate therapy due to adverse reactions, nor has any treatment not been completed. No individual has formed antibody to the trace amounts of murine and hamster proteins in the preparation.

One of the complications of treatment of hemophilia A with factor VIII is the development of a neutralizing antibody for factor VIII, so called inhibitor antibodies. Approximately 15% of hemophiliacs develop these antibodies when exposed to plasma-derived factor VIII[17]. This is a serious complication, and it often makes therapy with factor VIII difficult if not impossible. The pathogenesis for inhibitor formation is currently unknown as is the reason why some hemophiliacs develop this complication while others do not. We do know already that some individuals will form inhibitor antibodies with recombinant factor VIII, as is seen with plasma-derived factor VIII. Whether or not the frequency of inhibitor formation will be the same with recombinant factor VIII as with plasma-derived factor VIII is currently under study.

In summary, recombinant DNA technology offers new hope for hemophiliacs. This new technology is evolving rapidly, with clinical trials involving recombinant factor VIII now well advanced. This preparation will offer a new treatment modality for subjects with hemophilia A.

ACKNOWLEDGEMENT

Participating investigators of the international clinical research team for recombinant factor VIII are as follows: Charles F. Abildgaard, M.D. (University of California, Davis, CA); Louis M. Aledort, M.D., Steven Arkin, M.D. (Mt. Sinai Medical School, New York, NY); Arthur L. Bloom, M.D., Justin Harrison, M.D. (University of Wales, Cardiff, UK); Hans H. Brackmann, M.D., Johannes E. Egli, M.D. (University of Bonn, Bonn, FRG); Peter H. Levine, M.D., Doreen B. Brettler, M.D. (Memorial Hospital, Worcester, MA); Hiromu Fukui, M.D., Akira Yoshioka, M.D. (Nara Medical College, Nara, Japan); Michio Fujimaki, M.D. (Tokyo Medical College, Tokyo, Japan); Margaret W. Hilgartner, M.D. (Cornell University Medical Center, New York, NY); Martin J. Inwood, M.D. (St. Joseph's Hospital, London, Ontario, Canada); Carol Kasper, M.D. (Orthopaedic Hospital, Los Angeles, CA); Peter B.A. Kernoff, M.D., Paul L. Giangrande, M.D. (Royal Free Hospital and School of Medicine, London, UK); Jeanne Lusher, M.D., Indira Warrier, M.D. (Children's Hospital of Michigan, Detroit, MI); Pier M. Mannucci, M.D., Alessandro Gringeri, M.D., Loredana Simoni, M.D. (University of Milan, Milan, Italy); Shelby Dietrich, M.D. (Pasadena, CA); Inge Scharrer, M.D., Emel Aygören, M.D., Wolfhart Kreuz, M.D. (University Hospital, Frankfurt, FRG); CA: Joseph E. Addiego Jr., M.D. (Children's Hospital, Oakland, CA); Jennifer M. Pearce, M.D. (Albany Medical College, Albany, NY); Thomas Coyle, M.D. (State University at New York, Syracuse, NY); Leticia P. Valdez, M.D. (Children's Medical Center, Dayton, Ohio); Carl E. Krill, M.D. (Children's Hospital, Akron, Ohio); Data Collection, clinical study coordination, and analysis: North America - Cutter Biological, Miles Inc., Berkeley, CA: Richard S. Schwartz, M.D., Randy U. Allred, D.P.H., Mary A. MacKenzie, B.S., R.N., Pat Bailey, R.N., M.S., Stephen Dyla, Ph.D., Barbara Nielsen, B.A., Ralph H. Rousell, M.D., Stephen Coleman, M.D., James E. Pennington, M.D.; Europe - Bayer AG, Wuppertal, FRG: Dieter Maruhn, M.D., Peter Eckert, M.D., Jean Pierre Fallise, M.D.; Tropon - Cologne, FRG: Roswita Neumann, Ph.D.; Japan: Takeyoshi Minaga, M.D.

REFERENCES

1. Pool JG, Hershgold EJ, Pappenhagen AR: High-potency antihaemophilic factor concentrate prepared from cryoglobulin precipitate (letter). Nature 203:312, 1964

2. Levine PH: Efficacy of self-therapy in hemophilia: a study of 72 patients with hemophilia A and B. N Engl J Med 291:1381, 1974

3. Smith PS, Keyes NC, Forman EN: Socioeconomic evaluation of a state-funded comprehensive hemophilia-care program. N Engl J Med 306:575, 1982

4. Goldsmith MF: Hemophilia, beaten on one front, is beset on others. JAMA 256:3200, 1986

5. Brettler DB, Levine PH: Factor concentrates for treatment of hemophilia: which one to choose? Blood 73:2067, 1989

6. Pierce GF, Lusher JM, Brownstein SP, Goldsmith JC, Kessler CM: The use of purified clotting factor concentrates in hemophilia. Influence of viral safety, cost, and supply on therapy. JAMA 261:3434, 1989

7. Schwartz RS: Recombinant derived blood clotting factors including factor VIII and their potential, in Schild GC, Haseltine WA, Mann RD (eds): Applications of Biotechnology in Therapeutics. Carnforth, Parthenon Publishing, 1991 (in press)

8. Gitschier J, Wood WI, Goralka TM, Wion KL, Chen EY, Eaton DH, Vehar GA, Capon DJ, Lawn RW: Characterization of the human factor VIII gene. Nature 312:326, 1984

9. Wood WI, Capon DJ, Simonsen CC, Eaton DL, Gitschier J, Keyt B, Seeburg PH, Smith DH, Hollingshead P, Wion KL, Delwart E, Tuddenham EGD, Vehar GA, Lawn RM: Expression of active human factor VIII from recombinant DNA clones. Nature 312:330, 1984

10. Vehar GA, Keyt B, Eaton D, Rodriguez H, O'Brien DP, Rotblat F, Oppermann H, Keck R, Wood WI, Harkins RN, Tuddenham EGD, Lawn RM, Capon DJ: Structure of human factor VIII. Nature 312:337, 1984

11. Toole JJ, Knopf JL, Wozney JM, Sulzman LA, Buecker JL, Pittman D, Kaufman RJ, Brown E, Shoemaker C, Orr EC, Amphlett GW, Foster WB, Coe ML, Knutson GJ, Fass DN, Hewick RM: Molecular cloning of a cDNA encoding human antihaemophilic factor. Nature 312:342, 1984

12. White GC, Shoemaker CB: Factor VIII gene and hemophilia A. Blood 73:1, 1989

13. Pittman D, Kaufman R: Proteolytic requirements for thrombin activation of anti-hemophilic factor (factor VIII). Proc Natl Acad Sci 85:2429, 1988

14. Rotblat F, O'Brien DP, Middleton SM, Tuddenham EGD: Purification and characterisation of human FVIII:C. Thromb Haemost 50:108, 1983

15. Fass DN, Knutson GJ, Katzman JA: Monoclonal antibodies to porcine factor VIII coagulant and their use in the isolation of active coagulant protein. Blood 59:594, 1982

16. Eaton DL, Hass PE, Riddle L, Mather J, Wiebe Wiebe, Gregory T, Vehar GA: Characterization of recombinant human factor VIII. J Biol Chem 262:3285, 1987

17. Giles AR, Tinlin S, Hoogendoorn H, Fournel MA, Ng P, Pancham N. In vivo characterization of recombinant factor VIII in a canine model of hemophilia A (factor VIII deficiency). Blood 72:335, 1988

18. Gill FM. The natural history of factor VIII inhibitors in patients with hemophilia A, in Hoyer LW (ed): Factor VIII Inhibitors. New York, Liss, 1984, p 19

Clinical Trials of Recombinant Factor VIII

Gilbert C. White, II, Campbell W. McMillan, Edward D. Gomperts, Joseph E. Addiego, Jr., Hans H. Brackmann, Doreen B. Brettler, Joan C. Gill, Jonathon C. Goldsmith, W. Keith Hoots, Craig M. Kessler, C. Thomas Kisker, Y. Laurian, B. Lewis, Jeanne M. Lusher, Piero M. Mannucci, Messimo Morfini, Inge Scharrer, Sam Schulman, Martin L. Lee, Suzanne G. Courter, Henry S. Kingdon, and Harold R. Roberts

University of North Carolina School of Medicine, Children's Hospital of Los Angeles, Children's Medical Center of Northern California, Uversitate Bonn-Venusberg, Worcester Memorial Hospital, Great Lakes Hemophilia Foundation, University of Nebraska, M.D. Anderson Hospital, George Washington University, University of Iowa, Hopital de Bicentre, Alta Bates Hospital, Children's Hospital of Michigan, Milano, Ospedale Careggi, Frankfurt University Hospital, Karolinska Hospital, and Baxter Healthcare Corporation

INTRODUCTION

Preparations of plasma coagulation factors have been mainstays in the treatment of hemorrhagic disorders since their introduction in the middle 19th century. One of the first recorded uses of blood products to correct the hemorrhagic tendency in a hemophiliac was by Lane in 1840[1]. By the middle of the present century, the use of whole plasma[2] or various fractions derived from plasma[3] was accepted therapy in hemorrhagic disorders although the large volumes required and the lack of plasma availability limited the use of these materials to major bleeding episodes. The development in 1964 of cryoprecipitate[4], a cold-precipitated plasma fraction enriched in factor VIII, was a significant advance for patients with hemophilia. Not only was this used in the treatment of hemophilia for the next twenty years, but it was to be the starting material for many of the factor VIII concentrates to follow. In 1968, glycine-precipitated preparations of human factor VIII were first introduced for clinical use[5]. This procedure was amenable to large scale production of factor VIII and therefore to commercial preparation of factor VIII concentrates. By the early 1970s, factor VIII prepared in this way had become the major method for treatment of patients with hemophilia.

Despite this, plasma-derived clotting factors presented numerous problems. Foremost was the contamination of these products with viruses, including hepatitis B virus, hepatitis C virus, and, most importantly, human immunodeficiency virus (HIV-1)[6]. As a result, a number of techniques for the inactivation or removal of viruses were developed. These included heat treatment[7], solvent or detergent extraction[8], monoclonal antibody affinity chromatography, and pasteurization[9]. The ability to clone the genes for clotting factors introduced a new dimension in the preparation of viral-free proteins. Coupled with the development of methods for the high level expression of proteins in cell lines, this led to the ability to prepare large quantities of recombinant protein and resulted in 1987 in the production of the first recombinant clotting factor for clinical use, recombinant factor VIII[10]. In this chapter, the synthesis of recombinant factor VIII and the initial clinical studies with recombinant factor VIII will be presented. These studies provide preliminary evidence for the safety of recombinant clotting factors for patients with hemophilia and provide further support for the development of these products.

Recombinant Technology in Hemostasis and Thrombosis
Edited by L.W. Hoyer and W.N. Drohan, Plenum Press, New York, 1991

FACTOR VIII STRUCTURE AND FUNCTION.

Factor VIII is a large molecular weight glycoprotein which circulates in plasma bound non-covalently to multimeric von Willebrand factor. As shown in Fig. 1, analysis of the structure of factor VIII reveals several distinct features which may be important for its function in blood coagulation[11-13]. At the amino terminus of the protein are three non-contiguous A domain repeats, each approximately 350 amino acids in length with 30 percent homology between repeats. Similar A domain repeats are present in the copper-binding protein, ceruloplasmin, and are important for copper binding[14]. Based on this homology with ceruloplasmin, it has been suggested that one function of the A domains in factor VIII may be to bind divalent cations. Coagulation factor V, a protein that is structurally and functionally homologous with factor VIII, also contains homologous A domain repeats[15-16]. The first A domain repeat is at the amino terminus of the molecule and is separated from the second A domain repeat by a stretch of 36 predominantly acidic amino acids. In the middle of factor VIII is a large B domain, which is 983 amino acids in length and is encoded by a single exon. The B domain separates the second and third A domain repeats and together with the A1 and A2 domains constitutes the heavy chain of plasma factor VIII. The function of the B domain, if any, is unknown. Genetically engineered factor VIII lacking the B domain retains full procoagulant activity[17-20]. Recent studies suggest that the B domain may affect the efficient intracellular transit of factor VIII during synthesis[21]. A second stretch of 41 predominantly acidic amino acids is between the B domain and the third A domain repeat. This second acidic region contains a sulfated tyrosine residue and is the site of interaction with plasma von Willebrand factor[22]. Two C domain repeats, each approximately 150 amino acids in length, form the carboxy terminus of factor VIII. These repeats share approximately 30 percent homology and are homologous with a phospholipid and lectin binding sequence in discoidin I, a protein involved in cellular aggregation in the slime mold *Dictyostelium discoideum*. Based on a variety of studies, the C domain repeats have been implicated in lipid binding[23-25] and may therefore play an important role in the assembly of the factor X activating complex on the surface of platelets and other cells. Together with the third A domain repeat, the two C domain repeats constitute the light chain of plasma factor VIII.

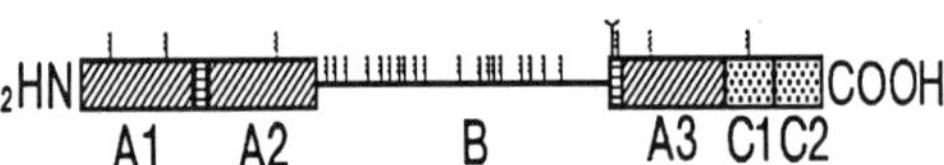

Fig. 1. **Factor VIII domain structure**. Potential *N*-linked glycosylation sites are shown by the curved lines. Y, tyrosine sulfate residue in the second acidic region. The A domain repeats (hatched boxes), B domain, C domain repeats (dotted boxes), and acidic regions (horizontal lined boxes) are as shown.

Factor VIII is synthesized primarily in hepatocytes as a single polypeptide chain and undergoes posttranslational modification with addition of O- and N-linked carbohydrates, sulfation of tyrosine residues, and proteolytic processing by one or more unknown proteases[26]. This intracellular processing results in removal of varying lengths of the B domain with the formation of a two chain, calcium-linked molecule consisting of a heavy chain which ranges in size from 90,000 to 210,000 daltons and a light chain of 78,000 or 80,000 daltons (Fig. 2). As a result, plasma factor VIII is structurally heterogeneous with multiple molecular weight forms between 170,000 and 300,000 daltons, all having the same specific clotting activity.

236

Factor VIII functions as a cofactor in the activation of factor X to factor Xa by activated factor IX in the presence of calcium ions and a phospholipid surface. The catalytic efficiency of factor VIII is enhanced nearly 10,000-fold following treatment with thrombin or factor Xa[27]. As illustrated in Fig. 2, this enhanced activity is accompanied by a sequence of proteolytic cleavages in factor VIII that correlate with the increased activity[28-32]. Thrombin cleaves Arg-Ser bonds at position 372, the junction between the first acidic region and the second A domain; at position 740, the junction between the second A domain and the B domain; and at position 1689, the junction between the second acidic region and the third A domain. As a result, the variable heavy chain is converted a uniform 90,000 dalton heavy chain by cleavage at Arg_{740} and to 50,000 and 43,000 dalton fragments by cleavage at Arg_{372}, while the 80,000 light chain is converted to a 73,000 dalton fragment by cleavage at Arg_{1689} and the consequent release of the acidic peptide from the amino terminus of the light chain. The cleavages at positions 372 and 1689 are essential for the increased activity of factor VIII but cleavage at position 740 is not[32].

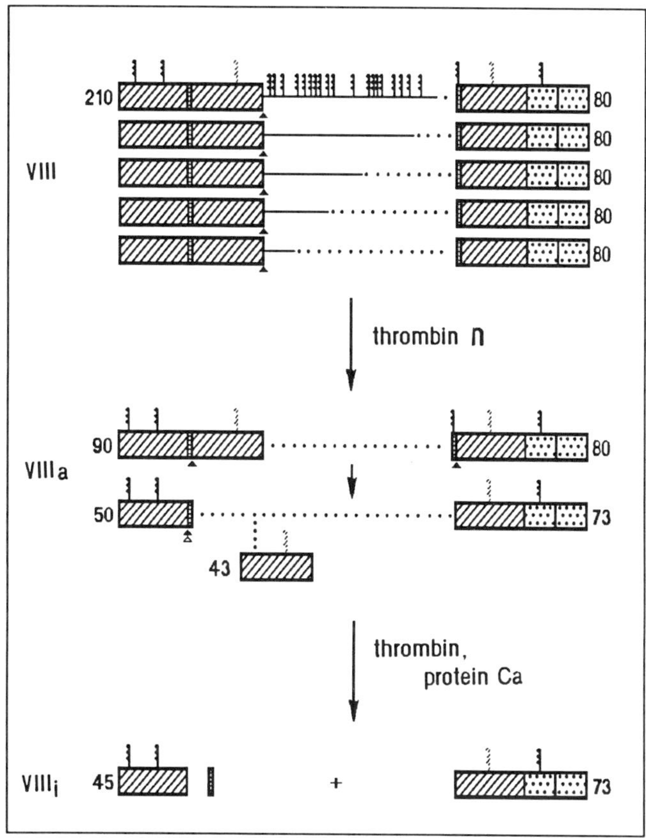

Fig 2. Factor VIII Activation and Inactivation. In plasma, factor VIII (VIII) is present as a family of proteins composed of a variable heavy chaim (90-210 kDa) and light chains (78-80 kDa). When activated by thrombin (VIII_a), the heavy chain is converted to a 90 kDa and then to 50 and 43 kDa fragments and the light chain to 75-73 kDa fragments by the cleavages indicated (▲). Further cleavage by thrombin or by activated protein C (protein Ca) (Δ) results in inactivation of factor VIII (VIII_i)

Factor VIII is inactivated by activated protein C as well as by further thrombin cleavage[33]. The proposed site of cleavage by activated protein C is the Arg-Ser bond at position 336, resulting in the release of the first acidic region from the 50,000 dalton heavy chain fragment of activated factor VIII.

RECOMBINANT FACTOR VIII

<u>Synthesis and formulation</u>. In general, recombinant hemostatic proteins such as factor VIII have been prepared using mammalian expression systems. This is in part because prokaryotic systems lack mechanisms for disulfide bond formation, *N*- and *O*-linked glycosylation, and sulfation of tyrosine residues, all important posttranslational modifications in factor VIII. In the studies described here, recombinant factor VIII has been prepared by a method developed by Dr. Randall Kaufman at Genetics Institute (Boston, MA), using Chinese hamster ovary cells as a host cell and a viral based plasmid expression vector that contains an SV40 enhancer, the adenovirus major late promoter, and an SV40 polyadenylation signal[34]. An adenovirus tripartite leader is included for efficient transcription of factor VIII. The CHO cells used for expression of factor VIII is a strain deficient in dihydrofolate reductase (DHFR) so that when cells are cotransfected with factor VIII cDNA and a DHFR minigene, DHFR positive cells (and therefore cells likely to also express factor VIII) can be selected by growth in nucleoside free medium followed by amplification of the DHFR positive cells in the presence of sequentially increasing concentrations of methotrexate, an inhibitor of DHFR. Initial studies by Kaufmann and coworkers showed that factor VIII levels of expression were low but were improved by coexpression of von Willebrand factor with factor VIII[35].

The factor VIII prepared in CHO cells is isolated from the cell medium by affinity chromatography using a monoclonal antibody to human factor VIII. The column is washed several times to remove loosely associated proteins and the bound factor VIII is eluted in 40% ethylene glycol. Further purification is obtained by ion exchange chromatography on QAE-Sepharose. The resulting factor VIII is pure and has a specific activity of 3000 units/mg.

<u>Comparison with plasma factor VIII</u>. Recombinant factor VIII has been compared structurally and functionally with purified plasma-derived factor VIII[10,26]. In CHO cells, recombinant factor VIII undergoes intracellular processing and proteolysis that is indistinguishable from that of plasma-derived factor VIII. Thus, like plasma derived factor VIII, recombinant factor VIII is isolated as a family of proteins with a variable heavy chain of 90,000 to 210,000 daltons and light chains of 78,000 and 80,000 daltons. The carbohydrate content of recombinant factor VIII is identical to plasma-derived factor VIII and the organization of the carbohydrates within the various domains of recombinant factor VIII appear to be the same as in plasma-derived factor VIII. Sulfation of tyrosine residues also occurs normally in recombinant factor VIII. Functionally, recombinant and plasma-derived factor VIII have equivalent specific activities. Recombinant factor VIII is activated by thrombin and factor Xa and inactivated by activated protein C in a manner similar to that of plasma-derived factor VIII and the proteolytic cleavages that occur during activation and inactivation are identical in the two proteins. Finally, plasma and recombinant factor VIII demonstrate similar antigenic specificity in mice suggesting that the two proteins are immunologically alike.

<u>*In vivo* response and toxicology</u>. A small phase I study with recombinant factor VIII was started in March 1987[10]. Recombinant factor VIII, prepared in CHO cells and purified and formulated as described above, was administered to two patients with severe classical hemophilia. Initial fall-off studies performed after administration of a dose of 50 units per kilogram were compared with fall-off studies after the same dose of heat-treated plasma-derived factor VIII. The recovery of recombinant factor VIII, expressed as a percentage of the recovery of plasma-derived factor VIII, was 188 and 102 percent in the two patients. The biological half-life of the recombinant factor VIII in the two patients was 16.1 ± 0.1 hours and was very close to the biological half-life of 15.3 ± 3.0 hours observed for plasma derived factor VIII. Thus, the biological response to recombinant factor VIII was equivalent to that to plasma-derived factor VIII.

After the initial half-life studies, each patient was placed on demand therapy with recombinant factor VIII for up to one year. Liver function tests, immunological studies, and inhibitor assays were performed monthly, and fall-off studies were performed at six and 12 months. No significant changes occurred in liver or immunologic function and there was no evidence of antibody formation by fall-off study or by Bethesda assay using either recombinant or plasma derived factor VIII as substrate. Clinically, both patients responded well to recombinant factor VIII with prompt control of joint and soft tissue hemorrhage. Thus, these initial studies provided preliminary evidence that recombinant factor VIII was tolerated well with no overt toxicity and was clinically efficacious.

CLINICAL STUDIES

To further evaluate the clinical effectiveness of recombinant factor VIII, a multicenter, phase II clinical trial was started in August 1988 using recombinant factor VIII prepared from CHO cells coexpressing factor VIII and von Willebrand factor[36]. The study is targeted for previously treated patients with severe classical hemophilia. Patients with inhibitors to factor VIII at the time of entry or with AIDS and ARC are excluded, but patients may be seropositive or seronegative for the human immunodeficiency virus (HIV). The study design is similar to that of the phase I study. After initial fall-off studies comparing monoclonal antibody-purified plasma-derived factor VIII with recombinant factor VIII, patients are started on demand home therapy with recombinant factor VIII with follow-up fall-off studies at three month intervals and liver function tests, immunological studies, and inhibitor assays at monthly intervals.

The study has been in progress for over 20 months. At the time of this analysis, 46 patients have been enrolled in the study. About two-thirds are HIV seropositive and nearly 90 percent are HBsAb positive. All have received previous treatment with plasma-derived factor VIII and most have been heavily treated with greater than 100 treatment-days of lifetime factor VIII exposure. This heavy lifetime exposure to factor VIII makes this group at low risk to develop inhibitors to factor VIII and would therefore make the development of inhibitors while on recombinant factor VIII noteworthy.

Clinically, the efficacy suggested in the phase I trial has been confirmed. Following the initial fall-off study, patients were started on home therapy with recombinant factor VIII and were asked to subjectively grade their response to treatment as excellent, good, fair, none, or worse. Of 558 evaluable episodes of bleeding treated so far with recombinant factor VIII, the response in 520 (93%) was judged to be excellent or good by the patients (Fig. 3). Another measure of the hemostatic effectiveness of recombinant factor VIII is the response to surgery. Two patients have undergone major surgery under coverage with recombinant factor VIII. The first is a six year old male who underwent bilateral ankle osteotomies for club feet. The second is a 32 year old male with with chronic hemophilic arthropathy and pain in the shoulder who underwent a thoracotomy for cervical sympathectomy. Both procedures were performed without complications.

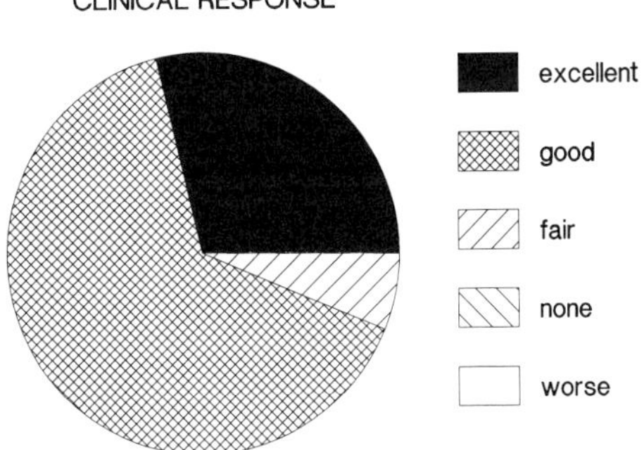

Fig. 3. Response to Recombinant Factor VIII. The response to treatment of bleeding episodes as assessed by patients themselves. Patients were asked to describe their response as excellent, good, fair, none, or worse.

There has been no evidence for inhibitor formation in any of the patients. The biological half-life and recovery of recombinant factor VIII, determined at three month intervals, is stable although the number of patients studied at nine and twelve months is small. Assessment of inhibitor formation *in vitro* by Bethesda assay showed a low level inhibitor (0.8 B.U.) in one patient, but the Bethesda assay was negative on subsequent follow-up and the biological half-life and recovery were normal in the patient. Both plasma-derived and recombinant factor VIII have been used as substrate in the Bethesda assay in order to detect antibodies specific for either protein. Liver function tests in patients on recombinant factor VIII have been unchanged from baseline. There has been no evidence for the development of antibodies to mouse monoclonal antibodies used in the purification of recombinant factor VIII.

Finally, there have been no complications of therapy. Patients with allergic responses to impure plasma-derived factor VIII concentrates have had no allergic responses to recombinant factor VIII. The recombinant product is reconstituted almost instantaneously, thus promoting more rapid treatment.

SUMMARY

Recombinant factor VIII is currently in the late stages of clinical trials. The available studies indicate that the product is safe and well-tolerated, and appears to be free of virus diseases such as HIV and hepatitis infections. Based on these early studies, recombinant coagulation factors appear to have enormous promise and potential for transfusion medicine. The synthesis of large quantities of safe material may lead to the development of techniques for daily administration of factor VIII aimed at the prevention of joint and soft tissue bleeds. There is also the promise of decreased cost as techniques for the efficient synthesis of recombinant proteins is refined further.

REFERENCES

1. Lane S. Haemorrhagic diathesis. Successful transfusion of blood. Lancet 1840;1:185-188.

2. Feissly R. Etudes sur l'hemophilie. Bull Mem Soc Mod Hop Paris 1923; 47:1778.

3. McMillan CW, Diamond LK, Surgenor DM. Treatment of classic hemophilia: The use of fibrinogen rich in factor VIII for hemorrhage and for surgery. New Eng J Med 1961;265:224-230,277-283.

4. Pool JG, Hershgold EJ, Pappenhagen MR. High potency antihaemophilic factor concentrate prepared from cryoglobulin. Nature 1964; 203:312.

5. Brinkhous KM, Shanbrom E, Roberts HR, Webster WP, Fekete R, Wagner RH. A new high-potency glycine precipitated antihemophilic factor (AHF) concentrate. J Am Med Assoc 1968; 205:613-617.

6. Horowitz B, Piet MPJ. Transmission of viral disease by plasma protein fractions. Plasma Ther Transfus Technol 1986;7:503-513.

7. Piszkiewicz D, Kingdon HS, Apfelzweig R, McDougal JS, Cort SP, Andrews J, Hope J, Cabridilla CD. Inactivation of HTLV-III/LAV during plasma fractionation. Lancet 1985;2:1188-1189.

8. Prince AM, Horowitz B, Brotman B. Sterilisation of hepatitis and HTLV-III viruses by exposure to tri(n-butyl)phosphate and sodium cholate. Lancet 1896;1:706-710.

9. Schimpf K, Brackmann HH, Kreuz W, Kraus B, Haschke F, Schramm W, Moesseler J, Auerswald G, Sutor AH, Keohler K, Hellstern P, Muntean W, Scharrer I. Absence of anti-human immunodeficiency virus types 1 and 2 seroconversion after the treatment of hemophilia A or von Willebrand's disease with pasteurized factor VIII concentrate. New Eng J Med 1989;321:1148-1152.

10. White GC II, McMillan CW, Kingdon HS, Shoemaker CB. Use of recombinant antihemophilic factor in the treatment of two patients with classic hemophilia. New Eng J Med 1989; 320:166-170.

11. Toole JJ, Knopf JL, Wozney JM, Sultzman LA, Buecker JL, Pittman DD, Kaufman RJ, Brown E, Shoemaker C, Orr EC, Amphlett GW, Foster B, Coe ML, Knutson GJ, Fass DN, Hewick RM. Molecular cloning of a cDNA encoding human antihaemophilic factor. Nature 1984;312:342-347.

12. Vehar GA, Keyt B, Eaton D, Rodriguez H, O'Brien DP, Rotblat F, Opperman H, Keck R, Wood WI, Harkins RN, Tuddenham EGD, Lawn RM, Capon DJ. Structure of human factor VIII. Nature 1984;312:337-342.

13. Truett MA, Blacher R, Burke RL, Caput D, Chu C, Dina D, Hartog K, Kuo CH, Masiarz FR, Merryweather JP, Najarian JR, Paschl C, Potter SJ, Pumma J, Quiroga M, Rall LB, Randolph A, Urdea M, Valenzuela P, Dahl HH, Favaloro J, Hansen J, Nordfang O, Ezban M. Characterization of the polypeptide composition of human factor VIII:C and the nucleotide sequence and expression of the human cDNA. DNA 1986;4:333-349.

14. Ryden L. Model of the active site in the blue oxidases based on the ceruloplasmin-plastocyanin homology. Proc Nat'l Acad Sci 1982;79:6767-6771.

15. Fass DN, Hewick RM, Knutson GJ, Nesheim ME, Mann KG. Internal duplication and sequence homology in factors V and VIII. Proc Nat'l Acad Sci 1985;82:1688-1691.

16. Kane WH, Davie EW. Cloning of a cDNA coding for human factor V, a blood coagulation factor homologous to factor VIII and ceruloplasmin. Proc Nat'l Acad Sci 1986;83:6800-6804.

17. Toole JJ, Pittman DD, Orr EC, Murtha P, Wasley LC, Kaufman RJ. A large region (95 kDa) of human factor VIII is dispensible for *in vitro* procoagulant activity. Proc Nat'l Acad Sci 1986;83:5939-5942.

18. Eaton DL, Wood WI, Eaton D, Hass PE, Hollingshead P, Wion K, Mather J, Lawn RM, Vehar GA, Gorman C. Construction and characterization of an active factor VIII variant lacking the central one-third of the molecule. Biochemistry 1986;25:8343-8347.

19. Pavirani A, Meulien P, Harrer H, Dott K, Mischler F, Wiesel M-L, Mazurier C, Cazenave J-P, Lecocq J-P. Two independent domains of factor VIII co-expressed using recombinant vaccinia viruses have procoagulant activity. Biochem Biophys Res Commun 1987;145:234-240.

20. Burke RL, Pachl C, Quiroga M, Rosenberg S, Haigwood N, Nordfang O, Ezban M. The functional domains of coagulation factor VIII:C. J Biol Chem 1986;261:12574-12578.

21. Pittman DD, Marquette KA, Kaufman RJ. Expression and characterization of factor VIII and factor V hybrid molecules produced in mammalian cells. Thrombos Haemostas 1989;62:34 (abstract).

22. Foster PA, Fulcher CA, Houghten RA, Zimmerman TS. An immunogenic region within residues Val_{1670}-Glu_{1684} of factor VIII light chain induces antibodies which inhibit binding of factor VIII to von Willebrand factor. J Biol Chem 1988;263:5230-5234.

23. Bloom JW. The interaction of rDNA factor VIII, factor $VIII_{des\ 979-1562}$ and factor $VIII_{des\ 979-1562}$-derived peptides with phospholipid. Thrombos Res 1987;48:439-448.

24. Arai M, Scandella D, Hoyer LW. Molecular basis of factor VIII inhibition by human antibodies. Antibodies that bind to the factor VIII light chain prevent the interaction of factor VIII with phospholipid. J Clin Invest 1989;83:1978-1984.

25. Foster PA, Fulcher CA, Houghten RA, Zimmerman TS. Synthetic factor VIII peptides with amino acid sequences contained within the C2 domain of factor VIII inhibit factor VIII binding to phosphatidylserine. Thrombos Haemostas 1989;62:34 (abstract).

26. Kaufman RJ, Wasley LC, Dorner AJ. Synthesis, processing, and secretion of recombinant human factor VIII expressed in mammalian cells. J Biol Chem 1988;263:6352-6362.

27. van Dieijen G, Tans G, Rosing J, Hemker HC. The role of phospholipid and factor VIIIa in the activation of bovine factor X. J Biol Chem 1987;256:3433-3442.

28. Vehar GA, Davie EW. Preparation and properties of bovine factor VIII. Biochemistry 1980;19:401-416.

29. Fulcher CA, Zimmerman TS. Characterization of the human factor VIII procoagulant protein with a heterologous precipitating antibody. Proc Nat'l Acad Sci 1982;79:1648-1652.

30. Hoyer LW, Trabold NC. The effect of thrombin on human factor VIII. Cleavage of the factor VIII procoagulant protein during activation. J Lab Clin Med 1981;97:50-64.

31. Hultin MB, Jesty J. The activation and inactivation of human factor VIII by thrombin: Effect of inhibitors of thrombin. Blood 1981;57:476-482.

32. Pittman DD, Kaufman RJ. Proteolytic requirements for thrombin activation of anti-hemophilic factor (factor VIII). Proc Nat'l Acad Sci 1988;85:2429-2433.

33. Fulcher CA, Gardiner JE, Griffin JH, Zimmerman TS. Proteolytic inactivation of human factor VIII procoagulant protein by activated protein C and its analogy with factor V. Blood 1984;63:486-489.

34. Kaufman RJ. High level expression of proteins in mammalian cells. In: Setlow JK ed. Genetic Engineering. Principles and Methods. Volume 9. New York: Plenum Press, 1987:155-198.

35. Kaufman RJ, Wasley LC, Davies MV, Wise RJ, Israel DI, Dorner AJ. Effect of von Willebrand factor coexpression on the synthesis and secretion of factor VIII in Chinese hamster ovary cells. Mol Cell Biol 1989;9:1233-1242.

36. White GC II, McMillan CW, Gomperts E, Addiego J Jr, Brackmann H, Brettler D, Gill J, Goldsmith J, Hoots K, Kessler C, Kisker T, Laurian Y, Lewis B, Lusher J, Mannucci P, Morfini M, Scharrer I, Schulman S, Lee M, Courter S, Kingdon H, Roberts H. Safety and efficacy of human factor VIII prepared by recombinant DNA techniques. Blood 1989;74(Suppl.1):55a.

CONCLUDING OVERVIEW

THE IMPACT OF RECOMBINANT TECHNOLOGIES IN UNDERSTANDING PLASMA PROTEINS

IMPORTANT FOR HEMOSTASIS AND THROMBOSIS

Kenneth G. Mann

Biochemistry Department
University of Vermont
Burlington, VT

It is clear that the power of molecular biology technology (1) has provided enormous access to information about molecular structure; (2) has provided access to biological problems which were formerly inaccessible; (3) has provided materials and resources for the management of human biology and the treatment of disease. Today, we truly have access to molecular pharmacology. We can modify molecules; we can create new molecules. We have the capacity for industrial scale production of otherwise scarce natural products. In the next decade the understanding of the transcriptional control of the gene and the translational control of message, as well as the trafficking of protein of cells, will become accessible through the technologies of molecular biology.

The dramatic consequences of the application of molecular biology techniques to the elucidation of primary structure bears some comparisons. For illustrative purposes let me consider human α-thrombin. Human α-thrombin is 308 amino acid residues divided into two chains of 49 and 259 amino acid residues. Ralph Butkowski, Michael Downing and Jacques Elion sequenced human α-thrombin in my laboratory during the period from 1974 to 1976. This study which used primarily sequential Edman degradation and spinning-cup sequencing technology, consumed about 100 mgs. of α-thrombin.[1] Inspection of table I illustrates a compelling problem which occurs when one considers a similar approach for molecules such as factor VIII. Factor VIII is nearly ten times the size of thrombin (2351 aa)[2,3] and it is present in blood at concentrations which are diminishingly small (0.01%[4]). Human factor VIII was sequenced at Genetic Institute with c-DNA technology using probes fashioned from knowledge of the amino acid sequence of fragments of porcine factor VIII obtained through the effort of David Fass and Rodney Hewick.[4,5] Genentech obtained peptide sequences of human factor VIII isolated by Tuddenham.[6] The two companies reported the formidable sequence of over 2,300 amino acids in back to back publications in 1984.

The isolation of factor VIII was preceded by the isolation of factor V, a protein which is highly homologous to factor VIII and which has served as a paradigm for Factor VIII isolation and subsequent structure/function studies.[7-10] However, although factor V has significant research value it has little economic value. Factor V deficiency unlike factor VIII deficiency, is autosomal and is represented by a very small population of individuals in the world. Microsequencing of human factor V led to the development of probes which were used by Richard Jenny, Randy Kaufman, and

Debbie Pittman at Genetics Institute to elucidate the human factor V sequence.[11] A contemporaneous study by Earl Davie and William Kane reported on the same sequence.[12] The gas phase sequencing techniques developed by Rod Hewick and Lee Hood were essential to develop sufficient amino acid sequence for nucleotide probe synthesis,[13] however the isolation of the proteins was a fundamental first step to execution of the sequence technology. Richard Jenny's Ph.D. thesis describes the sequence of factor V and peptide chemistry studies of its proteolytic fragmentation. Jenny's thesis topic would have been an inconceivable enterprise without the tools of modern molecular biology.

TABLE I

PLASMA PROTEINS OF THE VITAMIN K-DEPENDENT ENZYME COMPLEXES

| | | Average Plasma Concentration | |
Protein	Mw	μgm/ml	μM
Prothrombin	72,000	150	2.1
Factor X	59,000	10	0.17
Factor V	330,000	9	0.02
Factor IX	57,000	4	0.07
Factor VIII	285,000	0.12	0.0004
Factor VII	50,000	0.5	0.01
Protein C	62,000	4	0.06

There are two points of interest in the histories of the development of the factor V and factor VIII molecular structures which require reflection. The first is that the structures of both of these proteins were identified in studies conducted by virtue of collaborations between academic and industrial laboratories. The development of the techniques of molecular biology for sequencing and the generation of genetics companies were primarily through the efforts of enterprising academicians from university laboratories. The subsequent academic/industrial relationships have been important components in the progress which has been made in modern biological research. There was a natural foundation for establishing cooperation between academic and industrial laboratories because of the origins of the latter. I hope this sort of cooperation will continue into the future as it has made for tremendous breakthroughs in understanding molecular structure and human physiology. The second lesson to be learned is that the opportunities for talented individuals in well equipped industrial laboratories to participate in the solution of basic problems requires industrial support for these activities. Factor VIII and factor V represent a case in point. The frequency of hemophilia A or factor VIII deficiency is approximately one in twenty thousand. Factor VIII is a sex-linked gene product and inheritance of a single bad gene results in a bleeding disorder. Factor V is autosomal. If we assume that the mutation rate of the factor V gene is equivalent to that for the factor VIII gene we would expect the frequency of each gene mutation to be about 1/20,000, leading to absent or abnormal molecules. However, since 50% factor V levels would not produce a hemostatic defect and would be sufficient to avoid detection by a hemostasis laboratory, the observation of clinically relevant hemophilia associated with factor V deficiency would only result when two mutant genes were received. The frequency of this would be about one in 4 x 10^8. This sort of abnormal population is too small to merit an industrial market for factor V production. Thus the sequencing of factor V through the generous support of the Genetics Institute was certainly a "noncommercial" venture. We need to recognize the value of these collaborations and develop support models to encourage cooperation between the academic laboratory and the industrial laboratory to insure continued cooperation between these two enterprises.

The contributions of molecular biology technology to blood coagulation and thrombolytic pharmacology are impressive. However, a point not discussed at this meeting was the contributions of molecular biology to the area of growth factor technology. The hematopoietic blood cell system is immensely complicated and highly regulated. The available quantities of plasma proteins in coagulation depicted in Figure 1 are enormous relative to the amount of a growth factor required to regulate the hematopoietic lineage associated with red cells. Erythropoietin is a primary regulator of red blood cell mass. Compared to the 20,000 factor VIII deficient hemophiliacs with chronic need for replacement therapy there are 100,000 individuals undergoing chronic renal dialysis. These individuals represent just one segment of a much larger potential user base for this hematopoietic regulator. The industrial drive to produce growth factors in quantity has exceeded that to produce coagulation factors, partly because of this market demand. Erythropoietin is now produced by both Amgen and by Genetics Institute and it is approved for renal dialysis. The natural protein is produced in the kidney and influences maturation of committed precursor cells by a signalling process related to red cell mass. First associated with renal failure and anemia in 1836,[14] it wasn't until 1977 that Dr. Goldwasser's group purified human erythropoietin.[15] The protein was isolated, cloned, and was in clinical trials by 1985. The time interval between the expression of the protein and clinical approval for use was from 1983 to 1989. About six years to generate an approved product. This is very fast these days. I think that we need to encourage our colleagues in the regulatory agencies to be as responsive as possible in making recombinant products available in a timely manner. In many instances there is no other way to treat the disease. Other growth factors currently undergoing clinical trials include granulocyte colony stimulating factor, and granulocyte macrophage colony stimulating factor. These materials have been applied to cancer patients[16], bone marrow transplant recipients[17] and patients with anemia.[18] The regulator of megakaryocytopoesis, "thrombopoieten" has been much more elusive. Dr. Grant, a clinician-investigator who has been working in association with me has succeeded in isolating a factor from human aplastic urine which appears to fulfill the criteria of being the erythropoietin equivalent for platelets.[19]

The presentations at this conference did not deal with the issue of tissue plasminogen activator (t-PA), which has been a major player in the growth of the recombinant technology industry. This drug has had a major impact in the treatment of patients with myocardial infarction. t-PA is a very large market item. At a recent AHA Thrombosis Council meeting, Dr. Jack Hawiger reported that every year there are 800,000 first heart attacks, and 600,000 second heart attacks with a combined death total of 560,000. Intervention by thrombolytic therapy thus has a major potential for health management. Tissue plasminogen activator was isolated by Rikjin and Collen in 1981,[20] but it took eight years to get the product to market. A gestation phase of seven to nine years for approval of drugs, coupled with the complications of the number of programs that are initiated and fail, makes it important that we operate in such a fashion as to encourage exploitation of scientific discoveries rather than discourage them by making it economically unattractive to take chances. Much of the basic science of t-PA which has been conducted with mutated proteins produced by commercial enterprise. These have been made available to selected members of the scientific community, and these mutants have resulted in great progress in developing an understanding of the structure-function relationships of this molecule. This has occurred because of the cooperation between academics and industry.

I believe it is important that we continuously stress collaborative and cooperative research between industry and academe. This collaboration is under significant challenge in the political arena. Research cooperation typifies the relationship between the new biotechnology companies, and basic

science laboratories in academia over the past ten years. This behavior must
be reinforced and encouraged to expand. The new science of molecular biology
has flourished in an atmosphere of relatively free exchange. This
cooperation should be enabled and encouraged in every way possible.

SOME TECHNICAL NEEDS

Existing protein structural data obtained using both amino acid and nucleic
acid technology have been interpreted by pattern recognition to represent
domain features of molecules. However, we presently have an abundance of
physical information we are not adequately using because we do not have the
appropriate algorithms to reduce the information available for natural
proteins to manageable levels. We need better mathematical and computer
systems to utilize the information base available to us.

My final comments have to due with statistics. Most chemists do not worry
about statistics because they are protected by Avogadro's number (6.02 x
10^{23}). Chemists do not ordinarily use statistics because they work with very
large ensembles. Even a femptomolar solution (10^{-15}M) is a collection of 10^8
molecules per liter. The technology available in most molecular biology
laboratories does not provide such automatic protection by the inherent
statistics of large ensembles. Technology such as PCR can select for and
expand unusually small populations of RNA molecules. This approach has been
most useful for characterization of defective genes. While some concern
exists with respect to the fidelity of the expansion technique, an
additional problem is whether the single gene expanded is the basis for the
disease pathology.

We have studied, using peptide technology, one of the original
"hypoprothrombinemic" patients described by Armand J. Quick.[21] These studies
led to the observation that this deficient individual actually had two
defective prothrombins, each of which produced a defective thrombin. In one
of these molecules a cysteine replaces an arginine;[22] in the other, valine
replaces glycine.[23] Thus, in this patient there are two genetic defects each
of which give rise to a different abnormal thrombin. Each structural defect
is associated with a different functional defect. The Quick-2 enzyme has no
activity toward any thrombin substrate, while the Quick-1 enzyme has greatly
suppressed activity toward most substrates. In dealing with autosomal
congenital disease, caution in interpretation is required. Even in studies
of sex-linked proteins, when we are dealing with molecular defects in very
large molecules, we need to be cautious to be sure that we are indeed
identifying a defect associated with a specific physiologic lesion.

BIBLIOGRAPHY

1. Butkowski, R.M., Elion, J., Downing, M.R. and Mann, K.G.: 1977. J.
    Biol. Chem. 252:4942-4957, 1977.
2. Toole, J.J., Knopf, J.L., Wozney, J.M. Sultzman, L.A., Buecker, J.L.,
    Pittman, D.D., Kaufman, R.J., Brown, E., Shoemaker, C., Orr, E.C.,
    Amphlett, G.W., Foster, W.B., Coe, M.L., Knutson, G.J., Fass, D.N. and
    Hewick, R.M. 1984. Nature 312:342-47.
3. Gitschier, J., Wood, W.I., Goralka, T.M., Wion, K.L., Chen, E.Y.,
    Eaton, D.H., Vehar, G.A., Capon, D.J. and Lawn, R.M. 1984. Nature 312-
    326-30.
4. Hoyer, L.W. 1981. Blood 58:1-13.
5. Fass, D.N., Hewick, R., Knutson, G.J., Nesheim, M., Mann, K. 1983.
    Thromb. Haemostasis 50:255.
6. Rotblat, F., O'Brien, D.P., O'Brien, F., Goodall, A.H., Tuddenham,
    E.G.E. 1985. Biochemistry 24:4294-300.
7. Nesheim, M.E., Myrmel, K.H., Hibbard, L., Mann, K.G. 1979. J. Biol.
    Chem. 254:508-17.

8.   Esmon, C.T. 1979. _J. Biol. Chem._ **254**:964-73.
9.   Katzmann, J.A., Nesheim, M.E., Hibbard, L.S., Mann, K.G. 1981. _Proc. Natl. Acad. Sci._ USA **78**:162-66.
10. Kane, W.H., Majerus, P.W. 1981. _J. Biol. Chem._ **256**:1002-7.
11. Jenny, R.J., Pittman, D.D., Toole, J.J., Kriz, R.W., Aldape, R.A., et al. 1987. _Proc. Natl. Acad. Sci._ USA **84**:4846-50.
12. Kane, W.H., Ichinose, A., Hagen, F.S., and Davie, E.W. 1987. _Biochemistry._ **26**:6508-6514.
13. Hewick, R.M., Hunkapillar, M.W., Hood, L.E., and Bryer, W.J. 1981. _J. Biol. Chem._ **256**:7990.
14. Bright, R. 1836. _Guys Hospital Report._ **1**:338-379.
15. Mikake, T., Kung, C.K.H. and Goldwasser, E. 1977. _J. Biol. Chem._ **252**:558-5564.
16. Gabrilove, J.L., Jakubowski, A., Scher, H., Sternberg, C., Wong, G., Grous, J., Yagoda, A., Fain, K., Moore, A.S., Clarkson, B., Oettgen, H.F., Alton, K., Welte, K. and Souza, L. 1988. _N. Engl. J. Med._ **318**:1414-1422.
17. Brandt, S.J., Peters, W.P., Atwater, S.K., Kurtzberg, J., Borowitz, M.J., Jones, R.B., Shpall, E.J., Bast, R.C., Gilbert, C.J., Oette, D.H. 1988. _N. Engl. J. Med._ **318**:869-876.
18. Antin, J.H., Smith, B.R, Holmes, W., and Rosenthal, D.S. 1988. _Blood._ **72**:705-713.
19. Grant, B.W., Nichols, W.L., Solberg, L.A., Yachimiak, D.J. and Mann, K.G. 1987. _Blood._ **69**:1334-1339.
20. Rijken, D.C. and Collen, D. 1981. _J. Biol. Chem._ **256**:7035-7041.
21. Quick, A.J., Pisciotta, A.V., and Hussey, C.V. 1955. _Arch. Intern. Med._ **95**:2-14.
22. Henriksen, R.A. and Mann, K.G. 1988. _Biochemistry._ **27**:9160-9165
23. Henriksen, R.A. and Mann, K.G. 1989. _Biochemistry._ **28**:2078-2082.

CONTRIBUTORS

R.C. Austin
McMaster University Medical Center
Hamilton, Ontario, Canada

Kenneth A. Bauer
Harvard Medical School
Boston, MA

Michael N. Blackburn
Smith Kline-Beecham
King of Prussia, PA

M.A. Blajchman
McMaster University Medical Center
Hamilton, Ontario, Canada

Susan Clark Bock
Temple University School of Medicine
Philadelphia, PA

Paula R. Boerger
Smith Kline-Beecham
King of Prussia, PA

Andrew J. Dorner
Genetics Institute
Cambridge, MA

William Drohan
Holland Laboratory
American Red Cross
Rockville, MD

F. Fernandez-Rachubinski
McMaster University Medical Center
Hamilton, Ontario, Canada

Ruud D. Fonteijn
Central Laboratory of the
Netherlands Red Cross
Amsterdam, The Netherlands

Bruce Furie
Tufts University School of Medicine
Boston, MA

Barbara C. Furie
Tufts University School of Medicine
Boston, MA

Alan R. Giles
Queens University
Kingston, Ontario, Canada

Ulla Hedner
Novo Research Institute
Bagsvaerd, Denmark

Leon W. Hoyer
Holland Laboratory
American Red Cross
Rockville, MD

Randal J. Kaufman
Genetics Institute
Cambridge, MA

Anja Leyte
Central Laboratory of the
Netherlands Red Cross
Amsterdam, The Netherlands

George L. Long
University of Vermont
Burlington, VT

Morgan Lorio
Smith Kline-Beecham
King of Prussia, PA

Ross T.A. MacGillivray
University of British Columbia
Vancouver, BC, Canada

Kenneth G. Mann
University of Vermont
Burlington, VT

Koen Mertens
Central Laboratory of the
Netherlands Red Cross
Amsterdam, The Netherlands

Rekha Paleyanda
Holland Laboratory
American Red Cross
Rockville, MD

Hans Pannekoek
Central Laboratory of the
Netherlands Red Cross
Amsterdam, The Netherlands

D.B.C. Ritchie
University of British Columbia
Vancouver, BC, Canada

D.L. Robertson
University of British Columbia
Vancouver, BC, Canada

Richard S. Schwartz
Cutter Laboratories
Berkeley, CA

W.P. Sheffield
McMaster University Medical Center
Hamilton, Ontario, Canada

Arthur R. Thompson
Puget Sound Blood Center
Seattle, WA

Jan A. van Mourik
Central Laboratory of the
Netherlands Red Cross
Amsterdam, The Netherlands

Harm B. van Schijndel
Central Laboratory of the
Netherlands Red Cross
Amsterdam, The Netherlands

William Velander
Virginia Polytechnic Institute
Blacksburg, VA

Martin Ph. Verbeet
Central Laboratory of the
Netherlands Red Cross
Amsterdam, The Netherlands

Jan Voorberg
Central Laboratory of the
Netherlands Red Cross
Amsterdam, The Netherlands

Frederick J. Walker
American Red Cross Blood Services
Farmington, CT

Louise C. Wasley
Genetics Institute
Cambridge, MA

Gilbert C. White
University of North Carolina
School of Medicine
Chapel Hill, NC

Robert J. Wise
Genetics Institute
Cambridge, MA

Robert M. Wolcott
Smith Kline-Beecham
King of Prussia, PA

Janet Young
Holland Laboratory
American Red Cross
Rockville, MD